EBI Studies in Banking and Capital Markets Law

The European Banking Institute

The European Banking Institute based in Frankfurt is an international centre for banking studies resulting from the joint venture of Europe's preeminent academic institutions which have decided to share and coordinate their commitments and structure their research activities in order to provide the highest quality legal, economic and accounting studies in the field of banking regulation, banking supervision and banking resolution in Europe. The European Banking Institute is structured to promote the dialogue between scholars, regulators, supervisors, industry representatives and advisors in relation to issues concerning the regulation and supervision of financial institutions and financial markets from a legal, economic and any other related viewpoint.

As of May 2021, the Academic Members of the European Banking Institute are the following: Universiteit van Amsterdam, University of Antwerp, University of Piraeus, Alma Mater Studiorum–Università di Bologna, Universität Bonn, Academia de Studii Economice din București (ASE), Trinity College Dublin, University of Edinburgh, Goethe-Universität, Universiteit Gent, University of Helsinki, Universiteit Leiden, KU Leuven University, Universidade Católica Portuguesa, Universidade de Lisboa, University of Ljubljana, Queen Mary University of London, Université du Luxembourg, Universidad Autónoma Madrid, Universidad Carlos III de Madrid, Universidad Complutense, Madrid, Johannes Gutenberg University Mainz, University of Malta, Università Cattolica del Sacro Cuore, University of Cyprus, Radboud Universiteit, BI Norwegian Business School, Université Panthéon - Sorbonne (Paris 1), Université Panthéon-Assas (Paris 2), University of Stockholm, University of Tartu, University of Vienna, University of Wrocław, Universität Zürich.

Federica Agostini

The Law of Sustainable Securitisations

The Missing Piece in the Sustainable Finance Puzzle

palgrave
macmillan

Federica Agostini
International and European Law Department, and Centre
for Regulation and Enforcement in Europe (RENFORCE)
Utrecht University
UTRECHT, The Netherlands

ISSN 2730-9088 ISSN 2730-9096 (electronic)
EBI Studies in Banking and Capital Markets Law
ISBN 978-3-032-25045-2 ISBN 978-3-032-25046-9 (eBook)
https://doi.org/10.1007/978-3-032-25046-9

This Palgrave Macmillan imprint is published by the registered company Springer Nature Switzerland AG
The registered company address is: Gewerbestrasse 11, 6330 Cham, Switzerland

To my parents, Luca and Emanuela, and my grandfather Emilio,
for teaching me to work hard and make sacrifices for what really matters.
I hope this endeavour will do justice to the true values you have taught
me over the years.

Acknowledgements

This book builds on, but also substantially expands, the doctoral project I developed at the University of Glasgow between 2019 and 2024. I gratefully acknowledge the support of the University of Glasgow's Faculty of Law, whose academic environment and institutional backing made this research possible, and also to Fondazione Felice Gianani, which generously financed part of this research in the academic year 2021–2022. I owe special thanks to the many colleagues who, over the years, have provided insightful and constructive feedback that has significantly strengthened this work. In particular, I am deeply grateful to my supervisors, Prof. Iain MacNeil and Prof. Javier Solana, for their meticulous, constructive and at times challenging guidance, which nurtured my growth and development as a scholar. I am equally thankful to the fellow academics, peers, and friends whose thoughtful engagement with my research has helped me refine my ideas, particularly Bart Joosen, Chi Zhang, Giedre Jokubauskaite, Elia García Cerrato, Agnieszka Smoleńska, David Ramos Muñoz, Enrico Sartori, Camilla Faggioni, Pia Hüsch, Neofytos Sakellaridis–Mangouras, Walter Bruno, Lorenzo Soci, and Saskia Millmann (though too many others cannot be named here).

Many other exchanges with colleagues and students over the years, particularly through my experiences at the Bank of Italy, the European Bank for Reconstruction and Development, the European University Institute, and within the Centre for Regulation and Enforcement in

Europe (RENFORCE) at Utrecht University, have continuously nurtured my passion for financial regulation, sustainability matters, and EU law. Their influence is deeply embedded in this work, and I truly hope that the final output does justice to the unwavering support and guidance I have received.

Finally, I would like to thank those who have never ceased to support me through the highs and lows of these challenging years of professional and personal growth, especially my parents, Luca and Emanuela, and Davide. This would truly have never been possible without you.

Como, Italy Federica Agostini
December 2025

About the Book

The book examines the overlooked potential of securitisation as a technique to promote sustainability goals. Especially in the past decade, financial institutions (and, in residual cases, companies) have increasingly incorporated environmental, social, and governance (ESG) considerations into the design of securitisation transactions. In the early stages, and in the absence of ad hoc regulatory requirements, market actors independently characterised these transactions as 'sustainable', raising questions of legal certainty and potential 'greenwashing' risks. In contrast to other 'sustainable' asset classes, such as 'green' bonds and loans, the 'sustainable securitisation' market has also experienced a slower growth and less popularity.

Against this backdrop, the book contributes to the existing sustainable finance and securitisation scholarship by presenting a typology of the characteristics that render securitisations 'sustainable'. It also aims to offer the first comprehensive doctrinal analysis of the intersection between securitisation law and sustainability goals.

Drawing on theoretical inquiries into the inextricable nexus between law and finance, the book highlights how the development of the 'sustainable' securitisation market is fundamentally dependent on its legal framework. In particular, it identifies strategic areas where legal interventions could prompt sustainability-related practices. It also analyses and critically assesses the existing legal framework under the lens of such strategic areas. Such analysis first builds on examples of transactional documentation, which have been the primary *locus* for market actors to articulate the sustainability goals of the issuance. It brings to the fore the limited prescriptiveness of clauses related to sustainability matters ('private hard law'), as well as the lack of robust enforcement mechanisms.

In light of these shortcomings, two possible avenues for the coordination of 'sustainable' transactions lie in 'soft' legal tools, such as guidelines, metrics, and model provisions, or 'hard' legal tools, such as EU legislation. Thus, the book critically evaluates previous initiatives within these two realms, assessing their effectiveness in enabling and incentivising market actors to promote sustainability goals through their securitisations.

Overall, the book provides a timely and critical reflection on the gaps of the current sustainability and sustainable finance agenda. It also advances normative arguments around the opportunities for representatives of financial market actors, such as financial industry associations and EU regulators, to take action and address the existing shortcomings.

Contents

List of Figures

List of Tables

Abbreviations

ABA	American Bar Association
ABSs	Asset-Backed Securities
AFME	Association for Financial Markets in Europe
APLMA	Asia Pacific Loan Market Association
BCBS	Basel Committee on Banking Supervision
BREEAM	Building Research Establishment Environmental Assessment Method
CCSF	Commission de Surveillance du Secteur Financier (Luxembourg)
CLO	Collateralised Loan Obligation
CSDDD	Corporate Sustainability Due Diligence Directive
EBA	European Banking Authority
ECB	European Central Bank
EEMAP	Energy Efficiency Mortgages Action Plan
EIOPA	European Insurance and Occupational Pensions Authority
ELFA	European Leveraged Finance Association
ESMA	European Securities and Markets Authority
EUGB	EU Green Bond
EUGBR	EU Green Bond Regulation
EUGBS	EU Green Bond Standard
FCA	Financial Conduct Authority
GBPs	Green Bond Principles
GLPs	Green Loan Principles
ICMA	International Capital Markets Association
IFRS	International Financial Reporting Standards
IMA	Investment Association

IOSCO	International Organisation of Securities Commissions
ISDA	International Swaps and Derivatives Association
ISSB	International Sustainability Standards Board
KPIs	Key Performance Indicators
LEED	Leadership in Energy and Environmental Design
LMA	Loan Market Association
LSTA	Loan Syndications and Trading Association
NGFS	Network for Greening the Financial System
NIE	New-Institutionalist Economics
OECD	Organisation for Economic Co-operation and Development
OTC	Over-the-Counter
SFIG	Structured Finance Industry Group
SLBs	Sustainability-Linked Bond Principles
SLPs	Sustainability-Linked Loan Principles
SPTs	Sustainability Performance Targets
SPV	Special Purpose Vehicle
SSPEs	Special Purpose Entities
STS	Simple, Transparent and Standardised
TEU	Treaty on European Union
TFEU	Treaty on the Functioning of the European Union
TCFD	Task Force for Climate-Related Financial Disclosures
UNCTAD	United Nations Conference on Trade and Development
UNFCC	United Nations Framework Convention on Climate Change
UNPRI	United Nations Principles for Responsible Investment

CHAPTER 1

Introduction

1.1 Background and Aims

Especially since the 2010s, 'sustainable finance' and ESG (which stands for 'environmental, social, and governance') have become ubiquitous 'buzzwords' across academic and policy discussions on the role of financial markets, investors and companies.[1] While the notion of 'sustainable finance' remains nebulous and unsettled,[2] a compelling definition characterises it as 'any form of financial service with the objective of supporting the transition to a sustainable economy and society by integrating environmental, social, and governance factors into business and investment decisions'.[3]

[1] See, *ex multis*, Jennifer Acree, ESG: The popular corporate Buzzword and why leaders should care (15 March 2023, Forbes) https://www.forbes.com/councils/forbesbusinesscouncil/2023/03/15/esg-the-popular-corporate-buzzword-and-why-leaders-should-care/ accessed 5 December 2025.

[2] For example, the Commission's website only refers to sustainable finance as 'the process of taking environmental, social, and governance (ESG) considerations into account when making investment decisions in the financial sector, leading to more long-term investments in sustainable economic activities and projects', implying that it would only focus on investments. European Commission, 'Overview of sustainable finance', https://finance.ec.europa.eu/sustainable-finance/overview-sustainable-finance_en accessed 9 March 2025.
For a historical review of the meanings attributed to the notion see Andreas Dimmelmeier, 'Sustainable finance as a contested concept: tracing the evolution of five frames between 1998 and 2018' (2021) 13 *Journal of Sustainable Finance & Investment* 4, 1600 ff.

[3] See Swiss Sustainable Finance, 'Glossary' (2023) https://www.sustainablefinance.ch/en/resources/what-sustainable-finance/glossary.html accessed 10 December 2025.

F. Agostini, *The Law of Sustainable Securitisations*, EBI Studies in Banking and Capital Markets Law,
https://doi.org/10.1007/978-3-032-25046-9_1

The closely associated acronym 'ESG' encapsulates these three factors, while placing the emphasis on financial return and risk-management considerations.[4]

Several milestones have contributed to the unprecedented momentum for ESG issues. The UN Sustainable Development Goals (SDGs) and the Paris Agreement have raised awareness around the risks that climate change poses to the financial system and the need for large amounts of capital to meet 2030 sustainability targets.[5] The recovery from the COVID-19 pandemic and the energy crisis amplified by Russia's full-scale invasion of Ukraine have also urged regulators to think of ways to accelerate the 'green transition'.[6]

This context has intertwined with an evolving 'sustainable-finance' policy debate around potential measures to promote sustainability goals through financial markets and investments. In this regard, the EU, in particular, has acted as a forerunner on the global policy stage through a series of measures stemming from its 2018 Sustainable Finance Action Plan.[7] Its wide-ranging framework has resulted from the collaborative efforts among EU institutions, supervisory authorities, as well as stakeholders (involved in the Platform on Sustainable Finance[8]). The development of such a framework has, however, followed a fragmented logic, similar to the solution of a puzzle.

A notable piece of such a puzzle that has yet to receive comparable attention from policy-makers and market actors is securitisation. As I will further elaborate in Chap. 2, securitisation is essentially a financing technique which revolves around three core elements, namely (i) a bundle of

[4] On the emergence of the notion of ESG, see Irene-marié Esser, Iain MacNeil, 'From a Financial to an Entity Model of ESG' (2022) 23 *European Business Organisation* Law Review 9, 14–15.

[5] United Nations, 'Transforming our world: the 2030 Agenda for Sustainable Development' (2015) A/RES/70/1 https://undocs.org/en/A/RES/70/1 accessed 10 December 2025; United Nations Framework Convention on Climate Change (hereinafter UNFCC), 'Paris Agreement' (2015). https://unfccc.int/files/essential_background/convention/application/pdf/english_paris_agreement.pdf accessed 10 December 2025.

[6] See a speech by one of the members of the European Central Bank (hereinafter ECB)'s Executive Board, illustrating the challenges for policy-makers in the light of these historical developments. Isabel Schnabel, 'Looking through higher energy prices? Monetary policy and the green transition' (8 January 2022) American Finance Association 2022 Virtual Annual Meeting https://www.ecb.europa.eu/press/key/date/2022/html/ecb.sp220108~0425a24eb7.en.html accessed 10 December 2025.

[7] EU Commission, 'Action plan: financing sustainable growth' (2018) 97.

[8] Advising the EU Commission on the framework, See 5., ibid.

assets, such as loans and leases, (ii) an independent entity, called a special purpose vehicle (SPV), to which the assets are assigned, and (iii) the issuance of financial instruments, denominated asset-backed securities (ABSs), backed by the underlying asset pool in the SPV.

The intersection between securitisation and sustainable finance represents a promising development, which previous academic studies have examined only preliminarily and to a limited extent. The present book aims to bridge this gap by drawing on two noteworthy trends.

First, press coverage and a few academic studies have documented the growing use of securitisation as a means to promote environmental, social, and governance goals, particularly over the past fifteen years. For instance, one could mention Toyota, one of the frontrunners in Europe, which raised finance to expand its production of hybrid vehicles through the issuance of 'green' ABSs in 2014.[9] The example illustrates that market actors were the ones taking the lead to name securitisation operations as 'green' or 'sustainability-linked', in the absence of a legal definition.

Second, empirical data point to the dramatic rise in the popularity of financial instruments enabling market participants to further sustainability goals, such as 'green', 'social', and 'sustainability-linked' bonds and loans over the last decade.[10] While the development of the 'sustainable securitisation' market is situated within this broader trend, its growth has lagged behind that of comparable asset classes. For instance, the EBA reported only 13 cases of securitisation deals with 'green' or other 'sustainable' features[11] within the EU and UK market between 2016 and 2022.[12] Data further suggests that these transactions constitute a minuscule fraction of the EU's 'sustainable' debt landscape, amounting to around 1.9% of

[9] See Chap. 4, Sect. 4.3.

[10] For data on the issuance of 'sustainable' bonds and loans, especially between 2016 and 2024, see AFME, 'ESG Finance Report – European Sustainable Finance – Q4 2024' (2025) 3, 6 https://www.afme.eu/Portals/0/DispatchFeaturedImages/ESG%20Finance%20Report%204Q24.pdf accessed 5 December 2025. On the contrary, Q3 2025 witnessed a decline in the issuance of such instruments, though it is too early to draw any meaningful conclusions. AFME, 'ESG Finance Report – European Sustainable Finance – Q3 2025' (2025) 3, 4 https://www.afme.eu/publications/data-research/afme-esg-finance-report-q3-2025/ accessed 5 December 2025.

[11] For the distinction, see below Chap. 4, Sect. 4.4.

[12] European Banking Authority, 'Developing a Framework for Sustainable Securitisation' (2022) 20–21 https://www.eba.europa.eu/sites/default/documents/files/document_library/Publications/Reports/2022/1027593/EBA report on sustainable securitisation.pdf accessed 5 December 2025.

total issuance in 2024,[13] and an even smaller proportion of the EU securitisation market, where sustainability-oriented operations accounted for merely 6% in 2021.[14]

These trends raise a fundamental question, underexplored by previous academic scholarship: How can the legal framework governing these transactions, or the lack thereof, contribute to the sector's development? This question will be at the core of the present book. Addressing it requires investigating whether it is possible to formulate a systematic definition of sustainable securitisations based on the distinctive elements of previous transactions. The book also critically examines how, and to what extent, the law allows and incentivises market actors to incorporate sustainability considerations in the design and management of their securitisation deals, as well as the potential for legal reforms.

Overall, the research questions which the book seeks to address are the following:

- What makes securitisations 'sustainable'?
- What is the added value of securitisation for sustainable finance and for the sustainability agenda more broadly?
- How and to what extent does law support sustainability goals through securitisations?
- How could law most effectively advance sustainability goals through securitisations?

Before turning to the core analysis, the introductory chapter situates the work within the existing literature and identifies the gaps which it seeks to fill. It offers methodological remarks, focusing on the scope of the research and its limitations. It also provides an overview of the structure.

[13] Notably, 'sustainable' securitisations reached an overall volume of €5.5 bln in 2024, while the overall issuance of 'sustainable' bonds and loans amounted to €278 bln. AFME, 'ESG Finance Report– Q4 2024' (n. 10) 12. Previous reports have highlighted consistent trends, with securitisations accounting for around 2% of the total issuance of 'sustainable' instruments in 2021. AFME, 'ESG Finance Report – –European Sustainable Finance – –Q2 2021' (2022) 12–17 https://www.afme.eu/Portals/0/DispatchFeaturedImages/AFME%20Sustainable%20Finance%20Report%20-%20Q2%202021-%20v2.pdf accessed 5 December 2025.

[14] AFME, 'Q2 2021 Securitisation Report – –European Structured Finance' (2021) 6 https://www.afme.eu/Portals/0/DispatchFeaturedImages/AFME%20Q2%202021%20Securitisation%20Report1.pdf accessed 5 December 2025.

1.2 Relevance and Original Contributions

The book aims to make three relevant and original contributions.

First, it draws upon the academic scholarship on the law relating to financial markets[15] by investigating the applicability of recurring financial regulation techniques to the securitisation context.[16] More specifically, it aims to enrich the diverse literature on securitisation and its legal regime. On the one hand, commentators have praised its various functions: For instance, Schwarcz compared it to an 'alchemy' that 'turns base metals into gold'.[17] In other words, securitisation acts as a risk-management technique, allowing financial institutions to release part of their capital from their balance sheets, while also meeting legal requirements (for example, in terms of capital holdings), as well as serving as a liquidity-providing tool and as a substitute for bank lending.[18] On the other hand, many have criticised securitisation, especially after the 2007 Global Financial Crisis, for the market failure and financial stability risks it posed.[19] The aftermath of the crisis, and the ensuing policy reforms in the EU context, also sparked comparative analyses of securitisation legislation across EU Member States as well as positive and normative studies on the design of an optimal regime.[20] More nuanced accounts of securitisation have also started

[15] For example, John Armour, Dan Awrey, Paul Davies et al, *Principles of Financial Regulation* (Oxford University Press 2016).

[16] See further in Chap. 2, Sect. 2.6.

[17] Steven L Schwarcz, 'The Alchemy of Asset Securitisation' (1994) 133 *Stanford Journal of Law, Business & Finance* 133.

[18] Ibid., 3, 7, 8. Vincenzo Bavoso, 'Legal Issues on Financial Innovation and Structured Finance' (2011) 34 *The Company Lawyer* 5; Claire A Hill, 'Securitisation: A Low-Cost Sweetener for Lemons' (1997) 10 *Journal of Applied Corporate Finance* 64 1061, 1086–1098.

[19] For example, Atif Mian and Amir Sufi, *House of Debt: How They (and You) Caused the Great Recession, and How We Can Prevent It from Happening Again* (The University of Chicago Press 2014) 95–98; For literature reviews, see João Pinto and Paulo Alves, 'The Economics of Securitisation: Evidence from the European Markets' (2016) 13 *Investment Management and Financial Innovations* 112; Bank for International Settlements (hereinafterBIS), 'Securitisation is not that evil after all' (2011) BIS Working Paper 2–3 https://www.bis.org/publ/work341.pdf accessed 5 December 2025.

[20] See, *ex multis,* David Ramos Muñoz, Kevin Ingram, *The Law of Transnational Securitisation* (Oxford University Press, 2010); Vincenzo Bavoso, 'High Quality Securitisation and EU Capital Markets Union-Is It Possible?' (2017) 7 *Accounting, Economics and Law: A Convivium 3*, 1–29; see further Chap. 3, Sects. 3.2.–3.3.

to emerge,[21] advocating for targeted policies to revitalise the market.[22] The debate on the salience of securitisation has more recently intertwined with reflections on the progress of the Capital Markets Union (CMU), also thanks to the 2024 reports commissioned to Enrico Letta and Mario Draghi. They both underscored the untapped potential of securitisation for the growth and competitiveness of the EU market.[23]

Against this background, the present analysis aims to dispel the stigma and to bring an underexplored angle of securitisation to the fore. It particularly seeks to shed light on its potential to allocate investors' funds to sustainability-related goals and projects, as well as to bundle together loans which make a positive contribution to the sustainability agenda in turn.[24] These functions of securitisation also shift the narrative, placing the focus not on how securitisation transfers existing risks, but rather on how it contributes to *further* sustainability-related final goals. Therefore, building on and going beyond the above-mentioned contributions, the present analysis examines how law could intervene to prompt sustainability-related practices in securitisations.

Second, the book introduces a novel perspective to the burgeoning 'sustainable finance' scholarship by analysing the role of securitisation law. Financial, political economy and legal scholars have analysed the role of financial markets in furthering environmental, social, and governance goals from numerous angles, which cannot be replicated here due to space

[21] '*The broad policy implication of the paper is that the Securitisation of prime mortgages can function well*', BIS (n. 19), 6, 20–26.

[22] ECB, Bank of England, 'The case for a better functioning securitisation market' (2014) https://www.ecb.europa.eu/pub/pdf/other/ecb-boe_case_better_functioning_securitisation_marketen.pdf accessed 7 December 2025; more recently, see '*the available literature suggests that risk retention and higher prudential requirements have generally enhanced the resilience of securitisation markets*' and the broader analysis of the state of global securitisation markets following post-crisis reforms, Financial Stability Board, 'Evaluation of the Effects of the G20 Financial Regulatory Reforms on Securitisation – –Final report' (2025) pp. 36 ff. https://www.fsb.org/2025/01/evaluation-of-the-effects-of-the-g20-financial-regulatory-reforms-on-securitisation-final-report/accessed 5 December 2025.

[23] Within the EU context, Enrico Letta, 'Much more than a market' (2024) 32, 38, https://www.consilium.europa.eu/media/ny3j24sm/much-more-than-a-market-report-by-enrico-letta.pdf accessed 10 December 2025; Mario Draghi, 'The future of European Competitiveness – Part A: a competitive strategy for Europe' (2024) 64–65 https://commission.europa.eu/document/download/97e481fd-2dc3-412d-be4c-f152a8232961_en accessed 10 December 2025.

[24] See further Chap. 4, Sect. 4.3.

constraints.[25] Legal and political economy scholars have also increasingly delved into the legal techniques and optimal policy options to prompt market actors such as companies, banks and investors to implement sustainable practices. For instance, they have investigated the potential of disclosure[26] or discussed the role of the EU in the global sustainable finance landscape.[27] Others have analysed the legal framework[28] and clauses[29] for the issuance of 'sustainable' financial instruments. Another *leitmotif* has been the intersection of climate change policy goals with central banks' powers and the related boundaries.[30]

[25] For examples of financial scholarship, see, among others, various contributions in the 2022 special issue 'Sustainable Finance' Alex Edmans, Marcin Kacperczyk (eds), 'Sustainable Finance' (2022) 26 *Review of Finance* 6, https://doi.org/10.1093/rof/rfac069 accessed 10 December 2025. Other perspectives that have attracted considerable academic interest include the rise in investments driven by sustainability considerations, see Benjamin J Richardson, 'To Govern and Be Governed: The Governance Dimensions of SRI's Influence', in C Louche, T Hebb (eds), *Socially Responsible Investment in the twenty-first century: Does it Make a Difference for Society?*, Vol. 7 (Emerald Group Publishing Ltd 2014); on the pricing of 'sustainable' financial instruments, see, among others, Caroline Flammer, 'Corporate green bonds' (2021) 41 *Journal of Financial Economics* 2 499; Timo Busch, Peter Bruce-Clark et al, 'Impact investments: a call for (re) orientation' (2021) 1 SN Business & Economics 2; Willem Schamrade, 'Investing in the UN Sustainable Development Goals: Opportunities for Companies and Investors' (2017) 29 *Journal of Applied Corporate Finance 2*, 1.

[26] Tobias Hans Tröger, Sebastian Steuer, 'The Role of Disclosure in Green Finance' (2021) 8 *Journal of Financial regulation* 1. See more extensively in Chap. 1, Sect. 1.5.

[27] Linn Anker-Sørensen, Dirk A. Zetzsche, 'Regulating sustainable finance in the dark' (2022) 4 *European Business Organisation Law Review* 47; Stephen K, Park, 'Sustainable finance in global capital markets' (2023) in Iris Chiu, Iain MacNeil (eds.), *Research Handbook on Global Capital Markets Law (2023)* Edward Elgar Publishing 2025, 236 ff.

[28] Scott Breen, Catherine Campbell, 'Legal Considerations for a Skyrocketing Green Bond Market' (2016) 1 *Nat. Resources & Env't*; Stephen K Park, 'Investors as Regulators: Green Bonds and the Governance Challenges of the Sustainable Finance Revolution' (2018) 54 *Stanford Journal of International Law*.

[29] Federica Agostini, 'From 'Green Bond Principles' to 'Green Bond Clauses': Mitigating Greenwashing Through Contract Law' in Mads Andenas, Maren Heidemann (eds) *Quo vadis Commercial Contract?* (Springer 2023) 1 LCF Studies in Commercial and Financial Law 151; Mark C Weidemaier, Quinn Curtis, Mitu Gulati, 'Green Bonds, Empty Promises' (2023) 102 *North Carolina Law Review* 131; Agnieszka Smoleńska, European capitalisms in sustainability transition: the case of green bonds (2025) 33 *Journal of European Public Policy* 1572.

[30] For example, Javier Solana, 'The Power of the Eurosystem to Promote Environmental Protection' (2019) 30 *European Business Law Review* 4 547; Chiara Zilioli, Michael Ioannidis, 'Climate change and the mandate of the ECB: Potential and limits of monetary contribution to European green policies' (2022) 59 *Common Market Law Review 2* 363, 391 ff.

Despite the multifaceted sustainable finance scholarship, prior contributions have only tangentially explored the intersection between sustainability goals, securitisation and its legal framework or have done so only on a preliminary level. Petit and Schlosser have discussed the drivers, incentives and costs of relying on securitisation to further environmental goals. They have described such 'green' transactions as a subset of 'sustainable securitisations', exclusively contributing to environmental goals.[31] Gabor has expounded on the main challenges of utilising securitisation to promote economic and social growth across emerging markets.[32] *Cornut St-Pierre* and *Gareau-Lindsay* have analysed the documentation of four green securitisation deals to distil common elements in the structure and arrangements chosen by market participants.[33] Agliardi has developed theoretical models for banks to create 'green' portfolios.[34] Lastly, a few commentators have recently provided preliminary doctrinal analyses of selected aspects of the EU and UK framework for 'green securitisations'.[35]

Yet, the present work goes further by considering the potential of securitisation for all sustainability-related, not only environmental, goals. It also aims to provide a more comprehensive overview of the most relevant aspects of the law governing sustainability-related aspects of securitisations.[36]

[31] Christy Petit, Pierre Schlosser, 'Rationale, Potential and Pitfalls of Green Securitisation' (2020) RSCAS Working Paper 35 https://hdl.handle.net/1814/67018

[32] Daniela Gabor, 'Securitisation for Sustainability: Does it help achieve the Sustainable Development Goals?' (2019) Heinrich Boll Stiftung. https://us.boell.org/sites/default/files/gabor_finalised.pdf accessed 10 December 2025.

[33] Pascale Cornut St-Pierre, Jasmine Gareau-Lindsay, 'Green Securitisation, A Legal Structure Currently Unfit for Ecological Transition' (2026) 16 *Accounting, Economics, and Law: A Convivium 2 11.*

[34] Rossella Agliardi, 'Green Securitisation' (2022) 12 *Journal of Sustainable Finance and Investment* 1330.

[35] Julian Craughan, Andrew Carey et al', Talking about a revolution: making ESG Securitisations mainstream' (2022) 37 *Butterworths Journal of International Banking and Financial Law* 6; on a similar vein, Samuel Pinson, 'Revamping Green Securitisation Frameworks in the EU' (2024) 31 *University of Miami International and Comparative Law Review* 1, 202, only covering some of the EU initiatives discussed in Chap. 6, and not discussing other 'private law' aspects covered in Chaps. 4 and 5. With regards to the 'green auto' securitisation market, see also Carmelo Latino, Loriana Pelizzon, Max Riedel, 'How to green the European auto ABS market? A literature survey' (2025) 31 *European Financial Management* 628–631.

[36] On why such aspects of the law are strategic for sustainability purposes, see below, Sect. 1.4; Chap. 3, Sect. 3.3.

Third, the book also maps onto the broader scholarly discourse on the role of the law for financial markets. A large body of law and economics scholarship,[37] as well as a variety of financial-law scholars, have identified legislation as the optimal response to costs and 'market failures' arising across financial transactions, which could foster efficiency in turn.[38] In line with such theoretical insights, the present analysis will highlight the various shortcomings of the current status quo, where market forces are independently responsible for determining the legal obligations applicable to sustainable securitisations.[39] However, while previous accounts have predominantly focussed on the need for *state-designed* regulation to address inefficiencies, the book will juxtapose two levels of rules that have the potential to coordinate 'sustainable securitisations'. These encompass the complementary endeavour by financial-industry associations (what I will call 'private soft law') and by supranational authorities, such as EU institutions (what I will call 'public hard law').[40] Therefore, the book will highlight how not only state authorities, but also selected private actors, representing the interests of specific market segments, can foster consistent behaviours and practices across financial transactions, thereby acting in a rule-making capacity.

These contributions could also be relevant beyond the confines of academic discourse. They could support policymakers in the EU and globally in understanding how sustainable securitisations diverge from other sustainable financial instruments, such as 'green bonds', and from traditional securitisations, with a view to designing targeted measures. They may likewise inspire financial-industry associations, already deeply committed to standardising sustainable financial instruments, to devise further securitisation-specific initiatives. Lastly, they could raise awareness among companies, financial institutions and investors of the benefits of using securitisation for sustainability purposes. In doing so, they could encourage market participants to originate these transactions and to purchase the related securities.

[37] On the methodological value of the law-and-economics and law-and-finance scholarship in the present work, see below, Sect. 1.3.

[38] See, among others, the New institutionalist economics scholarship. Mary M Shirley, Claude Ménard, *Handbook of New Institutional Economics* (Springer 2008). See further Chap. 2, Sect. 2.2.2–2.2.3.

[39] See Chap. 4, Sect. 4.5. ff.

[40] See Chap. 2, Sect. 2.2.

1.3 Methodology

The book will provide a doctrinal analysis of securitisation and of its legal framework through the use of a wide range of primary and secondary sources, which overall allow for the 'rigorous analysis and creative synthesis of multiple doctrinal strands'.[41]

The theoretical framework, articulated in Chap. 1 and recurrently mentioned throughout the remaining chapters, draws upon selected theoretical inquiries into the nexus between law and finance. The insights from Legal Institutionalists and from *Pistor* point to the 'hybrid' nature of the law of financial markets, where market actors and state authorities concurrently design the legal obligations governing their financial transactions.[42] Similarly, the present analysis combines 'private law' (in Chaps. 4 and 5) and 'public law' (in Chap. 6). It also ties to selected accounts from law and economics scholarship, depicting state regulation as a response to the high transaction costs between private market actors and to 'market failures'.[43] While it is not the aim of this book to offer an analysis grounded in the 'Law and Economics' methodology, such academic scholarship frames the discussion. Further, it justifies the choice of a *legal* approach for investigating the intersection between securitisation and sustainability goals.

The doctrinal analysis has revolved around the existing legal framework as well as the 'social data'[44] on the subject. This has required reviewing primary sources, including EU treaties, EU legislation, selected national statutory provisions and judicial precedents on securitisation by EU courts and across EU countries.

Furthermore, the present work draws upon an extensive array of secondary sources, including textbooks, articles, policy papers and reports,

[41] On the concept and methodology of doctrinal legal research (DLT) see Bhat P Ishwara, 'Doctrinal Legal Research as a Means of Synthesizing Facts, Thoughts, and Legal Principles' in Bhat P Ishwara, *Ideas and Methods of Legal Research* (2020) Oxford Academic https://doi.org/10.1093/oso/9780199493098.003.0005 accessed 5 December 2025. In view of its legal methodology, the book does not offer a comprehensive account of the economic theory that supports securitisations. See, *inter alia*, Brent W. Ambrose, Michael LaCour-Little, Anthony B. Sanders, 'Does Regulatory Capital Arbitrage, Reputation, or Asymmetric Information Drive Securitisation?' (2025) 28 *Journal of Financial Services Research* 1 113; Peter M. DeMarzo, 'The Pooling and Tranching of Securities: A Model of Informed Intermediation' (2005) 18 *The Review of Financial Studies* 1, 1–35.

[42] See Chap. 2, Sect. 2.1, 2.2.

[43] As extensively described in Shirley, Ménard (n. 38). See further Chap. 2, Sect. 2.2.–2.3.

[44] Expression borrowed from Ishwara (n. 41) 156.

among others. These refine the findings and contextualise the analysis within the broader academic and policy debate.[45]

The analysis findings are also linked to the broader EU policy discourse, notably a number of proposed and ongoing initiatives, consultations and selected stakeholders' perspectives. These corroborate the normative recommendations concerning potential additional market-level initiatives and EU reforms, which conclude the analysis in Chaps. 4 and 5.

Moreover, the analysis has also drawn on 'social data', that is, current market practices for the design and management of 'traditional' securitisation programmes, other types of 'sustainable' financial instruments and 'sustainable securitisations' more specifically. In particular, Chap. 4 explores several examples of non-binding legal tools designed by financial-industry associations, such as principles and standard clauses, widely applied in the securitisation sector to draft the documentation for these operations.[46]

The analysis in Chaps. 3 and 4 also recurrently refers to examples of terms from previous 'sustainable securitisation' transactions, such as prospectuses, or to other documentation describing issuers' strategy on sustainability matters. It is important to emphasise that this book does not aim to provide a comprehensive empirical study of sustainability-related practices across all transactions, which would not have been feasible given the confidentiality of some of the agreements governing securitisations.[47] Instead, the selection of clauses and documentation was purposive, in order to ensure an objective assessment and focused on publicly available documentation. The references to such documentation throughout this work exclusively serve to exemplify and strengthen the arguments concerning the *potential* risks that may arise from market actors independently negotiating sustainability-related documentation on a case-by-case basis. Where available, the arguments are also corroborated by selected empirical studies.[48]

[45] On the function of secondary sources in doctrinal legal research, see 'Method of conducting DLR', ibid.

[46] On the rule-making function of standard clauses, see Chap. 5, Sect. 5.2.4.

[47] On the confidentiality of loan purchase agreements, see also Chap. 4, Sect. 4.5.1. On the exemptions to disclosure requirements ahead of the issuance of ABSs, see Chap. 3, Sect. 4.5.2., Chap. 6, Sect. 6.5.

[48] For example, see the comprehensive empirical study over green bond clauses by Weidemaier, Curtis, Gulati (n. 29), mentioned in Chap. 4, Sect. 4.5.2.

Overall, combining traditional primary and secondary sources with examples from transactional documentation aims to merge the in-depth theoretical inquiry of the law with the realities and practicalities of its application. In so doing, it offers a novel approach to the study of securitisation and its interaction with ESG goals that strengthens the original contribution of the analysis.

1.4 Scope of the Book

The present book provides an in-depth analysis of selected facets of securitisation *law*[49], which arguably act as strategic 'focal points' for sustainability-related purposes. In other words, these 'focal points' have been, and should arguably be, the target of sustainability-oriented legal strategies, since they could most effectively encourage market actors to integrate sustainability considerations in the management of their existing or prospective securitisation deals. Leveraging these areas also allows regulators and market actors to use securitisation deals as a tool to attain sustainability goals.[50]

Another methodological decision to narrow the scope of the present analysis was to focus on the binding legal framework under EU law. This choice stems from the EU's comprehensive approach to sustainable finance, unrivalled in the global policy arena. Yet, such an approach is also likely to serve as a blueprint for other policymakers, as they implement their own initiatives further down the line. Moreover, while every effort has been made to take stock of the dynamic EU legal developments, the book represents the state of the art as of December 2025.

1.5 Roadmap

The book is structured as follows. Chap. 2 provides the conceptual framework underpinning the entire analysis, by unpacking the nexus between law, sustainability objectives, and financial markets, with a specific focus on securitisation. It discusses how law advances sustainability goals from various perspectives, notably who introduces it, how binding it is and

[49] For example, Chap. 6, Sect. 6.5.3., mentions how macroprudential requirements and fiscal treatment are outside the scope of the analysis.

[50] On the criteria for the choice of 'focal points' see Chap. 2, Sect. 2.3.

how it operates, i.e. the functions it serves and the means by which it achieves such goals.

Chap. 3 shows how securitisation and its legal framework are tied to the sustainable finance discourse. It provides an overview of the functioning of securitisations and identifies key areas most amenable to further legal initiatives promoting environmental, social, and governance (ESG) goals. Under the lens of such 'focal points', the analysis dissect the law on sustainable securitisation in Chaps. 4, 5 and 6.

Chap. 4 delves into the relationship between sustainable finance and securitisation by providing an overview of how these two spheres have intertwined in market practice and in policy debate. It also turns to analyse the law that supports these transactions by investigating how market actors have independently set rules to qualify their securitisations as 'sustainable' (what I will term 'private hard law') and by assessing the adequacy of the current framework.

Chap. 5 discusses the extent to which selected market participants and notably financial-industry associations ('private soft law'), have contributed to the standardisation of sustainability-related elements of securitisations. It also makes recommendations on further legal initiatives.

Following a similar structure to Chap. 6, Chap. 7 explores the role of EU law in regulating sustainable securitisations. It specifically focuses on the effects of the current rules and of the ongoing legal initiatives in regulating sustainability-related elements of securitisations, as well as providing normative considerations on potential legal reforms.

The concluding chapter concludes the analysis by summarising the key findings and normative considerations concerning future initiatives. It also identifies promising additional research questions as well as current trends that may signal how the sustainable securitisation market may further develop.

References

Acree J, 'ESG: The Popular Corporate Buzzword and Why Leaders Should Care' (Forbes, 15 March 2023) https://www.forbes.com/councils/forbesbusinesscouncil/2023/03/15/esg-the-popular-corporate-buzzword-and-why-leaders-should-care/ accessed 3 June 2026.

AFME, 'ESG Finance Report – European Sustainable Finance – Q3 2025' (2025). https://www.afme.eu/publications/data-research/afme-esg-finance-report-q3-2025/accessed 3 June 2026.

AFME, 'ESG Finance Report – European Sustainable Finance – Q4 2024' (2024). https://www.afme.eu/Portals/0/DispatchFeaturedImages/ESG%20Finance%20Report%204Q24.pdf accessed 3 June 2026.

Agliardi R, 'Green Securitisation' (2022) 12 *Journal of Sustainable Finance and Investment* 1330.

Agostini F, 'From 'Green Bond Principles' to 'Green Bond Clauses': Mitigating Greenwashing Through Contract Law' in Andenas M and Heidemann M (eds), *Quo vadis Commercial Contract?* (Springer 2023) 151.

Ambrose BW, LaCour-Little M and Sanders AB, 'Does Regulatory Capital Arbitrage, Reputation, or Asymmetric Information Drive Securitisation?' (2005) 28 *Journal of Financial Services Research* 1 113.

Anker-Sørensen L and Zetzsche DA, 'Regulating Sustainable Finance in the Dark' (2022) 4 *European Business Organisation Law Review* 47.

Armour J, Awrey D, Davies P and others, *Principles of Financial Regulation* (OUP2016)

Bavoso V, 'High Quality Securitisation and EU Capital Markets Union: Is It Possible?' (2017) 7 *Accounting, Economics and Law: A Convivium* 3.

Bavoso V, 'Legal Issues on Financial Innovation and Structured Finance' (2011) 34 *Company Lawyer* 5.

Bank for International Settlements, 'Securitisation Is Not That Evil After All' (2011) 341 BIS Working Paper https://www.bis.org/publ/work341.pdf accessed 3 June 2026.

Breen S and Campbell C, 'Legal Considerations for a Skyrocketing Green Bond Market' (2016) 31 *Natural Resources and Environment 3 16, 18-20.*

Busch T, Bruce-Clark P and others, 'Impact Investments: A Call for (Re)orientation' (2021) 33 *SN Business and Economics 1 32.*

Cornut St-Pierre P and Gareau-Lindsay J, 'Green Securitisation: A Legal Structure Currently Unfit for Ecological Transition' (2026) 16 *Accounting, Economics and Law: A Convivium 2 211.*

Craughan J, Carey A and others, 'Talking About a Revolution: Making ESG Securitisations Mainstream' (2022) 37 *Butterworths Journal of International Banking and Financial Law*6

DeMarzo PM, 'The Pooling and Tranching of Securities: A Model of Informed Intermediation' (2005) 18 *Review of Financial Studies*1

European Commission, 'Overview of Sustainable Finance'https://finance.ec.europa.eu/sustainable-finance/overview-sustainable-finance_en accessed 3 June 2026.

Dimmelmeier A, 'Sustainable Finance as a Contested Concept: Tracing the Evolution of Five Frames between 1998 and 2018' (2021) 13 *Journal of Sustainable Finance and Investment* 1600

Draghi M, 'The Future of European Competitiveness – Part A: A Competitive Strategy for Europe' (2024) https://commission.europa.eu/document/download/97e481fd-2dc3-412d-be4c-f152a8232961_en accessed 3 June 2026.

European Central Bank and Bank of England, 'The Case for a Better Functioning Securitisation Market' (2014) https://www.ecb.europa.eu/pub/pdf/other/ecb-boe_case_better_functioning_securitisation_marketen.pdf accessed 3 June 2026.

EdmansA and KacperczykM (eds), 'Sustainable Finance' (2022) 26 *Review of Finance* 6

Esser IM and MacNeil I, 'From a Financial to an Entity Model of ESG' (2022) 23 *European Business Organisation Law Review* 9

European Commission, 'Action Plan: Financing Sustainable Growth' (2018) 097.

Financial Stability Board, 'Evaluation of the Effects of the G20 Financial Regulatory Reforms on Securitisation' (2025) https://www.fsb.org/2025/01/evaluation-of-the-effects-of-the-g20-financial-regulatory-reforms-on-securitisation-final-report/ accessed 3 June 2026.

Flammer C, 'Corporate Green Bonds' (2021) 41 *Journal of Financial Economics* 499

Gabor D, 'Securitisation for Sustainability: Does It Help Achieve the Sustainable Development Goals?' (2019) Heinrich Böll Stiftung https://us.boell.org/sites/default/files/gabor_finalised.pdf accessed 3 June 2026.

Hill CA, 'Securitisation: A Low-Cost Sweetener for Lemons' (1997) 10 *Journal of Applied Corporate Finance* 64

Ishwara BP, 'Doctrinal Legal Research as a Means of Synthesizing Facts, Thoughts, and Legal Principles' in IshwaraBP, *Ideas and Methods of Legal Research* (OUP2020)

Latino C, Pelizzon L and Riedel M, 'How to Green the European Auto ABS Market? A Literature Survey' (2025) 31 *European Financial Management* 628

Letta E, 'Much More Than a Market' (2024) https://www.consilium.europa.eu/media/ny3j24sm/much-more-than-a-market-report-by-enrico-letta.pdf accessed 3 June 2026.

Mian A and Sufi A, *House of Debt* (University of Chicago Press 2014)

Park SK, 'Investors as Regulators: Green Bonds and the Governance Challenges of the Sustainable Finance Revolution' (2018) 54 *Stanford Journal of International Law 1-47*

Park SK, 'Sustainable Finance in Global Capital Markets' in Chiu I and MacNeil I (eds), *Research Handbook on Global Capital Markets Law* (Edward Elgar2023)

Petit C and Schlosser P, 'Rationale, Potential and Pitfalls of Green Securitisation' (2020) RSCAS Working Paper 35 https://hdl.handle.net/1814/67018 accessed 3 June 2026.

Pinson S, 'Revamping Green Securitisation Frameworks in the EU' (2024) 31 *University of Miami International and Comparative Law Review* 202

Pinto J and Alves P, 'The Economics of Securitisation: Evidence from the European Markets' (2016) 13 *Investment Management and Financial Innovations* 112

Ramos Muñoz D and Ingram K, *The Law of Transnational Securitisation* (OUP 2010)
Richardson BJ, 'To Govern and Be Governed: The Governance Dimensions of SRI's Influence' in Louche C, Hebb T, *Socially Responsible Investment in the twenty-first century: does it make a difference for society?* (Emerald 2014)
Shirley MM and Ménard C (eds), *Handbook of New Institutional Economics* (Springer2008)
Schramade W, 'Investing in the UN Sustainable Development Goals: Opportunities for Companies and Investors' (2017) 29 *Journal of Applied Corporate Finance*1
Schnabel I, 'Looking Through Higher Energy Prices? Monetary Policy and the Green Transition' (8 January 2022) https://www.ecb.europa.eu/press/key/date/2022/html/ecb.sp220108~0425a24eb7.en.html accessed 3 June 2026.
Schwarcz SL, 'The Alchemy of Asset Securitisation' (1994) 1 *Stanford Law Review* 133
Smoleńska A, 'European Capitalisms in Sustainability Transition: The Case of Green Bonds' (2025) 33 *Journal of European Public Policy 6 1572-1598.*
Solana J, 'The Power of the Eurosystem to Promote Environmental Protection' (2019) 30 *European Business Law Review* 547
Swiss Sustainable Finance, 'Glossary' (2023) https://www.sustainablefinance.ch/en/resources/what-sustainable-finance/glossary.html accessed 3 June 2026.
Tröger TH and Steuer S, 'The Role of Disclosure in Green Finance' (2021) 8 *Journal of Financial Regulation*1
United Nations Framework Convention on Climate Change, 'Paris Agreement' (2015) https://unfccc.int/files/essential_background/convention/application/pdf/english_paris_agreement.pdf accessed 3 June 2026.
United Nations, 'Transforming Our World: The 2030 Agenda for Sustainable Development' (2015) UN Doc A/RES/70/1 https://undocs.org/en/A/RES/70/1 accessed 3 June 2026.
Weidemaier MC, Curtis Q and Gulati M, 'Green Bonds, Empty Promises' (2023) 102 *North Carolina Law Review* 131
Zilioli C and Ioannidis M, 'Climate Change and the Mandate of the ECB: Potential and Limits of Monetary Contribution to European Green Policies' (2022) 59*Common Market Law Review* 363

CHAPTER 2

Law as a Vehicle for Sustainability Objectives in Securitisations

2.1 Introduction

Analysing how law can advance objectives pertaining to environmental, social, and governance issues across securitisation transactions requires an understanding of the intricate interplay between law, sustainability goals, financial markets, and transactions. Accordingly, this chapter aims to introduce the conceptual framework underpinning the book's analysis and to map it onto the broader theoretical inquiries into the relationship between law and finance. While the intersection of these two areas has extensively engaged legal and law and economics scholars,[1] the analysis does not aim to replicate the extremely broad and multifaceted understanding of this relationship. Instead, it seeks to bring in selected angles of the debate that specifically shed light on how law could incorporate sustainability considerations in the design of securitisations.

The following chapter will unpack the three core criteria that steer the analysis of law in this book, namely the rule-making entities, their prescriptiveness and their functions. These criteria correspond to three core questions: *who* issues rules, *how prescriptive* they are, and *how* law operates to introduce environmental, social, and governance considerations within the design and management of securitisation operations.

[1] For the meaning of 'Law-and-Economics' for the purposes of the present work, see s. 2.2.

F. Agostini, *The Law of Sustainable Securitisations*, EBI Studies in Banking and Capital Markets Law,
https://doi.org/10.1007/978-3-032-25046-9_2

Throughout the subsequent chapters, I will examine how law currently shapes sustainability-related elements of securitisation through the lens of these three criteria. These will also inform 'normative' considerations proposed in the book.[2] In other words, I will draw from these criteria to reflect on potential legal actions (notably legal reforms) to strengthen the consideration of sustainability factors across securitisations.

Sect. 2 will present the matrix that drives the design of the conceptual framework. The following sections will expound on the various components that it entails. Sect. 3 will describe the multiple actors designing the rules which set sustainability objectives for securitisation. Sect. 4 will differentiate them based on their normative strength, that is, the extent to which they bind market participants. Sect. 5 will classify the different functions of law in contributing to sustainability objectives. It will also describe four selected strategies that arguably hold the greatest potential for furthering sustainability objectives across securitisation transactions. While rule-makers have already partially embraced such strategies in the design of rules for securitisation, there remains scope for additional legal initiatives based on similar approaches.

2.2 The Conceptual Framework

The starting point for the analysis in this book is the inextricable nexus between law and finance. The theory that can most persuasively describe this connection is what Pistor has described as the 'Legal Theory of Finance'.[3] In Pistor's words, 'Finance is not only regulated by law, it is *made in law*'.[4] Rules are endogenous to capital market transactions since law legitimises financial activities and articulates the rights over financial assets.[5]

While predominantly drawing upon this theory, the analysis will also engage with other theoretical endeavours to illustrate how law affects and could affect sustainable securitisation. It will specifically examine the role of law from three angles, summarised in the figure 2.1 below.

[2] See s. 2.6.

[3] Katharina Pistor, 'A Legal Theory of Finance' (2013) 41 *Journal of Comparative Economics* 2 315.

[4] (Emphasis added) K. Pistor, 'Regulating Financial Markets – An LTF Perspective' in Avgouleas E, Donald DC, *The political economy of financial regulation* (Cambridge University Press 2019) 19, 32.

[5] K. Pistor, 'Legal Theory of Finance' (n. 3) 321, 324.

Rule-Makers *Prescriptiveness*	*Market Actors*	*State and Supranational Authorities*
Binding	Private hard law (Chap. 3)	Public hard law (Chap. 5)
Non-binding	Private soft law (Chap. 4)	Public soft law (not covered)

Fig. 2.1 Scope of the Legal Analysis

The following sections will explain the three criteria that form this conceptual framework and justify why they are relevant for the law of sustainable securitisation. They will also situate the criteria within broader theoretical inquiries into the role of law for finance.

2.3 The Rule-Making Entities: Private Law and Public Law

The first criterion that informs my analysis relates to the nature of the entities involved in the design of rules for 'sustainable' securitisations. Specifically, the analysis draws the line between 'private law' (in Chaps. 4 and 5) and 'public law' (in Chap. 6), following the divide driving the traditional understanding of the law.[6] The distinction primarily serves an explanatory purpose and aims to make the analysis comprehensive. Simultaneously, it is also essential to acknowledge the limitations of viewing the modern regulatory environment and its various nuances under the lens of this demarcation.[7]

The question of who should issue rules and why is recurrent across a wide number of theoretical endeavours analysing law and legal institutions under the lens of economic principles (what I will define as 'law-and-economics' scholarship[8]). Among theorists and theories of regulation,

[6] On the distinction between private and public law, see Ralf Michaels, Nils Jansen, 'Private Law Beyond the State? Europeanisation, Globalisation, Privatisation' (2007) 54 *American Journal of Comparative Law* 843, 846–853; Harlow C, 'Public' and 'Private' Law: Definition without Distinction' (1980) 6 *The Modern Law Review* 951, 958.

[7] Harlow, ibid., 257; Peter Zumbansen, Osgoode Hall, 'The Law of Society: Governance Through Contract' (2007) 14 *Indiana Journal of Global Legal Studies* 2 191, 214; Lauren B Edelman, Mark C Suchman, 'The Legal Environments of Organisations' (1997) 23 *Annual Review of Sociology* 479, 480; Darren Sinclair, 'Self-Regulation versus Command and Control? Beyond False Dichotomies' (1997) 19 *Law and Policy* 529.

[8] i.e. the analyses which primarily draw from microeconomics and welfare economics, as defined by Nicholas Mercuro, Steven Medema, *Economics and the Law: From Posner to*

my work will draw closer attention to modern Legal Institutionalists and especially to Pistor.[9] They have understood financial markets as fundamentally hybrid,[10] i.e. neither fully private nor fully public. In other words, private contracts and state laws, market participants and the State, all play complementary roles in the design of market rules. This interpretation is also coherent with a large body of scholarship discussing the pluralistic architecture of international financial law.[11]

2.3.1 *Private Law*

Throughout the present work, 'private law' will capture the set of rules created at the market level by market participants. These rules operate *horizontally* between parties owing to the overlap between the entities making the rules and their respective addressees. Academic commentators have also described specific market-level rule-making processes in the financial context as 'self-regulation' or 'meta-regulation'. However, the terms have come to capture a broad spectrum of initiatives, sometimes

Post-Modernism and Beyond (2nd edn, Princeton University Press 2006) 1. Some have traced the origins of the Law-and-Economics debate back to the reflections on the impact of economic theories on state policies by American scholars before the Progressive era as well as to the work of Institutionalists or Legal Realists (the 'first great law and economics movement'), see Barbara H Fried, 'Preface' in *The Progressive Assault on Laissez Faire: Robert Hale and the First Law and Economics Movement* (1998) Vol. 56, vii, viii.; Herbert Hovenkamp, 'The First Great Law & Economics Movement' (1990) 42 *Stanford Law Review* 993, 999–1004. Others adopt a narrower approach and associate the term with Coase's work and with the Chicago School of law and economics scholarship, developed in 1970s (the 'second great law and economics movement'), ibid., 994; Jedediah Britton-Purdy and others, 'Building a Law-and-Political-Economy Framework: Beyond the twentieth-century Synthesis' (2020) 129 *Yale Law Journal* 1784, 1815.

[9] Legal Institutionalism is the line of thinking of those who emphasize the role of law along with state law and institutions for the modern capitalist economy. For this and other meanings of Legal Institutionalism, see Simon Deakin et al., 'Legal Institutionalism: Capitalism and the Constitutive Role of Law' (2017) 45 *Journal of Comparative Economics* 188–189.

[10] Pistor, 'Law in Finance' (2013) 41 *Journal of Comparative Economics* 2 311, 312. Pistor, 'Legal Theory' (n. 3) 322–323. Similarly, see also ibid., 198.

[11] Joanna Benjamin, David Rouch, 'The International Financial Markets as a Source of Global Law: The Privatisation of Rule-Making' (2008) 2 *Law & Fin Mkt Rev* 78; Joanne P Braithwaite, 'Standard Form Contracts as Transnational Law: Evidence from the Derivatives Markets' (2012) 75 *Modern Law Review* 5 779; Caroline Bradley, 'Private International Law-Making for the Financial Markets' (2005) *Fordham International law Journal* 127.

involving some level of state intervention and oversight.[12] For clarity, I will prioritise the use of the term 'private law'.

In the context of securitisation, and specifically 'sustainable securitisation', 'private law' primarily encompasses contractual and, in part, proprietary arrangements. The actors with a rule-making role are primarily the parties (i.e. originating entities, special purpose vehicles and investors) when designing the clauses and agreements governing these operations. Chap. 3, which will unfold the core architecture of these transactions, will further illustrate how the provisions of the documentation determine the allocation of proprietary rights over the portfolio or the securities and give rise to binding obligations for the parties.[13]

The scrutiny of private law will not only consider the role of parties at the individual deal level (in Chap. 4), but also the involvement of additional, more prominent, entities, representing a specific market segment (in Chap. 5). This angle of the project draws upon the wide-ranging scholarship on the multi-layered and pluralistic nature of the legal framework governing financial markets.[14] A persuasive model that will be used describes the law-making process as triangular due to the involvement of legal 'intermediaries'. These intermediaries stand between state or international rule-makers and private actors[15] and include, by way of example,

[12] For a systematic overview of various accounts of 'self-regulation', also involving a role for state authorities, see J Black, 'Decentreing Regulation: Understanding the Role of Regulation and Self-Regulation in a 'Post- Regulatory' World' (2001) 54 Current Legal Problems 103, 115–119. Sinclair (n. 7) 144, 146 and the sources cited therein. On 'meta-regulation', where state authorities delegate some level of rule-making power while retaining oversight and control, see Julia Black, 'Regulatory Styles and Supervisory Strategies' in Niamh Moloney, Eilís Ferran, Jennifer Payne (eds.) *The Oxford handbook of financial regulation* (2015 Oxford University Press); Robert Baldwin, Martin Cave, Martin Lodge, 'Self-Regulation, Meta-Regulation, and Regulatory Networks', in Robert Baldwin, Martin Cave (eds.) *Understanding Regulation: Theory, Strategy, and Practice*, Vol. 15 (2011) 146–148. Cary Coglianese, Evan Mendelson, 'Meta-Regulation and Self-Regulation' in *The Oxford Handbook of Regulation* (2010) 150. On the contrary, others have used to describe market-driven regulatory dynamics. See, inter alia, Rotem Medzini, David Levi-Faur, 'Self-Governance via Intermediaries: Credibility in Three Different Modes of Governance' (2023) 25 *Journal of Comparative Policy Analysis: Research and Practice*, 3, 323–345.

[13] See Chap. 3, s. 3.3.

[14] Eg, among others Black, 'Regulatory Styles' (n. 12); Benjamin, Rouch (n. 11); Bradley (n. 11) 131–135. On regulation addressing businesses, not only related to finance, see Kenneth Abbott, Duncan Snidal, 'Strengthening International Regulation Through Transnational New Governance' (2009) 42 *Vanderbilt Journal of Transnational Law* 501, 512–519.

[15] The authors name the components of the model 'regulators', 'regulatory intermediaries' and 'targets' or 'rule takers'. These terms could be misleading for the purposes of the present analysis

professional associations, transnational networks or non-governmental organisations (NGOs).[16] They contribute to the design, monitoring, and enforcement of the law.[17] In turn, their prestige and expertise also create a feedback effect, which affects the production of further state and international rules.[18] As others have already borrowed the model to describe regulatory dynamics in the financial context[19], it is reasonable to transpose it to sustainable securitisation deals. Accordingly, Chap. 5 will identify the 'intermediaries', such as financial industry associations, that act as rule-makers for sustainable securitisations. The difference between the 'private law' by market participants (analysed in Chap. 4) and by intermediaries (in Chap. 5) lies in the prescriptiveness of their rules, for the reasons outlined below.[20]

When delving into private law, the present book will engage with the theoretical accounts of those who have viewed contracts as legal tools, operating like state-designed rules. For instance, *Pistor*, who depicts the financial system as a web of contracts[21], underlined the proactive role of market participants in defining binding commitments and enforcing contractual breaches.[22]

Relatedly, contract theorists in economics and sociologists have highlighted the rule-making power of contractual clauses.[23] In this light, clauses are instrumental to the long-term planning between parties and to the *ex ante* protection against mutual risks.[24] They are also inextricably linked

due to the concurrent role of market actors in setting and designing private law rules. See Kenneth W Abbott, David Levi-Faur, Duncan Snidal, 'Theorizing Regulatory Intermediaries: The Rit Model' (2017) 670 *Annals of the American Academy of Political and Social Science* 14.

[16] Luc Brès, Sébastien Mena, Marie Laure Salles-Djelic, 'Exploring the Formal and Informal Roles of Regulatory Intermediaries in Transnational Multistakeholder Regulation' (2019) 13 *Regulation and Governance* 127, 128; Abbott, Levi-Faur, Snidal, ibid., 15, 29.

[17] Which would not necessarily depend on a formal delegation by the state, ibid., 21–23.

[18] Ibid., 28. See also Arno Kourula, Markus Paukku, Andrew Peterman, Mikko Koria, 'Intermediary Roles in Regulatory Programs: Towards a Role-Based Framework' (2019) 13 *Regulation and Governance* 140, 140–142; Brès, Mena, Salles-Djelic (n. 16) 127–128.

[19] Abbott, Levi-Faur, Snidal (n. 15) 15–17.

[20] See below, s. 2.3.

[21] Pistor, 'Legal Theory' (n3).

[22] Ibid., 315, 321, 324.

[23] See, inter alia, Oliver Hart, John Moore, 'Contracts as Reference Points' (2008) CXXIII Quarterly Journal of economics 1, 8. Iain R MacNeil, 'Contracts: adjustment of long-term economic relations under classical, neoclassical, and relational contract law' (1978) 72 *Northwestern University Law Review* 854.

[24] MacNeil, ibid., 854.

to property law by allowing parties to attribute mutual entitlements over assets.[25] Yet, scholars have also acknowledged the inherent incompleteness and unavoidable costs of private bargaining. These shortcomings justify the need for further legal initiatives to standardise market transactions.[26]

The accounts from Chicago School and new-institutionalist economics (*hereinafter*: NIE) scholars[27] will also lie in the background. They have brought the merits and demerits of market transactions to the fore by reflecting on the extent to which privately designed arrangements would be preferable to the law shaped by public institutions.[28] Similarly, Chaps. 4 and 5 will reflect on the costs and 'failures' emerging when market participants independently set their rules on the sustainability objectives of securitisation transactions.

2.3.2 Public Law

On the other hand, the present book will offer an analysis of 'public law' as the manifestation of the *vertical* coercive power of the state towards private citizens.[29] It gives rise to rules promulgated, adjudicated and enforced at the state or international level. I will prioritise the word 'public law' over 'public regulation' due to its broader meaning (also encompassing constitutional provisions[30]) and to the

[25] Hart, Moore (n. 23) 8.

[26] Ian Ayres, Robert Gertner, 'Filling Gaps in Incomplete Contracts: An Economic Theory of Default Rules' (1989) 99 *The Yale Law Journal* 87, 87; Michael C Jensen, William H Meckling, 'Theory of the firm: managerial behaviour, agency costs and ownership structure' (1976) 3 *Journal of Financial Economics* 305, 308–323; Michael Klausner, 'Corporations, Corporate Law, and Networks of Contracts' (1995) 81 *Virginia Law review* 757. For an overview of the theories, see Mercuro, Medema (n. 8) 138–144.

[27] For an analysis of Chicago school and NIE thinking, see Mercuro, Medema, ibid., 94–156, 241–284.

[28] Ronald H Coase, 'The Problem of Social Cost' (1960) 3 Journal of Law and Economics 1, 17–18; on different approaches to the efficiency of private contracts by NIE scholars, see Rudolf Richter, 'The Role of Law in the New Institutional Economics in Comparison With the Economic Analysis of Law' (2015) 26 *Washington University Journal of Law and Policy* 77, 25–30, Mercuro and Medema, ibid., 127.

[29] Harlow (n. 6) 258; Nils Jansen, Ralf Michaels, 'Beyond the State? Rethinking Private Law: Introduction to the Issue' (2019) 56 The American Journal of Comparative Law 527, 348–349.

[30] The term 'public law' has, however, a broader meaning, also encompassing constitutional provisions, Javier Solana, 'The Roles of Law in Finance' (2022) Working Paper, on file with the author, 5.

contentious meanings that commentators have given to the notion of 'regulation'.[31]

Throughout the present analysis, public law[32] encompasses company law and capital markets law rules governing the entities, assets, and investments connected to sustainable securitisation, as primarily shaped by EU law.[33]

The analysis of the demarcation between private and public law will build on the accounts of the Chicago School and NIE scholars, particularly regarding the role of state rules in addressing the high costs of private transactions or situations in which market actors fail to independently achieve efficient outcomes ('market failures').[34] While the term has acquired divergent meanings in the law and economics literature, 'transaction costs' generally capture the costs arising from market exchanges. These could arise from the conclusion of contracts or from situations when parties seek to establish and protect their property rights.[35] According to NIE scholars, institutional frameworks at market level (designed by actors such as firms, i.e. at the 'private' level, following the terminology of the present work) as well as institutional and enforcement frameworks at the state level (i.e. at the 'public' level) become necessary when such costs are significant, in order to facilitate market exchanges.[36] It is possible to transpose such conclusions to the context of the present analysis. Due to the complexity of securitisation transactions, their multi-layered documentation and the multitude of parties involved, they inherently give rise to multiple transaction costs, ranging from the drafting of the documentation, to the

[31] For instance, some have seen 'law' as one of the various manifestations of 'regulation', even though not all laws have a regulatory function, see Parker Christine and others (eds), 'Introduction' in Christine Parker and others (eds) *Regulating Law* (Oxford, 2004; Oxford Academic 22 Mar. 2012) 2–5 and the other contributions within the volume. There is also considerable scholarship on the notion of 'self-regulation' as a product of market participants, see Chap. 4, Sect. 4.2.

[32] Also in line with K Pistor, 'LTF Perspective' (n. 4) 20.

[33] For the reasons behind the choice of EU law, see Chap. 5.

[34] Coase (n. 28) 17–18; Marie M Shirley, Claude Ménard, *Handbook of New Institutional Economics* (Palgrave 2008) 43, 44, 209.

[35] Shirley, Ménard, ibid., 42, 43, 45; Allen compellingly highlighted the complementarities but also divergences between the two definitions, Douglas W. Allen, 'Transaction Costs' in Boudewijn Bouckaert, Gerrit De Geest et al (eds), *Encyclopedia of Law and Economics* (Elgar 2000) Vol. I, 901–902; 895–899.

[36] Mercuro, Medema (n. 8) 255–257; Ménard, Shirley, ibid., 42, 43, 45. The NIE theory builds on Coase's theorem, according to which private transactions are only efficient in the absence of transaction costs, ibid., 113–119.

monitoring and potential enforcement of the securities.[37] As the analysis in Chap. 3 will highlight[38], such costs are amplified if market actors also seek to achieve sustainability goals through these transactions. Building on these and other theoretical insights, I will map the sources of transaction costs potentially arising across 'sustainable securitisations'. This mapping exercise will be the starting point for discussing (in Chaps. 5 and 6) how standardised tools designed at market level, as well as EU legislation, could reduce these costs.

The notion of 'market failures' will also be in the background of the analysis of 'public law'. The latter includes, among others, 'information asymmetry' between parties when concluding transactions or 'negative externalities', i.e. the costs that certain actions engender, not fully reflected in market prices and not entirely borne by the respective actors.[39] In the specific context of financial markets, one example of negative externality is 'systemic risk'. This may derive from market failures in certain components or transactions, which may have a domino effect on other closely connected parts of the financial system.[40]

All in all, the traditional assumption has been that 'public law' can lower transaction costs and address market failures, promoting economic efficiency.[41] Similarly, Chaps. 5 and 6 will analyse how certain market actors, and EU law, have responded to the costs and failures of privately designed rules on 'sustainable securitisation'. They will discuss to what extent recent sustainability-related initiatives align or deviate from the traditional discourse.

[37] On information, negotiation, monitoring and enforcement costs in general, see Allen (n. 35) 899–906; Ayres, Gertner (n. 26) 92–93. Jensen, Meckling (n. 26) 338–342.

[38] For a mapping exercise of transaction costs, see Chap. 3, Sect. 3.4.6.

[39] John Armour, Dan Awrey, Paul Davies et al, *Principles of Financial Regulation* (Oxford University Press 2016) 52, 57.

[40] Steven L Schwarcz, 'Regulating Financial Change: A Functional Approach' (2016) 100 *Minnesota Law Review* 1441, 1446.

[41] On the relevance of efficiency for Chicago School and NIE scholars, see Mercuro, Medema (n. 8) 105–107; on the interplay between market failures, financial regulation and efficiency, Schwarcz, ibid., 1444–1446. Armour et al (n. 15) 51, 53, 61. On 'efficiency' as a normative criterion for financial regulation, see also Javier Solana, 'Market Efficiency as a Directive Principle of EU Monetary Policy' in Anna Chadwick, Eleonora Lozano-Rodríguez, Andrés Palacios-Lleras, Javier Solana (eds.), *Markets, Constitutions, and Inequality* (2022) Routledge, 6, 13, 14, 16.

2.4 Prescriptiveness: Hard Law and Soft Law

The other fundamental criterion underlying the present analysis relates to the normative strength of rules. Following another well-explored demarcation, I will differentiate between the 'hard law' and 'soft law' of sustainable securitisation. Much like the preceding juxtaposition of private law and public law, this distinction will primarily serve to provide a comprehensive explanation of the framework, while coming with several limitations in practice.[42]

Commentators have provided disparate definitions of 'hard law', focusing on factors like the precision of rules, the entities entrusted with enforcement powers,[43] or the design of rules[44]. For the purposes of my analysis, the demarcation will rely on the types of sanctions for non-compliance and the type of entity enforcing them. In other words, 'hard law' refers to legally binding rules, backed by formal sanctions imposed by state and international authorities.[45]

By contrast, 'soft law' encompasses a broad spectrum of initiatives, including standards, principles, codes,[46] which lack binding force. Non-compliance may still give rise to remedies or informal sanctions, but these do not involve formal enforcement by state or interinstitutional authorities, such as EU supervisory authorities.[47] The use of soft law to regulate financial markets has been pervasive, especially in the aftermath of the 2007 global financial crisis, culminating in a multitude of initiatives at the market, state or supranational levels.[48] in the specific context of

[42] '*Any definition is ultimately arbitrary at the margins*', Kenneth W Abbott and others, 'The Concept of Legalisation' (2018) 54 *International Organisation* 401, 401.

[43] Ibid., 401, 404, Kenneth W Abbott, Duncan Snidal, 'Hard and Soft Law in International Governance' (2000) 54 International Organisation 421, 421.

[44] For instance, some have depicted soft law as the product of non-state institutions (what I define 'Private law'). John J Kirton, Michael J Trebilcock, 'Introduction' and the other contributions within Kirton, Trebilcock (eds) *Hard Choices, Soft Law: Voluntary Standards in Global Trade, Environment and Social Governance* (Routledge 2016) 10.

[45] One of the distinctive criteria advanced by Abbott, Snidal (n. 43) 401.

[46] For the proposed classification of the types of soft law instruments for sustainable securitisation, see Chap. 5, Sect. 5.2.3.

[47] For a systematic analysis of soft law disciplining and enforcement mechanisms within the financial regulation arena, see Christopher Brummer, *Soft Law and the Global Financial System – -Rule-Making in the twenty-first century* (2nd ed., Cambridge University Press, 2020) 119–182.

[48] There is copious scholarly analysis on the role of soft law in the financial context, see, among others, ibid.; Roberta S Karmel, Claire R Kelly, 'The Hardening of Soft Law in Securities Regulation' (2009) 34 *Brooklyn Journal of International Law* 883.

securitisation and sustainable securitisations, hard and soft rules play a complementary role.

2.4.1 Interplay of the Criteria in Practice: 'Private' and 'Public' Hard Law

Following the previously mentioned hybridity of the financial markets, state or international authorities, as well as market actors, have concurrent powers in the design of hard law and soft law rules for sustainable securitisations.

On the 'private law' side, market actors can create legally binding commitments linked to sustainability objectives throughout the contractual agreements and other documentation for sustainable securitisations. If the commitments are not substantiated, market actors can rely on enforcement mechanisms, including litigation. The analysis of this 'private hard law' will be at the core of Chap. 4.[49]

Concurrently, EU initiatives have culminated, or should culminate, in rules for the regulation of 'sustainable securitisations', also introducing enforcement mechanisms for which EU and national authorities are responsible. The investigation of 'public hard law' will be the focus of Chap. 5.

2.4.2 Interplay of the Criteria in Practice (II): 'Private Soft Law'

As far as 'soft law' is concerned, Chap. 4 will explore how certain 'intermediaries'[50], specifically financial industry associations, have introduced various non-binding legal tools, ranging from principles to model contractual clauses. These initiatives are defined here as 'private soft law'. Parties of sustainable securitisation deals *may* decide to apply them throughout their operations or may deviate from them without bearing any legal consequences or risks of formal enforcement by the state. While a more detailed account of the notion of 'private soft law' will be provided later, one example may facilitate the preliminary understanding of the concept: investors in sustainable securitisations willing to show their commitment to sustainability objectives may adopt the UN Principles for Responsible Investment (UNPRI).[51] Not fully complying with the Principles may prevent investors

[49] See further Chap. 4, Sect. 4.2.

[50] See above Sect. 2.2.1.

[51] UNPRI, 'What are the principles for responsible investment?' (2026) https://www.unpri.org/about-us/what-are-the-principles-for-responsible-investment accessed 5 June 2026.

from becoming formal members of the UNPRI network, but does not expose them to formal state sanctions.[52]

It is important to note that 'private soft law' (in Chap. 5) and 'public hard law' (in Chap. 6), while designed by distinct rule-making entities and differing in normative strength, arguably share a common 'coordination' rationale. In other words, they both try to address the inconsistencies and shortcomings of legal obligations on sustainability matters designed by market participants, albeit from different standpoints. Therefore, I will define 'standardisation' as the end goal of 'private soft law' initiatives and 'harmonisation' as the end goal of 'public hard law' initiatives in order to underline their differences but also their common root.[53]

One may wonder whether there is room to coordinate sustainable securitisations at the 'public' level through 'soft law' too. While such initiatives are in principle possible, they are out of the scope of the present analysis, as also highlighted in Chap. 6.[54]

2.5 The Functions of Law

The other fundamental question that informs the analysis in Chaps. 4, 5 and 6 is *how* private hard law, private soft law and public hard law may, in practice, support sustainability objectives within securitisations. Two types of considerations follow from this question, andnamely what *functions* the law could serve and what steps should be taken at policy level (*'strategies'*) in order to advance these objectives.

Drawing from the classification proposed by *Solana* for financial markets in general[55], it is possible to identify four different functions of the law.

First, law has an 'enabling' role since it confers certain rights on a selective basis which, in turn, allow market actors to conduct financial activities or to take action in the financial context. This function is visible across the rules introducing positive rights or exemptions from the application of negative rules.[56] For instance, legislation 'enables' investors when empowering them to bring legal proceedings when securitisations fail to achieve the pre-announced sustainability goals.[57]

[52] See further Chap. 5, Sect. 5.7.1.–5.7.2.
[53] For the reasons behind these linguistic choices, see Chaps. 5 and 6.
[54] See Chap. 6, Sect. 6.2.5.
[55] Javier Solana, 'The Roles of Law' (n. 30).
[56] Ibid., 3–4.
[57] See further Chap. 2, Sect. 3.4.4.

Second, law has a 'framing' function. This function echoes *Pistor*'s account of law's power to set the 'rules of the game'[58], by drawing boundaries to the behaviours of market participants.[59] For the purposes of the present analysis, law can also ***constrain*** the types of assets and activities eligible for sustainable securitisations.[60]

Third, law has a 'programmatic' function insofar as it sets objectives and values, especially in the form of constitutional provisions, which should inspire legal action, like legal reforms.[61] For instance, it can introduce high-level principles, like EU treaty provisions on environmental objectives,[62] which should guide the design of regulatory measures targeting sustainable securitisations.

Fourth, law serves a 'constitutive' function for financial transactions. While other scholars have given a wide range of meanings to the notion,[63] the insights from *Pistor* and other modern Legal Institutionalists[64] are the most compelling. They have viewed law, both in its 'private' and in its 'public' form, as 'constitutive', since it represents a precondition for financial markets and transactions.[65] For instance, *Pistor* highlights how parties

[58] The expression was first coined by Douglass C North, *Institutions, Institutional change, and Economic Performance* (Cambridge University Press 1991).

[59] See ibid. Pistor gives a similar meaning to the term 'framing', see Pistor, 'LTF Perspective' (n. 4) 23–24. Black appears to depict financial regulation exclusively as a 'framing' device, see '*regulation is the set of rules, principles and norms allowing to manage risks and behaviours*', Black, 'Regulatory Styles' (n. 12) 218. It should also be noted that other theorists, such as Legal sociologists, have conceptualised the framing role of law differently, suggesting that law provides 'frames' that result in structural changes across organisations. Gwendolyn Leachman, Lauren B Edelman, Doug McAdam, 'On Law, Organisations, and Social Movements' (2010) 6 *Annual Review of Law and Social Science* 65, 662.

[60] In the same vein, see Solana, 'The Roles of Law' (n. 30) 3. This account of the law appears to resonate with the findings of socio-legal, especially organisational, theory, where scholars have depicted this function of law as 'regulatory'. See Edelman, Suchman (n. 7) 483; Edelman, Stryker, A Sociological Approach to Law and the Economy in Neil J. Smelser, Richard Swedberg (eds.), *The Handbook of Economic Sociology* (2010 Princeton University Press), 547.

[61] Ibid., 3–4. On the concept of 'legal action', see also Sect. 1.5.

[62] For the analysis of EU treaty provisions on the subject, see Chap. 5, Sect. 5.2.4.

[63] E.g. Socio-Legal Scholars have underlined how law provides legal constructs which legitimise certain economic practices and institutions. Among other things, it defines what corporate forms are acceptable and what promises give rise to a contractual commitment. Edelman, Suchman (n. 7) 503; Edelman, Stryker (n. 60) 24–25.

[64] Deakin and others (n. 9) 198.

[65] *'Firms (...) too have to be treated as creatures of the law'*, ibid. 194. *'Markets themselves, just like financial intermediaries, are constructed in law and do not exist outside it'*, see Pistor, 'Legal Theory of Finance' (n. 3) 317–321.

independently design and enforce contractual rights over financial assets, while state law acknowledges what kinds of contracts are legally enforceable, thus making financial transactions possible.[66] The introductory analysis of securitisation law, conducted in Chap. 3, will draw further attention to the 'constitutive' foundations of such a legal framework.[67]

At the same time, as the following analysis will highlight, the persistently embryonic state of the law governing sustainable securitisations leaves much more limited room for the law to play a 'constitutive' role.[68]

2.6 Legal Strategies

In principle, there are multiple avenues through which private hard law, private soft law, and public hard law could serve the above-mentioned functions and – –through them – –further sustainability objectives. Rule-making entities willing to advance sustainability goals are likely to see the law 'strategically'. To this end, they are likely to use 'legal tools', like laws and contractual provisions, and to deploy legal procedures (what have been defined as 'legal actions'[69]). For instance, policy-makers could revise specific rules of the securitisation regime to incentivise sustainable business practices. Likewise, parties designing these transactions could incorporate certain clauses within the documentation, creating binding legal obligations that could require the originating entity to comply with environmental regulations. I refer to the combination of hard and soft legal tools, as well as legal actions with the specific goal of bringing about transformational change, as 'legal strategies'.[70] Throughout the book, I will

[66] Ibid., 318, 321, 324, 326–327. To some extent, Pistor's insights bear similarities to Coase's transaction costs theory, which emphasizes that law, especially in the form of court judgments, serves to delimit rights between parties, see Coase (n. 28) 15–16.

[67] See especially Chap. 3, Sect. 3.2.

[68] Though others have more recently advocated for a potentially broader 'co-constitutive' role of the law in regulating 'sustainable' financial transactions, leading to the stretching of, differentiation from, existing legal concepts, Agnieszka Smoleńska, 'The Constitutive Role of Law in Sustainable Finance' (2025) *Industrial Law Journal* (advance publication), 27–30.

[69] Following the vocabulary proposed by Javier Solana, 'Systemic Lawyering' (2022) Working Paper. On file with the author. 4–5.

[70] Following *Solana*, who defines 'legal strategies' as a combination of legal tools and actions informed by the expectation to achieve a certain objective ('theory of change'). 'Legal actions' put legal tools into practice. ibid., 28. Other academic work also refers to 'legal' and 'regulatory strategies' when analysing financial regulation and private law, Armour et al

examine how rule-makers have previously used, and how they could further use, four selected legal strategies to foster sustainability-related practices in securitisations. Despite the widely discussed role of litigation in prompting climate-related commitments[71], and the potential for enforcement actions by entities like supervisory authorities[72], the actions at the core of this analysis will be legal reforms. These may include changes to the securitisation regime at the 'public' level or to the contractual agreements at the 'private' level. The analysis will highlight why legal reforms would arguably be preferable to litigation, given their *ex ante* effects and broader scope, which can ensure more uniform application across market actors. On the contrary, court judgments would act *ex post* and *primarily* target the litigants, with only limited impact on third parties. The analysis will also acknowledge the inherent limitations of legal changes, such as the time and costs of the law-making process, regulatory capture risks and path dependence mechanisms.[73]

The first discussed strategy is 'disclosure'. It manifests in the form of rules requiring or recommending[74] the dissemination of specific information on sustainability-related aspects of securitisation deals. Disclosure

(n. 39) 52–53; John Armour, Henry Hansmann and Reinier Kraakman, 'Agency Problems and Legal Strategies' (2017) 9 in Reinier Kraakman, John Armour, Paul Davies et al, *The Anatomy of Corporate Law: a comparative and functional approach* (3rd edn, Oxford University Press 2017) 31.

[71] There is abundant literature on 'climate litigation' as a 'strategic' or 'systemic' tool towards sustainability objectives. For a review see Jacqueline Peel, Hari M Osofsky, 'Climate Change Litigation' (2020) 16 Annual Review of Law and Social Science 21. For an overview of the main types of claims in financial markets see Javier Solana, 'Climate Litigation in Financial Markets: A Typology' (2020) 1 *Transnational Environmental Law* 103.

[72] E.g. the European Securities and Markets Authority (ESMA) has mapped examples of 'greenwashing' in the financial sector and is working towards the convergence of supervisory practices of national competent authorities in addressing it. ESMA, 'Final Report on greenwashing – Response to the European Commission's request for input on 'greenwashing risks and the supervision of sustainable finance policies' (2024) 2, 4 https://www.esma.europa.eu/sites/default/files/2024-06/ESMA36-287652198-2699_Final_Report_on_Greenwashing.pdf; The European Central Bank, in its supervisory capacity, has also imposed in 2025 the first periodic penalty payment to a bank that has allegedly failed to adequately manage climate-related and environmental risks within its portfolio. ECB, 'ECB imposes periodic penalty payments on ABANCA for failing to sufficiently identify climate risks' (2025) https://www.bankingsupervision.europa.eu/press/pr/date/2025/html/ssm.pr251110~3e0b6f579e.en.html accessed 7 December 2025.

[73] See Chaps. 5, Sect. 5.9., and 6, Sect. 6.4.–6.5.

[74] Depending on whether it acts as a 'hard law' or 'soft law' rule, see above Sect. 2.4.

obligations are widely used in the financial regulation remit[75] and have become ubiquitous in the sustainable finance landscape.[76] The EU initiatives[77] in this space provide a compelling example, illustrating a narrative that aligns with, yet also goes beyond, the narrative traditionally underpinning financial regulation.

In line with the prevailing regulatory approach,[78] EU preparatory documents and harmonisation efforts on sustainable finance[79] refer or implicitly allude to 'market failures'.[80] For instance, they recurrently mention the lack of a clear definition of what makes financial products 'sustainable' and the related information asymmetry among investors.[81] The direct consequence, often only implicit in the texts of recent initiatives, is the systematic underpricing of 'green' assets.[82] Accordingly, disclosure of sustainability-related information would improve 'price discovery'[83], i.e. the mechanism for the determination of the prices. Such a development could lead to a scenario where prices fully reflect the available information ('informational efficiency'[84]). Disclosure may also prevent potential

[75] Luca Enriques, Sergio Gilotta, 'Disclosure and Financial Market Regulation' in Niamh Moloney, Eilís Ferran, and Jennifer Payne, *The Oxford Handbook of Financial Regulation* [Oxford University Press 2015].

[76] Dimitri G Demekas, Pierpaolo Grippa, 'Walking the Tightrope: Financial Regulation, Climate Change, and the Transition to a Low-Carbon Economy' (2022) 8 Journal of Financial Regulation 2 203, 217. For an inquiry into the role of disclosure obligations in the sustainable finance debate, Tobias Hans Tröger, Sebastian Steuer, 'The Role of Disclosure in Green Finance' (2021) 8 *Journal of Financial regulation* 1.

[77] Stemming from the 2018 EU Action Plan, EU Commission, 'Financing sustainable growth' (2018) 097.

[78] See above Sect. 2.3.

[79] For the meaning and an overview of the legal tools towards 'harmonisation', see Chap. 5, Sect. 5.2.1–5.2.2.

[80] On market failures, see above Sect. 2.3.1.

[81] As highlighted, inter alia, by Para 2.2.3, Impact Assessment ahead of the EU Taxonomy Regulation, EU Commission, 'Impact Assessment – –Accompanying the Document 'Proposal for a Regulation of the European Parliament and of the Council on the Establishment of a Framework to Facilitate Sustainable Investment' (2018) 264. In relation to 'green finance' in general, see also Tröger, Steuer (n. 26) 35.

[82] Hugues Chenet, Josh Ryan-Collins, Frank Van Lerven, 'Finance, Climate-Change and Radical Uncertainty: Towards a Precautionary Approach to Financial Policy' (2021) 183 *Ecological Economics* 106957, 3; Tröger, Steuer (n. 26) 33.

[83] On the definition, see Dan Awrey, 'The Mechanisms of Derivatives Market Efficiency' (2016) 91 *New York University Law Review* 1104, 1123–1124.

[84] Armour et al (n. 39) 51, 66, 101; Enriques and Gilotta (n. 75) 519. The concept derives from the Efficient Capital Markets Hypothesis, first articulated by Fama, Eugene F Fama,

'negative externalities', like the systemic consequences that may derive from the incorrect quantification of climate-related risks.[85]

EU regulators and other intermediaries involved in the design of sustainable finance rules[86] also see sustainability-related information as a 'public good'. They have noted the scarce incentives for companies and financial institutions to report on sustainability matters on a voluntary basis, along with the challenges and costs of accessing such information for prospective investors. Disclosure-oriented rules would thus encourage consistent reporting, providing all investors with equal access to information.[87] An interrelated theme, which also recurrently came to the fore in the traditional financial regulation discourse[88], is the challenge of comparing the (sustainability-related) features of financial instruments.[89] Consequently, disclosure-oriented rules would promote the standardisation of data for potential investors.

Along with traditional goals, rule-making actors have shown interest in promoting *novel* objectives through rules grounded on disclosure. For instance, the EU regulator has alluded to the role of more granular and higher-quality information in mobilising capital flows towards sustainability objectives.[90] In other words, it should encourage market actors to increase their financing of 'green' and other sustainability-related

Efficient Capital Markets: A Review of Theory and Empirical Work (1970) 25 *The Journal of Finance* 2 383, 384–85.

[85] Tröger, Steuer (n. 26) 3. On systemic risks in general as drivers of financial regulation, see above, Sect. 2.2.2.

[86] E.g. industry associations, see Chap. 6. On 'intermediaries' see above Sect. 2.3.1.

[87] See, inter alia, Paras 2.1.1., 2.1.2. EU Commission (n. 119). more generally, Tröger, Steuer (n. 26) 32–33. On disclosure obligations as a source of 'positive externalities' and on information as a 'public good', see, *inter alia*, Enriques, Gilotta (n. 71) 7; John C Coffee, 'Market Failure and the Economic Case for a Mandatory Disclosure System' (1984) 70 *Virginia Law Review* 717, 725. Disclosure would thus lead to 'non-excludable' advantages for all market participants, Coffee, ibid., 725. Commentators have, however, questioned the equation between wider disclosure and better investment decisions on economic and other grounds, Tröger, Steuer (n. 26) 19–24, 31, Linn Anker-Sørensen, Dirk A Zetzsche, 'Regulating Sustainable Finance In The Dark' (2022) 23 *European Business Organisation Law Review* 47, 55.

[88] Enriques, Gilotta (n. 71) 524.

[89] For example, para 2.1.1., 2.2.3 EU Commission (n. 81). as also noted by Tröger, Steuer (n. 26) 33–35.

[90] One of the objectives of the EU sustainable finance action plan was 'reorienting capital flows towards sustainable investments', See 1., EU Commission (n. 77); Colaert has defined this policy objective 'sustainable finance', however the term could be misleading for the present analysis, where the term is also used to capture the interplay between sustainability considerations and financial products or services more broadly. See Veerle A Colaert, 'The

activities.[91] In economic terms, the reorientation of capital may still resonate with the traditional goal of allocating resources within the market to those who value them the most (what economists have traditionally defined as 'allocative efficiency'[92]). At the same time, it may be possible to interpret disclosure-based rules as a means to shift investors' preferences towards 'green' assets and to divert funds away from 'brown' assets. Therefore, there may arguably be an implied attempt to 'nudge' investors, i.e. to shift their decision-making processes and behaviours.[93] This perspective may represent a novelty in the traditional goals of financial policies.[94]

Complexity theory also offers valuable inroads into the role of disclosure. One could view the financial system as a complex adaptive system, encompassing various interrelated and interdependent points, with the ability to evolve and adapt to a changing environment.[95] Legal tools like legislation, even when their direct effects are marginal, have the potential to give rise to nonlinear feedback effects, triggering larger changes in the system's dynamics. For instance, commentators have highlighted how small changes to disclosure rules may considerably impact asset prices or investors' choices, by prompting them to decide, among others, to stop investing in specific sectors altogether.[96] For the purposes of the present analysis, even marginal adjustments to sustainability-related disclosure requirements for securitisation could have a far-reaching impact, triggering

Changing Nature of Financial Regulation. Sustainable Finance as a New Policy Objective' (2022) 59 *Common Market Law Review* 6 1669.

[91] 'Increased transparency and reduced search costs will create a more favourable investment environment, where investors can make sustainable investment decisions more easily. As a result, once additional building blocks will be put in place, the proposed initiatives will also help reorient capital flows towards achieving sustainable growth', 'General objectives', 4.1. EU Commission (n. 77).

[92] Armour and others, 'The Goals and Strategies of Financial Regulation' in Armour et al (n. 39) 53, 245. In an ideal perfect competition scenario, it would result in a situation where no alternative allocation of resources is possible without making one market participant worse off ('Pareto-efficiency'), ibid., 54.

[93] Anker-Sørensen, Zetzsche (n. 87) 59–61.

[94] However, Colaert argues that it still does not represent an autonomous policy objective in the current scenario, see Colaert (n. 90) 15, 28–30.

[95] On the definition of system, see Daniel H Kim, *Introduction to Systems Thinking* (Pegasus Communications ed, 1999). Various commentators have applied complexity theory to the analysis of the financial system, see, among others, Stefano Battiston and others, 'Complexity Theory and Financial Regulation' (2016) 351 *Science* 818.

[96] Acting as a 'sensitive intervention point', JD Farmer and others, 'Sensitive Intervention Points in the Post-Carbon Transition' (2019) 364 *Science* 132, 132–133.

substantial changes to sustainability-related practices across originating companies or stimulating investors' appetite for securitised products.

Building on these premises, the book analyses how private and public law rules revolving around disclosure have affected the 'sustainable' elements of securitisations. It discusses the need for further initiatives based on said strategy. It also highlights the limitations of this strategy, which may not, on its own, suffice to adjust the pricing of 'sustainable' assets, inevitably affected by other factors, like further non-correlated, changes to the policy framework.[97] The dissemination of further information also postulates the ability of addressees to process it and act upon it,[98] which should not be taken for granted for innovative financial products like 'sustainable' ones.

The second discussed strategy is 'shielding', a distinctive feature of all entities with a separate legal personality.[99] It captures the insulation of the assets of a given entity (giving rise to a 'shield'), which should protect them from external claims, like those brought by creditors.[100] Chap. 2 describes the various nuances of the 'shielding' of SPVs, which is one of the core characteristics of securitisation law.[101] The most relevant rules for the design of sustainability-oriented legal strategies, however, are those protecting SPVs' assets. The goal of such rules is to ensure that SPVs serve the primary purpose of the operation (which also to some extent relates to what some commentators have defined as 'affirmative asset partitioning'[102]).[103] The 'shielding' of securitisation assets could be functional to

[97] On how markets partially price policy developments, Tom Gosling, Iain Macneil, 'Can Investors Save the Planet? – NZAMI and Fiduciary Duty' (2023) 18 *Capital Markets Law Journal* 2 172, 182–184.

[98] Enriques, Gilotta (n. 71) 528.

[99] Like companies, but there have been analyses of 'entity shielding' in relation to partnerships and charitable trusts. See Henry Hansmann, Reinier Kraakman, Richard Squire, 'Law and the Rise of the Firm' (2006) 119 *Harvard Law Review* 1333, 1338, 1384.

[100] The concept was first conceptualised in a previous article, see Henry Hansmann, Reinier Kraakman, 'The Essential Role of Organisational Law' (2000) 110 *The Yale Law Journal* 1 387, 390–394.

[101] See Chap. 3, Sect. 3.3.2., 3.4.3.

[102] Hansmann, Kraakman, Squire (n. 99) 1338, 1339. Throughout the book, the preferred notion will be 'shielding', which draws attention to the insulation of the SPV and of its assets for the well-functioning of the operation. Conversely, the notion of 'asset partitioning' captures the priority *rights between the creditors of an entity over the personal creditors of its shareholders, or the priority of creditors' over shareholders' rights.* These aspects are less relevant for the present analysis, as SPVs tend not to have shareholders, and as they appear to be less strategic from a sustainability perspective.

[103] Hansmann, Kraakman, 'The essential role' (n. 100) 389, 392–393.

sustainability objectives since it could ensure the efficient use of SPVs' patrimony towards environmental, social, and governance goals, while preventing expenses or further uses that may thwart them. Thus, the book examines how the current framework promotes the 'shielding' of SPVs for sustainability purposes, and how future initiatives could promote it.

The third legal strategy discussed is 'Gatekeeping'. Extensive academic scholarship identifies 'gatekeepers' as intermediaries involved in financial transactions due to their expertise. Gatekeepers verify issuers, statements or financial deals and attest to their reliability to investors and market participants.[104] As a matter of principle, gatekeepers' statements should be inherently true due to the reputational risks that they face.[105] Examples of these actors include audit firms or credit rating agencies, assessing companies' financial statements or securities' creditworthiness.[106] Similarly, the development of 'sustainable' financial products has resulted in an ever-growing number of professional services verifying how firms and financial instruments contribute to sustainability objectives. For instance, organisations like Climate Bond initiative, or credit rating agencies like Fitch Ratings, offer, among others, 'external reviews' of green bonds, through which they verify and attest to their environmental features.[107] Therefore, the book will discuss how and to what extent market actors have effectively acted as 'gatekeepers' to ensure that the 'sustainable' component of securitisation deals is credible. It will also investigate to what extent recent EU initiatives on sustainable finance have introduced adequate gatekeeping mechanisms for sustainable securitisations.[108]

The fourth legal strategy comprises rules promoting 'voice' mechanisms. *Hirschman* has identified 'voice' as one of the techniques allowing investors to react when dissatisfied with investee companies' performance

[104] Reinier H Kraakman, 'Gatekeepers: The Anatomy of a Third-Party Enforcement Strategy' (1986) 2 Journal of Law, Economics, and Organisation 53; John C Coffee, *Gatekeepers: The professions and corporate governance* (Oxford University Press, 2006) 248.

[105] They *'pledge their reputational capital',* John C Coffee, 'Gatekeeper Failure and Reform: The Challenge of Fashioning Relevant Reforms' (2004) 237, 239, 249.

[106] Ibid., 249.

[107] For further examples, see Chap. 4, Sect. 4.6.1.3–4.7.2.3.

[108] For example, David Ramos Muñoz, Elia Cerrato Garcia, Marco Lamandini have proposed 'green trustees' acting as 'green gatekeepers', see 'The EU's 'Green' Finance. Can 'Exit', 'Voice' and 'Coercion' Be Enlisted to Aid Sustainability Goals?'(2021) EBI Working Paper Series 90, 29–32. https://papers.ssrn.com/sol3/papers.cfm?abstract_id=3824792 accessed 5 December 2025.

or with their decisions.[109] They attempt to 'change, rather than escape from, an objectionable state of affairs'.[110] Investors are only likely to prefer 'voice' over selling their shares ('exit') under limited circumstances and on the basis of a cost–benefit analysis.[111] Relevant considerations in this context will be, among others, their bargaining power, the costs of coordinating with other investors and the foreseeable impact on investee companies.[112] Modern academic scholarship has extensively drawn from *Hirschman*'s classification for mapping the various behaviours through which investors raise their 'voice' in corporate governance, such as meetings with the management, voting or bringing resolutions.[113] It is clear that the key source for such rights is the legal framework, which can empower investors to act, thereby serving an 'enabling' function.[114]

Even though the studies on 'voice' rights predominantly revolve around *shareholders*' relationships with companies,[115] it is also, in principle, possible to transpose the notion to other types of investors, like those holding bonds and securitised products.[116] Chap. 3 will highlight the limited rights of underwriters and investors in securitisation deals vis-à-vis originating companies and SPVs, as well as identifying some mechanisms to increase their representation.[117]

The most recent years have witnessed an unprecedented rise in examples of activist behaviours by investors on environmental, social, and governance matters, spanning filing and voting on resolutions, litigation, or collaborative initiatives, among others.[118] Investors and underwriters in ABSs

[109] Albert O Hirschman, 'Voice' in *Exit, Voice and Loyalty: responses to decline in firms, organisations and state* (1970). For the definition of 'voice', see ibid., 30–33.

[110] Ibid., 30.

[111] Ibid., 1, 2, 4.

[112] Ibid., 34–36, 38–41.

[113] See, inter alia, Marco Becht and others, 'Corporate Governance Through Exit and Voice' (2019) 633 ECGI Working Paper Series in Law, 7, 20–22. Bernard S Black, 'Agents Watching Agents: The Promise Of Institutional Investor Voice' (1992) 39 *UCLA Law Review* 830–844.

[114] See above Sect. 2.4.

[115] Hirschman focuses on 'customers' and 'members' of organisations. Investors fall in the latter category, Hirschman (n. 149) 4.

[116] For an application of the notion to bondholders, see George G Triantis, Ronald J Daniels, 'The Role of Debt in Interactive Corporate Governance' (1995) 83 *California Law Review* 1073; for 'voice' rights in the context of CLOs, see Suren Gomtsian, 'Debtholder Stewardship' (2023) 86 *Modern Law Review* 2 395, 37–39.

[117] See Chap. 3, Sect. 3.3.4.

[118] On voting on ESG matters across Italian and French listed companies, see, among others, Marco Fanari, Giuseppe Buscemi, Tiziana Di Stefano, 'La sostenibilità nelle assemblee

have a similar potential to drive corporate action on sustainability goals. For instance, the OECD has analysed how underwriters can exert pressure on companies from which they purchase securities on due diligence issues, like human rights.[119] the UNPRI has mapped various strategies for investors in bond and securitised products to engage with borrowers, like the right to have a say over contractual clauses on ESG matters and to renegotiate them.[120]

Against this background, the present book discusses the strengths and weaknesses of the mechanisms for investors and underwriters of sustainable securitised products to raise their 'voice' vis-à-vis originators and the SPV. It focuses on the rights shaped by the parties (through 'private law') and by EU law (through 'public hard law'). It also proposes new mechanisms to extend the spectrum of these rights.

The insights from Complexity Theory[121] are valuable for understanding the complementarity of the four above-mentioned strategies. They can only fully catalyse the transformation of sustainability practices linked to securitisation when operating in tandem. This becomes evident by observing the inherently complex nature of the financial system, consisting of interconnected nodes.[122] A rule targeting one component of the system can inevitably act as a 'shock' and affect the others. [123] For instance, actors not targeted by a rule may still act upon it through behaviour that contributes to advancing the very goal that the rule pursues (what has

societarie in Francia, Germania e Italia' (2024) 45 Banca d'Italia Working paper, https://www.bancaditalia.it/pubblicazioni/mercati-infrastrutture-e-sistemi-di-pagamento/approfondimenti/2024-045/N.45-MISP.pdf?language_id=1; on litigation, see Solana, 'Climate Litigation' (n. 71) 112–117, 121–122; on collaborative initiatives like the GFANZ and the Net Zero Asset Managers Initiative (NZAMI), see Gosling, MacNeil (n. 97) 2–3, 7–9. For a systematic and 'strategic' analysis of effective 'legal strategies' (which the authors depict as 'law tools') to unleash environmental objectives in the corporate law sphere, Jacqueline Peel, Anita Foerster et al, 'Governing the Energy Transition: The Role of Corporate Law Tools' (2019) 36 *Environmental and Planning Law Journal* 5 459.

119 OECD, 'Due Diligence for Responsible Corporate Lending and Securities Underwriting Key Considerations for Banks Implementing the OECD Guidelines for Multinational Enterprises' (2019) https://www.oecd.org/en/publications/due-diligence-for-responsible-corporate-lending-and-securities-underwriting_71fd2895-en.html accessed 5 December 2025.

120 UNPRI, 'ESG engagement for fixed income investors: managing risks, enhancing returns' (2018) 35–36 https://public.unpri.org/download?ac=4449 accessed 7 December 2025.

121 See above, Sect. 2.4.

122 See above Sect. 2.4.

123 Solana, 'Systemic Lawyering' (n. 69) 21–22.

been defined as 'reinforcing' or 'positive feedback').[124] Therefore, rules requiring originators to disclose the climate-related risks of the portfolio may also encourage market actors to support or withdraw from investing in certain companies. Due to 'positive feedback', each legal strategy may also yield the effects associated with *other* legal strategies. For example, disclosure-oriented rules should operate on market prices. At the same time, statutory provisions strengthening 'voice' mechanisms, like those introducing new judicial remedies, may also indirectly trigger a similar effect. For instance, legislation may empower investors to hold originating companies accountable for the truthfulness of their claims on environmental matters.[125] To prevent these rights from being exercised, companies may be incentivised to 'disclose' their commitment to these matters. The resulting transparency should also directly reflected in the prices for ABSs.

Building on these premises, Chap. 4 will highlight the various shortcomings of the current legal framework for sustainable securitisations, as shaped by market participants, and the need for legal change. Chaps. 5 and 6 will examine private soft law and public hard law initiatives grounded in the four strategies from a 'positive' and a 'normative' angle. In other words, they will discuss how law *is*, that is the extent to which these strategies have affected the framework on securitisation by encouraging or preventing sustainability-related practices. They will also reflect on how law *should be* and how legal reforms may most effectively trigger transformational changes in the design and management of these operations.[126]

References

Abbott KW and Snidal D, 'Hard and Soft Law in International Governance' (2000) 54 *International Organisation* 421

[124] Ibid., 21/22. On 'reinforcing feedback loops', see also Matthew D Stepp and others, 'Greenhouse Gas Mitigation Policies and the Transportation Sector: The Role of Feedback Effects on Policy Effectiveness' (2009) 37 *Energy Policy* 2774, 2776.

[125] See above Sect. 2.4.

[126] 'Positivistic' accounts of the law describe and explain the law or they provide a conceptualisation thereof. By contrast, 'Normative' accounts emphasize the coercive power of law and they often have prescriptive connotations by suggesting desirable forms of regulation. On the 'normativity' of law, see Andrei Marmor, Alexander Sarch, 'The Nature of Law' in Edward N Zalta (ed), *The Stanford Encyclopedia of Philosophy* (2019), 2.1.1, 2.1.2, 2.1.4, https://plato.stanford.edu/entries/lawphil-nature/; On the distinction between positive and normative accounts in legal-economic analysis, see Mercuro, Medema (n. 8) 47–50.

Abbott K and Snidal D, 'Strengthening International Regulation Through Transnational New Governance' (2009) 42 *Vanderbilt Journal of Transnational Law* 501
Abbott KW and others, 'The Concept of Legalisation' (2018) 54 *International Organisation* 401
Abbott KW, Levi-Faur D and Snidal D, 'Theorizing Regulatory Intermediaries: The RIT Model' (2017) 670 *Annals of the American Academy of Political and Social Science* 14
Allen DW, 'Transaction Costs' in Bouckaert B and De Geest G (eds), *Encyclopedia of Law and Economics*, vol I (Edward Elgar 2000)
Anker-Sørensen L and Zetzsche DA, 'Regulating Sustainable Finance in the Dark' (2022) 23 *European Business Organisation Law Review* 47
Armour J, Awrey D, Davies P and others, *Principles of Financial Regulation* (OUP2016)
Armour J, Hansmann H and Kraakman R, 'Agency Problems and Legal Strategies' in Kraakman R, Armour J, Davies P and others (eds), *The Anatomy of Corporate Law: A Comparative and Functional Approach* (3rd edn, OUP2017)
Awrey D, 'The Mechanisms of Derivatives Market Efficiency' (2016) 91 *New York University Law Review* 1104
Ayres I and Gertner R, 'Filling Gaps in Incomplete Contracts: An Economic Theory of Default Rules' (1989) 99 *Yale Law Journal* 87
Baldwin R, Cave M and Lodge M, 'Self-Regulation, Meta-Regulation and Regulatory Networks' in Baldwin R, Cave M (eds) *Understanding Regulation: Theory, Strategy and Practice* (2nd edn, OUP2011)
Battiston S, Farmer JD, Flache A and others, 'Complexity Theory and Financial Regulation' (2016) 351 *Science* 818
Becht M and others, 'Corporate Governance Through Exit and Voice' (2019) ECGI Law Working Paper No 633
Benjamin J and Rouch D, 'The International Financial Markets as a Source of Global Law: The Privatisation of Rule-Making' (2008) 2 *Law and Financial Markets Review* 78
Black B, 'Agents Watching Agents: The Promise of Institutional Investor Voice' (1992) 39 *UCLA Law Review* 830
Black J, 'Decentering Regulation: Understanding the Role of Regulation and Self-Regulation in a 'Post-Regulatory' World' (2001) 54 *Current Legal Problems* 103
Black J, 'Regulatory Styles and Supervisory Strategies' in Moloney N, Ferran E and Payne J (eds), *The Oxford Handbook of Financial Regulation* (OUP2015)
Bradley C, 'Private International Law-Making for the Financial Markets' (2005) 29 *Fordham International Law Journal* 127
Braithwaite JP, 'Standard Form Contracts as Transnational Law: Evidence from the Derivatives Markets' (2012) 75 *Modern Law Review* 779
Brès L, Mena S and Salles-Djelic ML, 'Exploring the Formal and Informal Roles of Regulatory Intermediaries in Transnational Multistakeholder Regulation' (2019) 13 *Regulation & Governance* 127

Britton-Purdy J and others, 'Building a Law-and-Political-Economy Framework' (2020) 129 *Yale Law Journal*1784
Brummer C, *Soft Law and the Global Financial System: Rule-Making in the twenty-first century* (2nd edn, Cambridge University Press 2020)
Chenét H, Ryan-Collins J and Van Lerven F, 'Finance, Climate Change and Radical Uncertainty' (2021) 183 *Ecological Economics* 106957
Coase RH, 'The Problem of Social Cost' (1960) 3 *Journal of Law and Economics* 1
Coffee JC, 'Market Failure and the Economic Case for a Mandatory Disclosure System' (1984) 70 *Virginia Law Review* 717
Coffee JC, 'Gatekeeper Failure and Reform' (2004) 84 *Boston University Law Review*301
Coffee JC, *Gatekeepers: The Professions and Corporate Governance* (OUP2006)
Coglianese C and Mendelson E, 'Meta-Regulation and Self-Regulation' in Baldwin R, Cave M and Lodge M (eds), *The Oxford Handbook of Regulation* (OUP2010)
Colaert V, 'The Changing Nature of Financial Regulation: Sustainable Finance as a New Policy Objective' (2022) 59 *Common Market Law Review* 1669
Deakin S and others, 'Legal Institutionalism: Capitalism and the Constitutive Role of Law' (2017) 45 *Journal of Comparative Economics* 188
Demekas DG and Grippa P, 'Walking the Tightrope: Financial Regulation, Climate Change, and the Transition to a Low-Carbon Economy' (2022) 8 *Journal of Financial Regulation* 203
Edelman LB and Suchman MC, 'The Legal Environments of Organisations' (1997) 23 *Annual Review of Sociology*479
Edelman LB and Stryker R, 'A Sociological Approach to Law and the Economy' in Smelser NJ and Swedberg R (eds), *The Handbook of Economic Sociology* (Princeton University Press 2010)
Enriques L and Gilotta S, 'Disclosure and Financial Market Regulation' in Moloney N, Ferran E and Payne J (eds), *The Oxford Handbook of Financial Regulation* (OUP2015)
European Central Bank, 'ECB Imposes Periodic Penalty Payments on ABANCA for Failing to Sufficiently Identify Climate Risks' (2025) https://www.bankingsupervision.europa.eu/press/pr/date/2025/html/ssm.pr251110~3e0b6f579e.en.html
European Commission, *Financing Sustainable Growth* COM (2018) 97 final
European Commission, *Impact Assessment Accompanying the Proposal for a Regulation on the Establishment of a Framework to Facilitate Sustainable Investment* SWD (2018) 264 final
European Securities and Markets Authority, *Final Report on Greenwashing* (2024) https://www.esma.europa.eu/sites/default/files/2024-06/ESMA36-287652198-2699_Final_Report_on_Greenwashing.pdf
Fama EF, 'Efficient Capital Markets' (1970) 25 *Journal of Finance* 383
Fanari M, Buscemi G and Di Stefano T, 'La sostenibilità nelle assemblee societarie in Francia, Germania e Italia' (2024) *Banca d'Italia Working Paper* No 45

https://www.bancaditalia.it/pubblicazioni/mercati-infrastrutture-e-sistemi-di-pagamento/approfondimenti/2024-045/N.45-MISP.pdf?language_id=1
Farmer JD and others, 'Sensitive Intervention Points in the Post-Carbon Transition' (2019) 364 *Science* 132
Fried BH, 'Preface' in Fried BH, *The Progressive Assault on Laissez Faire: Robert Hale and the First Law and Economics Movement* (Harvard University Press 1998)
Gomtsian S, 'Debtholder Stewardship' (2023) 86 *Modern Law Review* 395
Hansmann H and Kraakman R, 'The Essential Role of Organisational Law' (2000) 110 *Yale Law Journal* 387
Hansmann H, Kraakman R and Squire R, 'Law and the Rise of the Firm' (2006) 119 *Harvard Law Review* 1333
Harlow C, 'Public' and 'Private' Law: Definition without Distinction' (1980) 6 *Modern Law Review* 951
Hart O and Moore J, 'Contracts as Reference Points' (2008) 123 *Quarterly Journal of Economics* 1
Hirschman AO, *Exit, Voice and Loyalty* (Harvard University Press 1970)
Hovenkamp H, 'The First Great Law & Economics Movement' (1990) 42 *Stanford Law Review* 993
Jansen N and Michaels R, 'Beyond the State? Rethinking Private Law' (2019) 56 *American Journal of Comparative Law* 527
Jensen MC and Meckling WH, 'Theory of the Firm' (1976) 3 *Journal of Financial Economics* 305
Karmel RS and Kelly CR, 'The Hardening of Soft Law in Securities Regulation' (2009) 34 *Brooklyn Journal of International Law* 883
Kim DH, *Introduction to Systems Thinking* (Pegasus Communications1999)
Kirton JJ and Trebilcock MJ (eds), *Hard Choices, Soft Law* (Routledge 2016)
Klausner M, 'Corporations, Corporate Law, and Networks of Contracts' (1995) 81 *Virginia Law Review* 757
Kourula A, Paukku M, Peterman A and Koria M, 'Intermediary Roles in Regulatory Programs' (2019) 13 *Regulation & Governance* 140
Kraakman RH, 'Gatekeepers: The Anatomy of a Third-Party Enforcement Strategy' (1986) 2 *Journal of Law, Economics, and Organisation* 53
Macneil I, 'Contracts: Adjustment of Long-Term Economic Relations' (1978) 72 *Northwestern University Law Review* 854
Marmor A and Sarch A, 'The Nature of Law' in ZaltaEN (ed), *Stanford Encyclopedia of Philosophy* (2019) https://plato.stanford.edu/entries/lawphil-nature/
Medzini R and Levi-Faur D, 'Self-Governance via Intermediaries' (2023) 25 *Journal of Comparative Policy Analysis* 323
Mercuro N and Medema S, *Economics and the Law* (2nd edn, Princeton University Press 2006)
Michaels R and Jansen N, 'Private Law Beyond the State?' (2007) 54 *American Journal of Comparative Law* 843

North DC, *Institutions, institutional change and Economic Performance* (Cambridge University Press 1991)

OECD, *Due Diligence for Responsible Corporate Lending and Securities Underwriting*: Key Considerations for Banks Implementing the OECD Guidelines for Multinational Enterprises' (2019) https://www.oecd.org/en/publications/due-diligence-for-responsible-corporate-lending-and-securities-underwriting_71fd2895-en.html

Parker C and others (eds), *Regulating Law* (OUP2004)

Peel J and Osofsky HM, 'Climate Change Litigation' (2020) 16 *Annual Review of Law and Social Science* 21

Peel J, Foerster A and others, 'Governing the Energy Transition' (2019) 36 *Environmental and Planning Law Journal*459

Pistor K, 'Law in Finance' (2013) 41 *Journal of Comparative Economics* 311

Pistor K, 'A Legal Theory of Finance' (2013) 41 *Journal of Comparative Economics* 315

Pistor K, 'Regulating Financial Markets – An LTF Perspective' in Avgouleas E and Donald DC (eds), *The Political Economy of Financial Regulation* (CUP2019)

Ramos Muñoz D, Cerrato Garcia E and Lamandini M, 'The EU's 'Green' Finance' (2021) EBI Working Paper No 90, https://papers.ssrn.com/sol3/papers.cfm?abstract_id=3824792

Richter R, 'The Role of Law in the New Institutional Economics' (2015) 26 *Washington University Journal of Law and Policy* 77

Schwarcz SL, 'Regulating Financial Change: A Functional Approach' (2016) 100 *Minnesota Law Review*1441

Shirley MM and Ménard C (eds), *Handbook of New Institutional Economics* (Palgrave Macmillan2008)

Sinclair D, 'Self-Regulation versus Command and Control?' (1997) 19 *Law & Policy*529

Smoleńska A, 'The Constitutive Role of Law in Sustainable Finance' (2025) *Industrial Law Journal (advance publication). https://doi.org/10.1093/indlaw/dwaf026*

Solana J, 'Climate Litigation in Financial Markets' (2020) 1 *Transnational Environmental Law* 103

Solana J, 'Market Efficiency as a Directive Principle of EU Monetary Policy' in Chadwick A, Lozano-Rodríguez E, Palacios-Lleras A and Solana J (eds), *Markets, Constitutions, and Inequality* (Routledge2022)

Solana J, 'Systemic Lawyering' (2022) Working Paper, on file with the author

Solana J, 'The Roles of Law in Finance' (2022) Working Paper, on file with the author

Stepp MD and others, 'Greenhouse Gas Mitigation Policies and the Transportation Sector' (2009) 37 *Energy Policy* 2774

Triantis GG and Daniels RJ, 'The Role of Debt in Interactive Corporate Governance' (1995) 83 *California Law Review* 1073
Tröger TH and Steuer S, 'The Role of Disclosure in Green Finance' (2021) 8 *Journal of Financial Regulation* 1
UNPRI, '*ESG Engagement for Fixed Income Investors: Managing risks, enhancing returns*' (2018) https://public.unpri.org/download?ac=4449
UNPRI, '*What Are the Principles for Responsible Investment?*' (2026) https://www.unpri.org/about-us/what-are-the-principles-for-responsible-investment
Zumbansen P, 'The Law of Society: Governance Through Contract' (2007) 14 *Indiana Journal of Global Legal Studies* 191

CHAPTER 3

The 'Anatomy' of Securitisation Law

3.1 Introduction

Before investigating how law could embed sustainability considerations in the structuring and management of securitisations, it is pivotal to grasp how law shapes securitisation operations *in general.*

Securitisation is a financing technique that essentially relies on credit claims, such as loans, leases or mortgages. The entities putting the transaction in motion ('originators' or 'sponsors') pool them together and entrust independent entities to manage them and to issue tradable financial instruments, which generally take the form of bonds or notes. Through the purchase of these securities, investors provide upfront liquidity and thereby allow the the credit claims to be 'refinanced', while the investors are repaid from the cash flows of the underlying assets. The rights over the securities (commonly referred to as 'asset-backed securities', *hereinafter*: ABSs) are also generally divided into different 'tranches', which determine the distribution of losses and risks across investors.

The present chapter aims to provide an 'anatomy'[1] of the legal framework applicable to securitisations and to introduce the aspects which could directly affect sustainability-related practices. The previous remarks

[1] Ie, an analysis of the most salient aspects. The term is borrowed from Reinier Kraakman, John Armour, Paul Davies, *The Anatomy of Corporate Law: A comparative and functional approach* (3rd edn, Oxford University Press 2017).

F. Agostini, *The Law of Sustainable Securitisations*, EBI Studies in Banking and Capital Markets Law,
https://doi.org/10.1007/978-3-032-25046-9_3

on the 'constitutive' role of law[2] are relevant to contextualise the role of securitisation law. In particular, following *Pistor*'s insights into financial markets in general, securitisation is 'made in law', since it presupposes a web of rules which delineate the proprietary and contractual rights over the various assets of the operation. As the following chapters will illustrate, securitisation rules advancing sustainability goals also serve a concurrent 'framing', 'enabling', and 'programmatic' function.[3]

Following *Pistor*'s theoretical framework, securitisation law is 'hybrid', since its legal framework is neither fully 'private' nor fully 'public'.[4] In other words, market participants, state, and international authorities concurrently design the rules for these operations. Therefore, the focus will be on the legal regime under EU law and selected examples of legislation across European countries (like the UK, which had initially retained the EU securitisation regime after its exit from the EU[5]). The chapter will also mention selected examples of standard clauses that market actors tend to replicate when drafting provisions for their own securitisation operations.[6]

As a result, the present chapter will have a twofold goal. First, it will *describe* how law shapes the foundational aspects of securitisations. In particular, Section 1 will explain what securitisations are and who is involved in these transactions. Section 2 will provide an overview of the foundational aspects of securitisation law that lie in the background of the analysis. Second, Sect. 3 of the chapter will *discuss* which areas of securitisation law could act as 'focal points' for legal initiatives to promote sustainability objectives. It will also highlight the peculiarities of the distribution of costs among parties when utilising securitisations for sustainability purposes. Such 'strategic' discussion will serve as background for Chaps. 4, 5 and 6, which will focus on how market actors, state, and international authorities have intervened, in their rule-making capacity, on these focal points.

[2] See Chap. 2, Sect. 2.4.

[3] On the various functions of law for finance, see Chap. 2, Sect. 2.4.

[4] Katharina Pistor, 'Law in Finance' (2013) 41 *Journal of Comparative Economics* 311, 312; ibid., 'A Legal Theory of Finance' (2013) 41 *Journal of Comparative Economics* 2 315, 322–323. See Chap. 2, Sect. 2.3.

[5] See, however, Securitisation Regulations 2024 (UK).

[6] On the role of standard provisions, see Chap. 5, Sect. 5.2.3.

3.2 Securitisation and Securitisations: A Bird's Eye View over Core Elements and Actors

Securitisation (which is sometimes referred to as 'structured finance'[7]) is a financing mechanism which is mostly recurrent across larger-sized entities, including banks and companies.

Pistor traces the origins of securitisation back to the creation of cooperatives at the end of eighteenth century in Prussia. The creation of independent structures, jointly responsible for their debts, helped indebted aristocrats in the wake of the Seven Years' War.[8] Other commentators have referred to the 1970s, when US government-sponsored agencies bought bundles of residential mortgages in order to boost the market. As a means to finance the purchases, they started to issue certificates, giving investors the right to the principal and to the interest of the underlying mortgages.[9]

These historical developments have common traits: they involve a bundle of long-term, illiquid credit claims, and the intervention of external entities, *other* than those originating the claims. These latter entities take

[7] It is possible to understand 'Structured finance' as an overarching category capturing traditional securitisations, 'synthetic' ones as well as 'resecuritisations' (also referred to as transactions for the 'repackaging' of bonds), see H Wagner, A Heinzmann, 'Structured finance in Luxembourg: a new dawn?' in Bourin, P., & Thieltgen, N. (Eds.), *Droit bancaire et financier au Luxembourg* (Larcier 2014) Vol. IV, 2231–2233, 2240–2244. Under the EU framework, the (Sect.equently amended) Credit Rating Agencies Regulation used the term 'structured finance instruments' to capture securitisation instruments. See art 3 (1)(l) Regulation 1060/2009 of the European Parliament and of the Council of 16 September 2009 on credit rating agencies (2009) OJ L 302/1. More recent examples of EU legislation refer to 'securitisation' instead, such as Regulation 2017/2402 of the European Parliament and of the Council of 12 December 2017 laying down a general framework for securitisation and creating a specific framework for simple, transparent and standardised securitisation, and amending Directive 2017/35 (2017) OJ L 347/35 (hereinafterSecuritisation Regulation). In market practice, 'structured finance' and 'securitisation' are often used interchangeably. See Fitch Ratings, 'Representations, Warranties and Enforcement Mechanisms in Global Structured Finance Transactions' (2016) -warranties-enforcement-mechanisms-in-global-structured-finance-transactions-31-05-2016 accessed 5 December 2025.

[8] Katharina Pistor, *The Code of Capital: How the Law Creates Wealth and Inequality* (Princeton University Press, 2019) 93–94.

[9] *inter alia,* Vincenzo Bavoso, 'Legal Issues on Financial Innovation and Structured Finance' (2012) 34 The Company Lawyer 1, 3–4.; Jonathan C Lipson, 'Defining Securitisation' (2012) 85 *Southern California Law Review* 1229, 1247. On the history of these transactions, see also Thomas E Plank, 'Regulation and Reform of the Mortgage Market and the Nature of Mortgage Loans: Lessons from Fannie Mae and Freddie Mac' (2009) 60 *South Carolina Law Review* 3 9, 796–799.

the credit risk (as in the case of the Prussian cooperatives) or transfer the risk thereof to investors (as in the case of US agencies). Modern securitisations follow a similar mechanism: in line with the US example, an external actor is formally part of the operation and the funding for the credit claims comes from capital markets.[10] The integration of such additional entities ('structures') in the operation also shows why they fall within the broader category of 'structured finance'.

Lipson persuasively identified three indispensable elements to qualify financial transactions as 'securitisations'. First, they derive from an 'input' which consists of a pool of assets producing a regular and predictable flow of funds, such as loans, mortgages or leases.[11] Second, they require a legally distinct 'entity' which formally holds or takes the credit risk of the assets[12], generally called a special purpose vehicle (SPV). Third, they lead to an 'output', i.e. the issuance of securities to raise finance, generally called ABSs.[13] These provide upfront liquidity to SPVs, and indirectly benefit the entities that have arranged the deal ('originators' when they are the direct creditors of the claims, 'sponsors' when they purchase them for the operation from third parties). Credit claims 'back' the operation in that they enable the repayment of investors.[14]

An example to illustrate this structure, which I will delve into in greater detail later in the following chapters, involves a transaction where Toyota Finance (the 'originator') pooled together a series of car leases ('input') and assigned them to an independent 'entity'. Subsequently, the latter issued ABSs which investors could purchase, directly providing liquidity to Toyota Finance. Such investors received repayment when the underlying leases matured.[15]

Beyond these three distinctive traits, the structure of securitisation transactions tends to be very deal-specific. A recurrent element is the use of two external ratings, which generally evaluate various aspects, with a specific focus on the likelihood of default of the portfolio. Rating agencies tend to publish ratings even if unsolicited by the parties involved in

[10] Lipson, ibid., 1240–1241.
[11] On the types of assets which can be securitised, see Sect. 3.3.1.–3.4.1.
[12] On the formal rights of SPVs over the securitised assets, see below Sect. 3.3.2.
[13] Lipson (n. 9) 1271–1275. *Schwarcz* proposes a similar definition, albeit broader. See Steven L Schwarcz, 'What Is Securitisation? And for What Purpose?' (2012) 85 *Southern California Law Review* 1283.
[14] For the different 'pass-through' and 'pay-through' structures, see Sect. 3.3.3.1.
[15] See Chap. 4, Sect. 4.4.

the deals, by relying on publicly available information. Entities putting the transaction in motion are also likely to introduce 'credit enhancement mechanisms' which should safeguard the repayment to investors. These could, in practice, consist of contractual clauses where originators bind themselves to repay in case of default or deterioration of quality of the securitised assets. Alternatively, they may provide letters of credit or 'over-collateralise', i.e. add more assets to the portfolio than would be necessary to repay investors. All these measures should improve the marketability of the operation and improve the ratings of the securities.[16] Moreover, another common feature is the presence of a 'waterfall scheme', determining the order of repayment of ABSs: in other words, the issuance is structured around various 'tranches', which distribute the risks and losses among investors.[17] Such tranches, different in their pricing, should also cater to various types of investors with different risk tolerances. Generally, the 'senior' tranche will be the first in the repayment order. As it promises safer returns, it may attract investors like government-sponsored entities. There will then be one or more 'mezzanine' or 'junior' tranches. By offering higher returns for higher risks, these could target investment funds or hedge funds.[18] Lastly, a 'residual interest' tranche should provide a right to the outstanding balance after other investors have been repaid.[19] In some cases, the originator may retain it for risk-management and regulatory purposes.[20]

The following diagram (Fig. 3.1) provides a simplistic illustration of the structure of these deals, while acknowledging that there may be significant deal-to-deal variations in practice.

Financial innovation has fostered several variations of this original structure. One trend, which especially gained traction during the 2007 global financial crisis, encompasses cases where SPVs do not purchase credit claims but rather *other* asset-backed securities. [21] These operations (also referred to as 'resecuritisations') lead to the issuance of securities

[16] David Ramos Muñoz, *The Law of Transnational Securitisation* (Oxford University Press, 2010) 5. Jacob Bonavita, 'Asset Securitisation in Germany: Risk Transfer or Legal Transformation?' (2016) 17 *European Business Organisation Law Review* 15, 20.

[17] Definition in line with (1), Art 2(1) Securitisation Regulation.

[18] Ramos Muñoz (n. 16) 3, 5. For an informative example of different tranches and investors, see Pistor, *The Code of Capital* (n. 8), 84–85.

[19] Ramos Muñoz (n. 16) 4–5.

[20] Ibid., 3, 5.

[21] The term is widely used across international standards and EU legislation, see Art 8 Regulation 2017/2402.

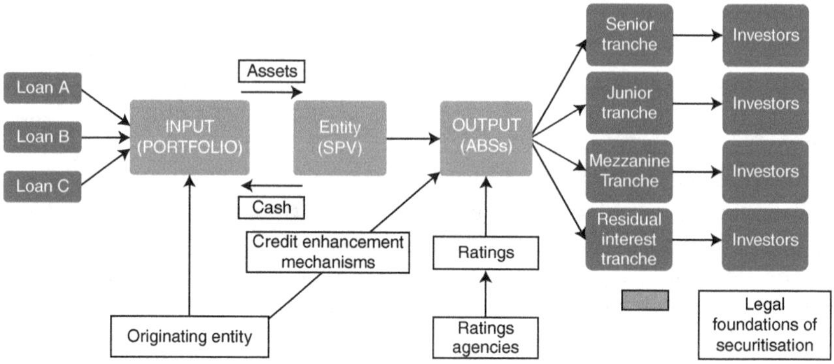

Fig. 3.1 A Simplistic Overview of Securitisation Operations

such as collateralised debt obligations (CDOs). Their peculiarity lies in the use of *other* financial instruments traded on capital markets to back the repayment.[22]

Alternatively, 'synthetic' securitisations[23] enable the transfer the risks of the underlying exposures (an objective which may in itself give rise to regulatory advantages[24]) while originating entities retain them on their balance sheet. The transfer of risk is generally operationalised through a 'credit protection agreement', giving rise to a guarantee or to credit derivatives[25], like credit default swaps or credit-linked notes.[26] Originators

[22] Lipson (n. 9) 1251; Alastair Hudson, *The Law of Finance* (Sweet & Maxwell ed, 2nd edn, 2009) 1304.

[23] The present work will use the term 'synthetic' securitisations to capture transactions not involving the assignment of the assets to SPVs, in line with EU legislation (11), art 1 (10) Securitisation Regulation, as amended by Regulation (EU) 2021/557 (2021) OJ L 116/1, and also in line with some commentators, Bonavita (n. 16) 20–21; Vincenzo Bavoso, 'Filling the Accountability Gap in Structured Finance Transactions: The Case for a Broader Fiduciary Obligation' (2016) 23 *Columbia Journal of European Law* 5. However, other commentators use 'synthetic securitisation' as an umbrella term which could capture both transactions based on derivatives and CDOs. see, among others, '*CDOs (. . .) are a specific and notorious example of the broader phenomenon of 'synthetic' Securitisation'*, Lipson (n. 9) 1234.

[24] For an overview see Ramos Muñoz (n. 16) 324–350.

[25] Art 2 (10), 26b (5) Regulation 2017/2402.

[26] The latter form has become way less prominent in Europe after the 2007 Global financial crisis, see Para 3, Report from The Commission to the European Parliament and the Council on the creation of a specific framework for simple, transparent and standardised synthetic securitisation, limited to balance-sheet synthetic securitisation (2020) 284. On the use of derivatives across synthetic securitisations see Lipson (n. 9) 1234; Bavoso, 'Legal Issues' (n.

undertake to pay an amount ('credit protection premium') to investors, while the latter commit to pay an amount ('credit protection payment') to originators in case of contractually defined 'credit events'.[27] Due to the central role of the risk transfer mechanism, the involvement of an external 'entity' (i.e. the SPV) is not strictly necessary.[28]

Following the aforementioned three-pronged definition of securitisation, the present work does not examine covered bonds despite their similarities. The issuance of covered bonds follows the use of a similar 'input', since it is also linked to a portfolio of assets. However, the operation differs in practice since investors have a security interest over the asset pool, in addition to a contractual claim against the originator.[29] Unlike securitisation, covered bonds do not arise from 'tranching', so investors' claims will always correspond to the entire amount of 'covered' assets. Besides, an external entity is not involved since those who originate the credit claims are generally also those issuing covered bonds.[30] These peculiarities also result in a widely different legal treatment.[31] Correspondingly, even though sustainable finance has also affected the issuance of covered bonds (and fostered 'green' and more broadly 'sustainable' covered bonds[32]), their legal framework is outside the scope of the present analysis.

9) 7. Ramos Muñoz (n. 16) 8–9; D Ramos Muñoz, 'In Praise of Small Things: Securitisation and Governance Structure' (2010) 5 Capital Markets Law Journal 363, 365–366.

[27] Art 2(26) Securitisation Regulation.

[28] Para 2., Report from the Commission (n. 26).

[29] See For example, The Regulated Covered Bonds Regulations (2008) SI 346, rFor example, 1 (UK); Art 3(1) Directive 2019/2162 of the European Parliament and of the Council of 27 November 2019 on the issue of covered bonds (2019); Sascha Kullig, Luca Bertalot and Jennifer Johnson, 'European Covered Bond Fact Book 2024' (2024) 33–34 https://hypo.org/app/uploads/sites/3/2024/08/FactBook-2024_web.pdf accessed 3 June 2026. Pierre De Gioia Carabellese, Camilla Della Giustina, *The law of securitisations: from crisis to techno-sustainability* (Routledge 2023), 66.

[30] In certain instances, they may also purchase the assets from third parties, yet there would not be a further formal assignment to a distinct entity. See Paolo Bontempi, *Diritto bancario e finanziario* (Giuffrè, 6th edn 2019), 204–206.

[31] E.g. there are special rules on the authorisation and supervision of issuers to safeguard their creditworthiness, see Art 52 (4) Directive 2009/65 of the European Parliament and of the Council of 13 July 2009 on the coordination of laws, regulations and administrative provisions relating to undertakings for collective investment in transferable securities (2009) OJ 302/32; Art 4, Directive (EU) 2019/2162 of the European Parliament and of the Council of 27 November 2019 on the issue of covered bonds and covered bond public supervision and amending Directives 2009/65/EC and 2014/59/EU.

[32] Climate Bonds Initiative, 'Green Securitisation: Unlocking Finance for Small-Scale Low Carbon Projects' (2018) 3. https://www.climatebonds.net/data-insights/publications/

Due to space constraints, the present chapter will not focus on all the manifold areas where the law shapes the functioning of securitisations. Instead, the following section will provide a succinct analysis of the 'foundational' aspects that lie in the background of the subsequent 'strategic' analysis.

3.3 The Legal Foundations of Securitisation

A premise for analysing securitisation under a legal lens is the 'constitutive' power of law, andnamely its ability to vindicate the rights over financial assets, as described by Legal Institutionalists such as Pistor.[33] Securitisation is possible thanks to a series of rules which delineate the proprietary and contractual rights over the various assets of the operation.

As mentioned in Chap. 1, the law governing financial markets is 'hybrid', i.e. neither fully 'private' nor fully 'public'. Its legal framework is the result of the complementary effort of market actors, state and international authorities (i.e. the two spheres which I have defined as 'private law' and 'public law'[34]).

First, market actors establish legally binding obligations and clauses throughout securitisation documentation.[35] These cover, among other things, the rights over the portfolio assets and the securities, as well as the mutual obligations between the different parties.

The following diagram (Fig. 3.2), which builds on the one in Sect. 3.2, summarises in orange the key documents comprising securitisation

green-securitisation-unlocking-finance-small-scale-low-carbon-projects accessed 5 December 2025; Frank Damerow, Sean Kidney, Stuart Clenaghan, 'How Covered Bond Markets Can Be Adapted for Renewable Energy Finance and How This Could Catalyse Innovation in Low-Carbon Capital Markets – Unlocking Bank Lending in an Era of Capital Constraint and Limited Public Budgets' (2021) Climate Bonds Initiative Discussion paper http://ecosystemservices.co.uk/_docs/Climate%20Bonds%20RE%20covered%20bonds%2022May2012%20%28published%29.pdf accessed 3 June 2026. for an overview of their distribution see Kullig, Bertalot, Johnson (n. 29) 41–77.

[33] Pistor, 'Legal Theory of Finance' (n. 4) 317–321; Simon Deakin et al., 'Legal Institutionalism: Capitalism and the Constitutive Role of Law' (2017) 45 Journal of Comparative Economics 194–195. On different insights into the 'constitutive' role of law, see Chap. 2, Sect. 2.4.

[34] See Chap. 2, Sect. 2.3.

[35] Like the articles of association or prospectuses, see Sect. 3.3.2.–3.3.3. see also Lipson (n. 9) 1234.

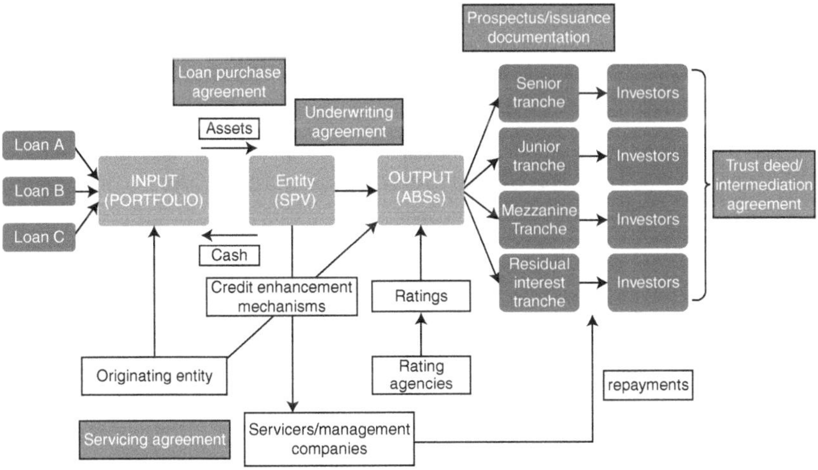

Fig. 3.2 An Overview of Securitisation Documentation

documentation, which I will recurrently mention throughout the subsequent chapters.

Second, national and EU provisions intervene to recognise and legitimise the various phases of these transactions.[36] For instance, legislation lays out the requirements for assigning credit claims, as well as the rules for determining who has the 'entitlement', i.e. the right over certain assets, in conflicting situations.[37] Statutory provisions also articulate contractual rights between the parties by providing 'mandatory rules', which automatically apply to all transactions, or 'default rules', which allow parties to opt out.[38]

Before examining the aspects of securitisation law with a connection to sustainability objectives, it is essential to draw attention to the foundational

[36] Legal institutionalists have also acknowledged the role of state legislation in vindicating and legitimising the rights over financial assets. see Simon Deakin et al., 'Legal Institutionalism: Capitalism and the Constitutive Role of Law' (2017) 45 *Journal of Comparative Economics* 194–195.

[37] On property law as a source of 'entitlements', see Guido Calabresi, AD Melamed, 'Property Rules, Liability Rules, and Inalienability: One view of the Cathedral' (1972) 85 *Harvard Law Review* 6 1089, 1090, 1092.

[38] Ayres I, Gertner R, 'Filling Gaps in Incomplete Contracts: An Economic Theory of Default Rules' (1989) 99 The Yale Law Journal 87, 92.

aspects which lie in the background of the analysis. Following the three-pronged definition of securitisation provided above21, I will differentiate the rules targeting the 'input', the legally distinct entities (SPVs), and the 'output' of these operations.

3.3.1 The 'Input': Rights over the Portfolio

The core concern for market actors putting securitisations into motion ('originators') is to convert their credit claims into cash. Thus, the primary legal question relates to the creation and to the assignment of a portfolio of claims. A number of legal issues, cutting across property and contract law, arise during such a phase. Clearly establishing who holds the rights over the portfolio assets and the type of proprietary right is a crucial part of the securitisation process[39], also holding broader regulatory implications in contexts such as bankruptcy law[40], accounting law, and capital requirements, which are outside the scope of the present analysis.[41] Depending on the jurisdiction, SPVs may have proprietary or contractual rights over the assets.24

The traditional structure of securitisation deals relies on a 'true sale', with SPVs acquiring full title over the asset pool.[42] Correspondingly, EU rules qualify traditional securitisations as those involving the 'transfer of ownership of exposures'.[43] Technically, the transaction acts as an assignment, as the wording of some national statutory provisions across

[39] On the significance of certainty over legal rights for securitisations, see Pistor, *The Code of Capital* (n. 8) 95–98.

[40] See For example, Ramos Muñoz, 'In praise of small things' (n. 26) 36–39, 68–70.

[41] Ibid., 300–350; Alessandro Scopelliti, 'Securitisation, Bank Capital and Financial Regulation: Evidence from European Banks', *ECB Forum* (2016) https://www.ecb.europa.eu/pub/conferences/ecbforum/shared/pdf/2016/scopelliti_paper.pdf accessed 8 December 2025; for an overview of the advantages under the US framework, see Gary B Gorton, Andrew Metrick, 'Securitised Banking and the Run on Repo' (2012) 104 *Journal of Financial Economics* 3 425, 430–432.

[42] Hugh Beale, Michael Bridge et al, *The Law of Security and Title-Based Financing* (3rd edn, Oxford University Press 2018) paras 2.12; 2.26; 7.76. For a comparative analysis see Georg Graf, Monika Hinteregger, Manuela Weissenbacher et al, 'Case 11: Securitisation' in Lionel Smith, Michele Graziadei, Ugo Mattei (eds) *Commercial Trusts in European Private Law* (Cambridge University Press 2005). 513, 525, 529; Latham & Watkins, *The International Comparative Legal Guide to: Securitisation 2013* (2013) Global Legal Group; about England, Thomas W Albrecht, Sarah J Smith, 'Corporate Loan Securitisation: Selected Legal and Regulatory' (1998) 8 *Duke Journal of comparative & international law* 411, 431–447.

[43] Art 2 (9) Securitisation Regulation.

EU countries demonstrates.[44] This means that the transaction follows the regime for obligations and contracts, so that it may be sufficient for the transferee to acquire possession of the assets without a written contract.[45] Different rules, predominantly jurisdiction-specific, determine the requirements for a valid assignment and ensure that the agreement between the parties is enforceable.[46] For instance, originating entities may have to give notice to the debtors of the claims composing the portfolio.[47] Legislation may also require the keeping of written record as evidence of the assignment[48] or lay out the criteria to identify the assets falling within the portfolio, especially when the assignment takes place before certain claims come into existence.[49] Therefore, the assignment is likely to entail costs and some level of publicity which may run counter to the interests of originators that prioritise simpler and more confidential deals. However, various market actors are still likely to opt for this proprietary arrangement in order to ensure the full insulation of the assets in SPVs for bankruptcy and other regulatory purposes.[50]

In some jurisdictions, it is also possible to assign the beneficial title over the asset pool to SPVs, such as the 'equitable assignment' in common law countries.[51] This method relieves originating entities from the duty to give prior notice to debtors.[52] However, debtors will continue to discharge their obligations (i.e. make the payments) only against the originator.[53] Leaving aside the priority and bankruptcy implications, which are beyond

[44] '*Abtretung*' (assignment) under German Law, Latham & Watkins (n. 42) 148, 213. 'cessione' under Italian Law, Art 1264 cc.; art 7.1 c 6 L 130/1999 on securitisation (IT);

[45] For example, in Italy under art 1262 Civil Code (cc).

[46] For an overview of formal requirements across various EU countries see Ramos Muñoz, *Transnational Securitisation* (n. 16) 49–51. For a more detailed comparative analysis see ibid.

[47] Additional registration requirements or notice requirements may also exclusively apply in relation to certain types of assets, see under Italian law, Art 4(1)(2) L 130/1999 (IT); in England, see S 136 Law of Property Act 1925 (UK), *Dearle v Hall* (1828) 3 Russ 1; SS 4, 29, 32 Land Registration Act 2002.

[48] For instance, EU rules require originating entities to make securitisation documentation available to competent authorities and to investors upon request. Art 7 (b)(ii) Securitisation Regulation.

[49] See Ramos Muñoz, 'In praise of small things' (n. 26) 56–57.

[50] See below Sect. 3.4.1.

[51] In England, Albrecht, Smith (n. 42) 434. Similar proprietary structures are also common in other countries like Ireland and Australia, see Latham & Watkins (n. 42) 46, 192–194.

[52] S 136 Law of Property Act 1925 (UK).

[53] Albrecht, Smith (n. 42) 435–437.

the scope of the present analysis[54], the convenience of this arrangement is the possibility of keeping the assignment confidential to debtors.[55]

At the same time, alternative solutions ensuring the partial or total separation of SPVs' assets from those of originators include the creation of security interests or trusts over the portfolio or its proceeds.[56] For instance, the clauses of securitisation documentation may specify that SPVs are temporarily secured creditors for all payments on the securitised assets, while originators retain ownership. They may then prescribe that SPVs acquire title over the assets if originators go insolvent. This arrangement is common in Germany.[57]

SPVs can also act as trustees of the asset portfolio thanks to a combination of statutory and contractual arrangements.[58] On the one hand, legislation can *allow* SPVs to take the form of trusts or funds, especially in civil law countries such as France, Spain and Portugal, where other trust-like structures are unlikely or not allowed. This would mean that SPVs are the formal owners of the securitised assets, while investors are the 'beneficial' owners.[59] Alternatively, a deed may also designate the SPV as trustee and create a trust or a 'master trust'.[60] The main requirements for these trusts to be valid are generally the evidence of a certain intention, the identification of the assets and of the beneficiaries.[61] All in all, parties may prefer

[54] *Dearle V Hall* (1828) 3 Russ 1, Beale, Bridge et al (n. 42) 7.93, 7.139.

[55] Graf, Hinteregger et al (n. 42) 509, 531.

[56] Under certain limited circumstances, this may result in lower formalities for the creation of a security interest as a result of the Directive 2002/47/EC of the European Parliament and of the Council of 6 June 2002 on financial collateral arrangements (2002) OJ L 168/43. See Chambers and Partners, 'Securitisation 2023 – UK law and practice' (2023) 6–7 https://www.lw.com/admin/upload/SiteAttachments/Chambers%20Securitisation%202023%20UK.pdf accessed 3 June 2026.

[57] Bonavita (n. 16) 31.

[58] Ramos Muñoz, *Transnational Securitisation* (n. 16) 154. On the complementarity of property and trust in finance and in the context of securitisation, see Pistor, *The Code of Capital* (n. 8).44–48, 80–86.

[59] For some examples, see Ramos Muñoz, 'In praise of small things' (n. 26) 372.

[60] The latter model is especially popular in US, see ibid., 363, 372, 375–376; Ramos Muñoz, *Transnational Securitisation* (n. 16) 52, 53, 93. For its use in England, see ACPR – Banque de France, 'Le marché de la titrisation en Europe: caractéristiques et perspectives' (2014) 31 Analyses et syntheses, p. 15 https://acpr.banque-france.fr/sites/default/files/medias/documents/20140701-marche-titrisation-en-europe-caracteristiques-et-perspectives2.pdf accessed 8 December 2025.

[61] *Knight v Knight* (1855) 49 ER 58; *Re London Wine Co Ltd [1986]* PCC 121; *Morice v Bishop of Durham* (1805) (UK); Ramos Muñoz, 'In praise of small things' (n. 26) 54.

these arrangements over outright assignments to avoid formalities and maintain a direct relationship with borrowers of the underlying claims. At the same time, they may opt for outright assignments since these have clearer consequences in a bankruptcy scenario and are less exposed to the risk that courts recharacterise the transaction.[62]

Alongside the statutory provisions that set the validity requirements for these arrangements, the other key legal tool is the documentation giving SPVs legal rights over the portfolio. This may have different denominations, such as 'receivables purchase agreement' when the assets are assigned, or 'trust deed' (also referred to as 'declaration of trust'[63]) when SPVs act as trustees.[64]

As an alternative to the outright assignment of the exposures, originators may also conclude 'sub-participation agreements' with SPVs. These would transfer exclusively the contractual rights to the payments, in exchange for a financial contribution.[65] EU law recognises such agreements as valid for securitising assets, as long as the ABSs no longer constitute payment obligations of the originator.[66]

Transactional documentation acquires an even more prominent role in synthetic securitisations, which transfer only the risks associated with the exposures.[67] In the absence of a formal assignment of the assets, the 'credit protection agreement' becomes the key instrument for determining how risks are allocated between originators and protection buyers (i.e. SPVs, investors, or third parties). The agreement can give rise to a guarantee or, less commonly, to a credit derivative.[68] This agreement

[62] The present work does not aim to provide an in-depth analysis of the bankruptcy implications of security interests and trusts or of the respective re-characterisation risks, see Bonavita (n. 16) 31. Graf, Hinteregger (n. 42) 509. For re-characterisation risks, see Ramos Muñoz, *Transnational Securitisation* (n. 16) 36–39, 68–70.

[63] See the EU Securitisation Regulation, which lists 'any relevant *declaration of trust*' as part of the transactional documentation, see art 7 (b)(ii).

[64] See Fig. 2.1 above. For various examples of agreements and their clauses, see Chap. 5, Sect. 5.4. ff.

[65] For an example see For example, (22)-(24) *Szef Krajowej Administracji Skarbowej V O. Fundusz Inwestycyjny Zamknięty reprezentowany przez O SA* (2022) CJEU C-250/21. On the legal configuration of 'sub-participation agreements' under UK law see also Albrecht, Smith (n. 42) 442–443.

[66] Art 2 (9) Securitisation Regulation.

[67] See above Sect. 3.2.

[68] In 2023, 96% of the synthetic securitisation transactions complying with the 'simple, transparent and standardised' regime relied on a financial guarantee, while only 4% on a derivative. ESMA, 'TRV risk analysis – EU securitisation markets overview' (2023) 12 https://

generally identifies the 'Credit Events', i.e. the circumstances in which the protection buyer will have to make payments, such as the restructuring of the underlying exposures or claims over the underlying exposures remaining outstanding for over ninety days.[69] It also stipulates whether the protection buyer will pay upfront, giving rise to a 'funded' protection agreement, which is the dominant structure for EU synthetic securitisations[70], or only upon the occurrence of Credit Events (as in 'unfunded' transactions).

3.3.2 *The Intermediating Entities (Special Purpose Vehicles, SPVs)*

Defining the rights over the portfolio is contextual or follows the creation of SPVs, i.e. legal entities specifically devoted to the issuance of one or multiple securitisation deals.[71] Next to holding the portfolio assets, these entities are responsible for the issuance of ABSs and for the overall management of the repayment to investors, except in certain cases which I will illustrate below.[72] Similar entities, commonly used across 'resecuritisations'[73], are 'structured investment vehicles' (SIVs) (or 'conduits').25 They tend to purchase only longer-term debt securities, such as ABSs, and to only issue short-term securities.[74] two Two relevant legal questions that

www.esma.europa.eu/sites/default/files/2023-09/ESMA50-524821-2908_TRV_risk_analysis_-_EU_securitisation_markets_overview.pdf accessed 8 December 2025.

[69] European Banking Authority (*hereinafter*EBA), Report on synthetic securitisation (2015) 26 https://www.eba.europa.eu/sites/default/files/documents/10180/983359/03ed077f-cdf9-4f6c-901b-fb7fda0e44a3/EBA-Op-2015-26%20EBA%20report%20on%20synthetic%20securitisation.pdf accessed 3 June 2026.

[70] ESMA (n. 68) ibid. Albrecht and Smith (n. 42) 443, 444.

[71] Also referred to as 'special purpose entity' (SPE) or 'securitisation special purpose entity', see European Banking Authority, 'Developing a Framework for Sustainable Securitisation' (2022) 5, 6 https://www.eba.europa.eu/sites/default/files/document_library/Publications/Reports/2022/1027593/EBA%20report%20on%20sustainable%20securitisation.pdf . Accessed 3 June 2026.

[72] See Sect. 3.4.5.

[73] See Sects. 3.1.–3.2.

[74] Like asset-backed commercial paper (ABCP), see below Sect. 3.3.3. Basel Committee on Banking Supervision, Board of the International Organisation of Securities Commissions, 'Criteria for Identifying Simple, Transparent and Comparable Securitisations' 5 https://www.iosco.org/library/pubdocs/pdf/IOSCOPD494.pdf. accessed 8 December 2025. Daniel I Castro, 'Structured Investment Vehicles (SIVs)' in Gary Strumeyer, Sarah Swammy (eds), *Capital Markets: Evolution of the Financial Ecosystem* (John Wiley 2017) 381, 386.

come into play relate to the modalities for the creation of SPVs and to the insulation of the related assets.

3.3.2.1 Legal Issues Concerning the Creation of SPVs

Across EU countries, it is up to national legislation to determine the forms and structures which SPVs can take. For instance, national rules may qualify them as companies and subject them to the company law regime. [75] Statutory provisions are also likely to introduce the explicit option to set up SPVs as trusts or as investment funds.[76] For instance, Luxembourg law gives the option to create SPVs as companies or as 'funds': in the latter case, they can follow the law for collective investment undertakings or the trust and fiduciary law regime.[77] Especially when created as 'funds', the law may also allow or require the delegation of decision-making powers from SPVs to external management companies.[78]

Depending on the legal construction of SPVs, there are also likely to be corresponding contractual or quasi-contractual agreements, which make the creation of the entities effective and articulate their rights and purpose. These could be management regulations, when SPVs are set up as investment funds[79], or articles of association when incorporated as companies.[80]

3.3.2.2 The 'Shielding' of SPVs

Another fundamental function of statutory provisions and of the terms of the securitisation documentation is to ensure the insulation ('shielding') of SPVs' assets.[81]

[75] Ramos Muñoz, *Transational Securitisation* (n. 16) 53–55. Ramos Muñoz, 'In praise of small things' (n. 24) 372.

[76] This latter option is possible in Luxembourg, Italy and France, see art 2 Loi du 22 mars 2004 relative à la titrisation (LU) https://www.cssf.lu/wp-content/uploads/L_220304_titrisation.pdf accessed 8 December 2025; Art 7 (1)(b) L 130/1999 (Italy); Article L214-180 Code Monétaire et financier (FR).

[77] See Art 2, 6 (1) ibid.

[78] Ramos Muñoz, 'In praise of small things' (n. 24) 372–373. See For example, in Luxembourg and in Spain, art 2 Loi du 22 mars 2004 relative à la titrisation (LU), Art 25-26 Ley 5/2015 de fomento de la financiación empresarial (2015) (ES) https://www.boe.es/buscar/act.php?id=BOE-A-2015-4607 accessed 8 December 2025.

[79] In Luxembourg, management regulations are required for all SPVs created as funds, see Art 6 (1) ibid.

[80] In Luxembourg, see For example, art 5, 20 ibid. In Italy, see the provision referring to the elements that contracts and articles of association for the creation of SPVs must include, art 7.1 (5) L 130/1999.

[81] On the 'shielding' of SPVs as the consequence of legislation and contractual law provisions, see Bonavita (n. 16) 26.

As anticipated in Chap. 2, the 'shielding' of the assets also arguably constitutes a strategic area which could contribute to the attainment of sustainability goals.[82] The notion was first developed by organisational law scholars[83] to illustrate one of the most distinctive consequences of separate legal personality for entities like companies, partnerships and trusts. The concept encapsulates the idea that only the creditors of a given entity can enforce the respective assets, and not external creditors or personal creditors of the entity's shareholders.[84]

Acknowledging SPVs as legal entities, distinct from their originators, also means that they have their own patrimony. [85] They will use it to repay investors and for any outstanding projects or purposes, as long as the transactional documentation allows it.[86] In turn, the SPVs' autonomy may also positively affect the rating and increase the marketability of the securities. There could also be statutory provisions, especially across civil law countries, that explicitly acknowledge the separation between SPVs' and originators' patrimonies.[87] For instance, SPVs created as investment funds under Luxembourg law lack legal personality: for this reason, legislation specifically acknowledges that they have their own patrimony.[88]

Furthermore, SPVs' 'shielding' is also a result of the rules ensuring the exclusion of these entities from bankruptcy proceedings ('bankruptcy-remoteness').[89] It is visible on two fronts. First, legislation and contractual

[82] On SPVs' goals, see below Sect. 3.4.3.1.

[83] The concept was first conceptualised by Hansmann and Kraakman in Henry Hansmann, Reinier Kraakman, 'The Essential Role of Organisational Law' (2000) 110 *The Yale Law Journal* 387, 390–394.

[84] Ibid. see Hansmann, Kraakman, 'Anatomy of corporate law', 3–4.

[85] What has also been referred to as 'asset partitioning', see also Chap. 2, Sect. 2.6. ibid., 'Law and rise of the firm' (2006) 119 *Harvard Law Review* 1333.1338–1339. Hansmann and Kraakman, 'Organisational law' (n. n 83). 389, 392–393.

[86] See below Sect. 1.3.2.

[87] In the civil law legal doctrine, the equivalent of 'shielding' could be the notion of 'separate patrimonies', which captures a group of assets set aside for a specific purpose, Hansmann, Kraakman, ibid., 'Law and rise of the firm', 1338.

[88] Art 6(2) Loi du 22 mars 2004 relative à la titrisation (LU).

[89] What Hansmann and Kraakman have labelled 'strong entity shielding', see Hansmann, Kraakman, 'Law and rise of the firm', 1338–1339. In the context of securitisation, see David Ramos Muñoz, 'Bankruptcy-Remote Transactions and Bankruptcy Law-a Comparative Approach (Part 1): Changing the Focus on Vehicle Shielding' (2015) 10 *Capital Markets Law Journal* 239, 233–234.

provisions tend to ensure that SPVs count as distinct legal entities in the event of originators' bankruptcy. In doing so, they put securitisation assets, i.e. the cash flows from the claims within the portfolio and from investors, beyond the reach of originators' creditors.[90] However, it must be noted that SPVs' exposure to originators' bankruptcy varies in practice depending on the arrangements between originating companies and SPVs. For instance, the outright assignment of the full title to the assets should in principle ensure the complete separation of SPVs. There would not be a question of exposure to originators' insolvency when SPVs only have contractual obligations regarding the securitised assets, as in the case of synthetic transactions.[91] By contrast, when the documentation only designates SPVs as secured creditors, the portfolio assets may still be part of originators' insolvency proceedings.[92]

Second, some commentators have described SPVs' status as one of 'enhanced bankruptcy-remoteness', as national legislation across EU countries tends to exclude these entities from bankruptcy altogether,[93] or they only make alternative liquidation proceedings applicable.[94] This exclusion could also derive from contractual bargaining following the introduction of 'non-petition clauses', which prevent investors, originators or other entities[95] from filing for SPVs' bankruptcy.[96] The articles of association may also curtail the ability of SPVs' directors to file for insolvency by requiring a unanimous decision of the board or giving one

[90] Ibid.

[91] See above Sect. 3.2.

[92] On how the type of proprietary rights affects SPVs' exposure to insolvency, and on a German case where SPVs' assets were held not to be bankruptcy-remote since the entity was merely a secured creditor, see Bonavita (n. 16) 31–33. Due to space constraints, the present analysis will not illustrate the circumstances where judges have or are more likely to re-characterise the assignment of assets or the nature of SPVs in order for it to participate in originators' insolvency. See further in Ramos Muñoz, 'In praise of small things' (n. 24). 22, 23, 85, 88–90; ibid., Ramos Muñoz, 'Vehicle shielding' (n. 89) 257.

[93] Bonavita (n. 16) 33. Ramos Muñoz, *Transnational Securitisation* (n. 16) 5, 243.

[94] For example, under art 2447-novies cc (IT); Art L 214-43 Code monétaire et financier (FR).

[95] Like underwriting entities or servicers. On their respective function, see Sect.s. 3.4.2.1.–3.4.5.

[96] Bonavita (n. 16) 34; Ramos Muñoz, *Transnational Securitisation* (n. 16) 96, 98–100; Phillip Wood, 'Securitisation structures' in *Project finance, securitisations and subordinated debt* (3rd edn, Sweet & Maxwell 2019) Vol. 7, para 6-024. On the validity of such clauses, see *Highberry Ltd v Colt Telecom Group (No 2)* [2002] EWHC 2815 (UK). Legislation may also expressly recognise these clauses as valid, see in Luxembourg art 64 (3) Loi du 25 février 2022 portant 1° modification de la loi modifiée du 22 mars 2004 relative à la titrisation.

independent director veto rights.[97] As I will illustrate below, transactional documents for the issue of ABSs may also introduce ad hoc enforcement procedures based on specific events ('Events of Default') which *de facto* allow avoidance of bankruptcy proceedings.[98]

3.3.3 The 'Output': The Issuance of ABSs

The third key pillar for securitisations is fund-raising activities which involve the issuance of ABSs by SPVs.[99] ABSs tend to receive similar legal treatment even though they also take different denominations depending on the composition of the portfolio, as I will also discuss below.[100] An exception is asset-backed commercial paper (ABCP), which tends to have a distinct status due to its shorter maturity. For instance, EU Securitisation legislation and Basel Committee prudential requirements provide for different rules for securities with a maturity of up to one year.[101]

The legal nature of ABSs is peculiar. In some jurisdictions, SPVs have the option to issue ABSs as 'equity' or 'debt'.[102] However, tax and regulatory complications generally make debt securities more appealing and widely used. In line with this qualification, investors have a right to be repaid at maturity or in certain circumstances. It also affects the rights of ABS holders: unlike equity, which makes shareholders effective members of the investee companies, debt securities only provide limited participation rights in the decision-making processes of SPVs (what I have defined as 'voice' rights[103]).[104]

The issuance of ABSs generally occurs on capital markets. At the same time, a noteworthy trend is the rise in 'private placements', which entails the sale of the securities to a smaller pool of investors outside the remit of publicly regulated markets. They protect the confidentiality of the transaction,[105] while avoiding certain regulatory requirements[106] and the

[97] Ramos Muñoz, *Transnational Securitisation* (n. 16) 96.
[98] On 'early amortisation events', see below Sect.ection 3.4.2.1.–3.4.4.
[99] See securitisation operations as defined by (1), Art 2(1) Securitisation Regulation.
[100] See below Sect. 3.4.1.
[101] Article 2(7) Securitisation Regulation; Para 40.8, CRE40-Securitisation, BCBS, 'General Provisions, Basel Framework'(2023) https://www.bis.org/basel_framework/chapter/CRE/40?inforce=&published=20201126 accessed 8 December 2025.
[102] For example, under French, Portuguese and Luxembourg law, Ramos Muñoz, 'In praise of small things' (n. 24) 40. Wagner, Heinzmann (n. 7) 2253–2255.
[103] See Chap. 2, Sect. 2.6.
[104] See below Sect. 2.3.5.
[105] As also highlighted by recital (13) Securitisation Regulation.
[106] Like disclosure obligations, see Bavoso, 'Legal Issues' (n. 9) 9.

need for ratings. In the EU, these arrangements have experienced a dramatic rise, to the point that three-quarters of the transactions under the 'Simple, Transparent and Standardised' (STS) regime, which I will cover in Chap. 5[107], were private placements in 2020.[108] As Chap. 4 will illustrate, market actors have previously issued 'sustainable' ABSs both across publicly regulated markets and through private placements. [109]

3.3.3.1 The Offer of ABSs

Before issuing ABSs, prospective issuers generally issue a 'prospectus', or other similar documents, outlining the purpose of the offer and the key characteristics of the securities.[110] They will also articulate investors' rights over ABSs based on their 'tranches' therein.[111] These rights are generally contractual in nature, meaning that investors have a claim once the credit claims mature ('pass-through' structure) or automatically, regardless of the underlying repayment ('pay-through' structure).[112]

At times, investors may also have proprietary rights over ABSs. For instance, when SPVs are trusts or collective investment funds, the issuance creates 'co-ownership' rights in the entity.[113] When SPVs are 'master trusts', they also issue 'trust certificates' in favour of end investors, which

[107] See Chap. 6, Sect. 6.3.

[108] Report from the Commission to the European Parliament and the Council On the functioning of the Securitisation Regulation (2022) 5, 21. https://eur-lex.europa.eu/legal-content/EN/TXT/PDF/?uri=CELEX:52022DC0517 accessed 5 December 2025. The ECB and ESMA also estimated that around the 43% of outstanding securitisations as of 30 December 2023 was not public. ESMA, 'Consultation Paper – –Consultation Paper on the revision of the disclosure framework for private securitisation under Article 7 of the Securitisation Regulation' (2025) ESMA12-2121844265-4462, https://www.esma.europa.eu/press-news/esma-news/esma-consults-revised-disclosure-requirements-private-securitisations accessed 8 December 2025.

[109] See Chap. 4, Sect. 4.3.

[110] For the main functions and items of a prospectus under EU law, see recitals (7)-(26), Art 6 Regulation 2017/1129. For other examples of issuance documentation see below Sect. 2.3.2.

[111] See above Sect. 2.1. Bonavita (n. 16) 20.; art 2 (6) Securitisation Regulation.

[112] Federal Reserve Bulletin, 'Asset Securitization: A supervisory perspective' (1989) 661-662 https://fraser.stlouisfed.org/files/docs/publications/FRB/pages/1985-1989/32502_1985-1989.pdf. Bavoso, ' Accountability Gap' (n. 23) 11. Ramos Muñoz, 'In praise of small things' (n. 24) 6, 7, 40.

[113] This is the case, among others, under Luxembourg law ('fonds copropriete'). See Art 6 (1) Loi du 22 mars 2004 relative à la titrisation; Alain Couret, Alain Stechen, André Prüm et al, *Titrisation* (Anthemis 2008) 49.

evidence an undivided interest in the trust assets to the extent specified in the respective trust deed.[114]

3.3.3.2 The Management of the Issuance

Legislation and the transactional documentation will design ad hoc procedures to oversee the circulation of cash flows from the portfolio and vis-à-vis investors. These arrangements are highly deal-specific and jurisdiction-specific. In some jurisdictions, such Italy, the law designates SPVs as default entities for the management of the issuance.[115] However, it is more common in market practice to only vest them with limited decision-making powers,[116] and to involve additional 'servicers' or 'management companies'.[117] The main legal sources outlining the duties of these entities include legislation and the agreements for their appointment ('servicing' or 'administration agreements'[118]). These generally require them to collect the payments and bring enforcement actions on the portfolio,[119] or to monitor the use of funds, as also illustrated later.[120]

3.4 A 'Strategic' Discussion of Securitisation: Focal Points for Sustainability-Oriented Legal Initiatives

The foundational aspects of securitisation law provide a valuable background which should help the reader navigate the book. Throughout the remaining chapters, the focus will be on selected aspects of securitisation

[114] Steven L Schwarcz, 'Commercial Trusts as Business Organisations: Unraveling the Mystery' (2003) 58 *Business Lawyer* 559, 5–6; Gorton, Metrick (n. 41) 11–12. Federal Reserve Bulletin (n. 112) 10. On the blurring of the line between proprietary and contractual rights, see Bonavita (n. 16) 38. Jongchul Kim, 'Property-Ization: The Process by Which Financial Corporate Power Has Risen' (2018) 3 *Review of Capital as Power* 1 58, 60, 66–69.
[115] 'the amounts paid by the assigned debtor or debtors are applied by the assignee only in satisfaction of the notes issued'. Art 1, c 1 b) L 130/1999 (IT). This is the default position, but the appointment of an additional servicers is also possible. Art 2, c 3)c) ibid.
[116] See Sect. 3.4.3.
[117] 'Management companies' are also generally responsible for the selection of portfolio assets. Ramos Muñoz 'In praise of small things' (n. 24) 383. See below Sect. 2.3.5.
[118] See Fig. 2.1, Sect. 2.2. The EU Securitisation Regulation also mentions the servicing agreement as part of the documentation to be made available to investors, See 2.5, Art 7(1)(b)(iv) Regulation 2017/2042. For its main provisions, see Phillip Wood, 'Transfer and Administration Agreement' in *Project Finance, Securitisations* (n. 96) paras 7–003.
[119] Ramos Muñoz, *Transnational Securitisation* (n. 16) 4, 234.
[120] See Sect. 3.4.5.

law (that I will define as 'focal points'), which are arguably particularly amenable to sustainability-oriented legal initiatives. The distinctive feature is their potential *transformative impact*. In other words, their functions within the securitisation process make them the ideal starting point for adjustments to promote sustainability-related practices at the transactional level (as discussed in Chap. 5) and the statutory level (as explored in Chap. 6). Initiatives based on these focal points can arguably encourage more companies and financial institutions to engage in 'sustainable securitisations' or promote a more efficient design and management of existing securitisations which already present a 'sustainable' profile.[121]

Moreover, the analysis of the legal framework in the following chapters aims to highlight the various costs of utilising securitisation to achieve sustainability goals, as well as how legal initiatives address, and could address, such costs.

Accordingly, the following subsections will present the five focal points and discuss their relevance for the sustainability agenda. They will also map the various costs of integrating sustainability considerations in the design and management of securitisation transactions. Subsequently, I will discuss the extent to which securitisation rules have advanced sustainability goals in these areas and lowered the costs throughout Chaps. 4, 5 and 6. I will also provide a series of recommendations on possible legal changes that leverage these points.

3.4.1 The Composition of the Portfolio

As previously mentioned, the minimum requirement for the creation of a securitisation portfolio is a bundle of assets producing a future stream of income, which generally consists of loans, leases or mortgages, among others. These could range from student loans to aircraft leases, trade and credit card receivables.[122] Those who put the transaction in motion can either directly originate the credit claims (e.g. by extending the loans) or purchase them from third parties.[123] Such 'originators' are the key actors

[121] For the different circumstances where market actors have so far defined securitisation operations as 'sustainable', see Chap. 4, Sect. 4.3.–4.4.

[122] For other examples, see Art 1 (a) Commission Delegated Regulation 2019/1851 of 28 May 2019 supplementing Regulation (EU) 2017/2402 of the European Parliament and of the Council with regard to regulatory technical standards on the homogeneity of the underlying exposures in securitisation.

[123] Under EU law, the 'sponsor' is the one arranging the deal, see Art 2 (5)(b), Art 20 (7) Securitisation Regulation. However, the notion of 'sponsors' is sometimes used to

in the management of the deal: For this reason, they are generally responsible for ensuring compliance with the securitisation framework, such as the drafting and publication of reports.[124]

Bundling together the assets for the creation of the portfolio is such a fundamental step that the types of credit claims also drive the denomination of the securities issued at the end of the process. In other words, we generally differentiate ABSs on the basis of the characteristics of the portfolio: Examples include residential mortgage-backed securities (RMBS), backed by residential mortgages, commercial mortgage-backed securities (CMBS), backed by commercial mortgages, auto asset-backed securities, backed by automobile loans.[125]

The underlying credit claims could also be mixed types of loans, driven by different purposes and extended to various kinds of obligors, such as different types of bank loans. Market actors have generally depicted the resulting securities as collateralised loan obligations (CLOs).[126]

Besides, the composition of the asset pool may also change during the course of the transaction, such as in the case of a 'revolving' portfolio. This pattern, particularly common across CLOs and ABCP programmes[127], sees parties only identifying general criteria for the characteristics of eligible assets and postponing the exact identification thereof. Throughout the duration of the programme, a third party is generally in charge of the periodic selection and replacement of the assets ('collateral manager').[128] Lastly, the previous discussion has highlighted how other securities, or other ABSs may also be pooled together, thereby becoming portfolios for 'resecuritisations'.[129]

capture those who purchase assets from third parties. See ICMA, 'Sustainable Securitisation – –Related questions' (2022) 3 https://www.icmagroup.org/assets/documents/Sustainable-finance/2022-updates/Sustainable-Securitisation-QAs_CLEAN-and-FINAL_2022-06-24-280622.pdf accessed 8 December 2025. For clarity, I will use the term 'originator' to capture actors setting up the deal.

[124] Ibid.

[125] See, among others, the terminology used by Gorton, Metrick (n. 41) 26–27.

[126] Ibid.

[127] See above, Sect. 3.2.

[128] Ramos Muñoz (n. 16) 7, 9, 137 ff. Ramos Muñoz, 'In praise of small things' (n. 24) 366, 370.

[129] See above Sect. 3.2.

Regulating the types of assets that can be securitised, as well as introducing incentives to securitise certain assets, hold untapped potential for embedding sustainability goals in these deals.

3.4.1.1 Preliminary Strategic Considerations (I): Direct Requirements on Portfolios

Laws and contractual provisions on the composition of the portfolio serve multiple functions that point to their strategic potential.

First, legislation can provide a 'frame' by defining the perimeter of the types of assets eligible for inclusion in the portfolio. For instance, EU and national provisions may prevent the securitisation of certain receivables, like those that have not yet come into existence.[130] On a slightly different note, some national frameworks also enable securitisation over *the risks* related to economic activities or over assets other than credit claims, such as IP rights.[131] Like legislation, the documentation for the assignment or the offer of ABSs[132] often contains clauses listing the 'eligibility criteria' that the underlying assets must meet.[133] It may also determine that a change in asset quality, such as a deterioration of their risk profile, automatically triggers the repayment of the securities.[134]

Laws concerning the composition of the portfolio can also encourage or deter investors from purchasing securitised products, thereby affecting liquidity. For instance, several economic studies view information asymmetry on the features of financial assets as well as the high search costs to assess them, as barriers to liquidity.[135] Therefore, legislation, particularly in the form of disclosure requirements,[136] may require originators to provide

[130] For some examples in Germany see Ramos Muñoz (n. 16) 56–57; for the limitations under French law, see

Latham & Watkins (n. 42) 137. The EU Securitisation regulation excludes 'specialised lending exposures' under Art 2 (1)(c) Regulation 2017/2402.

[131] Like Luxembourg. For an in-depth analysis of securitisation over economic activities and IP rights, see Alexandre Quiquerez, *La titrisation des actifs intellectuels* (Larcier 2013), especially 75–77.

[132] See below Sect. 1.3.3.

[133] For some examples, see Chap. 3.

[134] i.e. the 'amortisation of the tranches', especially for 'revolving' deals, see For example, Art 21 (6) Securitisation Regulation.

[135] ECB, Bank of England, 'The Case for a Better Functioning Securitisation Market in the European Union' (2014) 29 https://www.ecb.europa.eu/pub/pdf/other/ecb-boe_case_better_functioning_securitisation_marketen.pdf accessed 20 December 2025.

[136] On the function of disclosure rules, see Chap. 2, Sect. 2.5

investors with or publish more information about the qualities of the portfolio assets. By lowering information asymmetries, these measures could encourage investments in ABSs.

3.4.1.2 Preliminary Strategic Considerations (II): Indirect Incentives

Law may also act on incentives to originate or pool together certain types of assets. For instance, it can introduce restrictions which increase costs, formalities, or complicate the process for the securitisation of certain assets. This may in turn disincentivise originators from using them, or from engaging in securitisation operations altogether. An example is 'resecuritisation' of ABSs, which is only possible under EU law subject to stringent conditions[137] and is thus unlikely.

The composition of the portfolio may also determine the eligibility of ABSs for transactions with central banks. The latter generally set requirements to accept assets, including securitised assets, as collateral. Financial institutions will naturally tend to originate or to purchase from third parties and securitise assets meeting such criteria with a view to accessing central bank liquidity.[138] In other words, as *Nyborg* argued, 'if central bank money were only available against (...) igloo-backed securities, igloos would be built'.[139] Indeed, one of the drivers of the growth of the securitisation market during the 2007 financial crisis was the use of ABSs as collateral for repo transactions with central banks (causing a real 'run on repos'[140]).[141]

For the purposes of the present analysis, the Eurosystem's collateral framework introduces detailed requirements for the ABSs that the European Central Bank (ECB) accepts as collateral.[142] These also restrict ('frame') the types of eligible assets and introduce qualitative criteria on the homogeneity of the portfolio.[143]

[137] Art 8 (2) – (3) Securitisation Regulation.

[138] Ibid.; Scopelliti (n. 39) 27–30.

[139] Kjell G Nyborg, 'Central Bank Collateral Frameworks' (2017) 76 Journal of Banking and Finance 198, 204.

[140] A repurchase agreement ('repo') is, in essence, a sale of securities with an agreement to repurchase them at a later stage. It is essential for short-term liquidity. See John Armour and others, *Principles of Financial Regulation* (2016) 317.

[141] Gorton and Metrick (n. 41) 2–3. The ECB has also documented the steady increase in the use of ABS as collateral for repos between 2004 and 2013, Nyborg (n. 139) 206.

[142] Annex VIII, Guideline 2015/510 of the European Central Bank of 19 December 2014 on the implementation of the Eurosystem monetary policy framework (General Documentation Guideline) 2014 (OJ L 91) 5.

[143] Specifically art 62–73 ibid.

Other types of incentives, albeit less popular among regulators, include preferential capital requirements. For instance, the EU regime for STS securitisations rewards market actors with a lower capital surcharge when they combine high-quality and 'homogeneous' assets.[144] In instances where *ad hoc* capital requirements or differentiated prudential treatments stretch existing legal concepts or establish a distinct treatment for certain assets, the law could also play a more explicitly 'constitutive' role.[145]

3.4.1.3 *Implications*

The law governing the composition of portfolios performs multiple functions that are central to the design of sustainability-oriented legal initiatives.

First, legal requirements may directly 'frame' the use of certain assets with negative environmental or social impacts by prohibiting securitisation operations that include them.

Legislation, as well as central banks' frameworks, could also indirectly achieve a similar outcome by preventing or disincentivising the use of 'browner' assets, or by incentivising the use of 'greener' assets, through preferential fiscal treatment or capital requirements. These measures could also have a trickle-down effect across the entire operation: for instance, legislation requiring disclosure of the impact of the portfolio on environmental, social, and governance factors may encourage or deter investors from purchasing 'sustainable' ABSs.

Second, central banks and supervisors that introduce differentiated treatments for 'sustainable' assets may contribute to 'co-constitutive' dynamics, in which 'sustainable' securitisations come to be distinguished from traditional securitisations.

3.4.2 *The Terms of Issue of ABSs*

The issuance of ABSs is preceded by a negotiation and underwriting phase, during which the terms of the securities are defined and the securities are offered to investors. Prior to the issuance, it is also common practice, or required by law, to disclose extensive information regarding

[144] Art 20 (8) Securitisation Regulation, Commission Delegated Regulation 2019/1851. On the STS regime, see also Chap. 6, Sect. 6.4.

[145] Agnieszka Smoleńska, 'The Constitutive Role of Law in Sustainable Finance' (2025) Industrial Law Journal, 24, 27–30.

the characteristics of the ABSs. Both phases offer meaningful avenues for advancing sustainability goals.

3.4.2.1 The Negotiation and Underwriting of ABSs

The negotiation of the terms of prospectuses, 'offering circulars' and other pre-issuance documentation[146] constitutes a strategic phase prior to 'public' issuances and 'private placements', which can also shape originating entities' and SPVs' commitment to sustainability objectives.

The law does not prescribe who should intervene in the drafting of securitisation documentation: Therefore the involvement of prospective investors will inevitably depend on their contractual and bargaining power. However, it is common practice for large and influential market actors, such as investment or commercial banks, to directly interact with originating entities before the issuance and act as 'underwriters' thereafter. In other words, one actor will sign an 'underwriting' (or 'subscription') agreement, where it undertakes to purchase all the securities and to market them further to investors. There is often a clause compelling underwriters to retain all the unsold securities.[147]

The OECD has exhaustively described underwriters' potential to influence originating companies on sustainability objectives. This could translate, among other things, into 'due diligence' checks to investigate the environmental and social practices of originating entities across their value chains.[148] EU law does not formally regulate underwriters' role, and neither do national frameworks of EU countries such as France, Luxembourg and Italy. However, it is common practice for market actors to set out mutual rights and obligations between originating entities and

[146] See above Sect. 3.3.3.

[147] Phillip Wood, 'Marketing and Distribution of Bonds', *International Loans, Bonds, Guarantees, Legal Opinions*, Vol. 5 (2019) para 25–040; Thomas GF Fuller, *The law and practice of international capital markets* (LexisNexis Butterworths, 3rd edn 2012) 325–330.

[148] 'Due diligence' involves *'a process conducted by the banks, with the assistance of external experts (typically including financial and legal experts) of due and careful investigation, taking into account the specific circumstances, into the affairs of the company and its business in the broadest sense, where the results of such investigation will be reflected (as appropriate) in the (...) prospectus'* Danny Busch and others, *Prospectus Regulation and Prospectus Liability* (Oxford EU, Oxford University Press 2020) para 2.30; on 'due diligence' on environmental and social matters, see OECD, 'Due Diligence for Responsible Corporate Lending and Securities Underwriting – –Key considerations for Banks implementing the OECD Guidelines for Multinational enterprises' (2019) https://mneguidelines.oecd.org/due-diligence-for-responsible-corporate-lending-and-securities-underwriting.pdf accessed 5 December 2025.

underwriters in clauses drafted on the basis of standard agreements.[149] As I will explain in the following Chap., these clauses may be the starting point to crystallise the necessary pre-conditions that originating entities and the securitisation operation itself must meet to make it suitable to underwriting. For instance, 'underwriting agreements' may also go as far as to allow underwriters to follow up with originating entities or to avoid subscribing to the securities when dissatisfied with their sustainability-related commitments.[150]

3.4.2.2 The Issuance Documentation

Legislation, especially based on EU provisions, and the contractual documentation introduce prerequisites which 'enable'[151] the offer of ABSs on regulated markets. As previously mentioned, these consist in most cases of a mandatory prospectus, which must be filed with and approved by a public authority, especially in cases of issuances on capital markets.[152] However, it is generally in originators' interests to circulate some information about the securities anyway, for transparency, reputational purposes, and marketing purposes. Therefore, entities often choose to publish a document describing the key characteristics of the securities even for operations that would not be subject to prospectus requirements under EU rules otherwise.[153] Similarly, while 'private placements' are in principle exempt from such requirements (except for 'STS-compliant' deals, which I shall analyse in Chap. 6[154]), it is still common practice to draft ad hoc documentation, though often more succinct, such as more general 'terms and conditions'[155] or 'offering circulars'.[156]

[149] See Chap. 5, Sect. 5.4.

[150] Ibid.

[151] For the 'enabling' function of law, see Chap. 2, Sect. 2.4.

[152] For the main functions and items of a prospectus under EU law, see recitals (7)-(26), Art 6 Regulation 2017/1129.

[153] Exemptions include, among others, issuances solely addressed to qualified investors, with a denomination per unit amounts of at least 100 000 €, with a total consideration of less than 1 000 000 € within a 12-month period, see art 1 (3)(4) ibid. see further in Chap. 6, Sect. 6.5.1.

[154] If originators decide to comply with the STS regime, they have to make available to investors prospectus or other 'final offering document'(s), Art 7(1)(b) Securitisation Regulation. See Chap. 5, Sect. 6.5.1.

[155] For some examples of transactional documentation, see Chap. 4.

[156] Phillip Wood, 'Securitisation Structures', in *Project Finance, Securitisations* (n. 96) paras 6–040.

Three main functions of issuance documentation make it strategic for sustainability purposes.

First, prospectuses or similar documents generally describe the key features of the securities, which make them particularly amenable to sustainability-related references. The following example, which will find its exhaustive analysis in Chap. 4[157], will illustrate the point. Prospectuses may include clauses, generally referred to as 'covenants', which may compel originating entities or issuers to meet certain requirements (e.g. related to their financial conditions) or to submit certain documents. Through these clauses, market actors may, among other things, set out additional legal obligations in the environmental, social, and governance sphere,[158] for instance, they may require originators to achieve a certain greenhouse gas (GHG) emission reduction target by a certain date.

Second, issuance documentation enhances issuers' accountability for the information included. In other words, national competent authorities scrutinise prospectuses and can issue administrative sanctions vis-à-vis issuers, suspend or prevent the offer in the case of misstatements or omissions.[159] Moreover, prospectuses and other voluntary issuance documents[160] act as contractual offers. As such, EU national frameworks provide for civil liability remedies enabling investors to seek compensation for false and misleading statements negatively affecting their investment decisions.[161] The double litigation risk at public and private level (i.e. by national authorities and investors) should deter issuers from including false or misleading claims in prospectuses, including on sustainability-related matters (colloquially defined as 'greenwashing' or 'sustainability washing' practices).[162]

Third, the documentation also lists the circumstances which trigger investors' power to enforce ABSs, i.e. to request the early repayment of their securities. These are often named 'Events of Default', 'Early

[157] See Chap. 4, Sect. 4.4.1.

[158] Jouffray et al make a similar point in relation to the potential of covenants, which could encourage seafood industries to implement sustainability-related measures. Jean Baptiste Jouffray and others, 'Leverage Points in the Financial Sector for Seafood Sustainability' (2019) 5 *Science Advances* 1, 4.

[159] Under EU law, Art 20, 32, 37,38 Regulation 2017/1129.

[160] See above Sect. 1.1.4.

[161] On the EU prospectus regime, see Art 11 Regulation 2017/1129.

[162] See, among others, Magali A Delmas, Vanessa Cuerel Burbano, 'The Drivers of Greenwashing' (2011) 54 *California Management Review* 64. Further on 'greenwashing', see Chap. 6.

Amortisation Events' [163] or 'Credit Events', especially across synthetic transactions[164]. As I will illustrate below, the right to act on these clauses often rests with one representative of all investors.[165] The scope of these clauses tends to be wide-ranging. Failure to comply with certain clauses, including the above-mentioned 'covenants', may constitute an 'Event of Default'. Thus, the above-mentioned clauses may go as far as to require originators to act in the environmental, social, and governance sphere, and to link non-compliance to the early repayment of the securities. In principle, this could provide a very powerful threat, incentivising originators to comply.

3.4.3 *The Governance of SPVs*

Discussing the 'governance', i.e. the 'system of rules, practices and processes by which a company is directed and controlled', and its purpose[166] may appear counterintuitive in the context of SPVs, which are often perceived as 'empty shells'. The traditional assumption in the academic literature is that the entities setting up and effectively managing the transaction are originators, with marginal or little role left for SPVs.[167]

A legal analysis of securitisation under a sustainability lens, however, requires partially challenge such an assumption. There are statutory and contractual provisions which effectively delineate the perimeter and purpose of SPVs' activities as well as the duties of its decision-making bodies.

3.4.3.1 The Specialisation of SPVs' Object

The law tends to conceive SPVs as specialised entities which should predominantly or exclusively focus on securitisation operations.[168] This is

[163] They have an 'early' nature since they avoid the liquidation of the entity, in line with the bankruptcy-remoteness of SPVs. Ramos Muñoz, *Transnational Securitisation* (n. 16) 109. 'Events of default' may be a more widely used term across standard bond issuances, see S 6.01. ABA (n. 190).

[164] In line with the standard documentation for derivatives, see Ramos Muñoz, *Transnational Securitisation* (n. 16) 104.

[165] See Sect. 2.3.5, Chap. 4, Sect. 4.6.1.

[166] See the definition of corporate governance by the Chartered Governance Institute UK and Ireland, CGI, *What is corporate governance?(2026)* https://www.cgi.org.uk/about-us/policy/what-is-corporate-governance accessed 3 June 2026; see also Colin Mayer, *Prosperity: Better Business Makes the Greater Good* (Oxford University Press ed, 2019) 159.

[167] See, among others, Bonavita (n. 16) 24.

[168] On the 'Principle of specialisation', see also the Q&A by the Luxembourg financial authority, Commission de Surveillance du Secteur Financier (hereinafterCCSF), 'Frequently Asked

another implication of the 'shielding' of SPVs alongside the previously mentioned unavailability of their patrimony to originators and other creditors, and alongside their 'bankruptcy-remoteness'.[169]

As part of this specialisation, EU and national legislation often specify and thereby 'frame' the 'object' that SPVs pursue.[170] For instance, EU legislation clarifies that their purpose is 'carrying out one or more securitisations', restricting the scope of their activities to those appropriate to accomplishing that objective.[171] Luxembourg legislation clarifies that the object of management companies (which effectively have decision-making powers when SPVs are trust funds[172]) is to manage securitisation funds and to act as fiduciaries.[173] Statutory provisions may introduce further restrictions to SPVs' ability to undertake other debt obligations or to create security interests.[174] Such rules should prevent SPVs from reducing their assets, thus safeguarding the repayments.

Similarly, SPVs' constitutional documents (such as the articles of association or management regulations[175]) are likely to contain object clauses conferring very limited powers to SPVs. They may specify that these entities are only meant to manage the portfolio and raise liquidity from capital markets. They may also prevent them from engaging in other business activities, hiring employees,[176] or appointing independent directors[177], who may interfere with SPVs' main activities.

Questions – Securitisation' (2013) 11 https://www.cssf.lu/en/Document/faq-securitisation/ accessed 9 December 2025.

[169] See above Sect. 3.3.2.

[170] On the role of management companies in pursuing SPVs' objects, see above Sect. 3.2.2.

[171] Art 2 (2) Securitisation Regulation.

[172] See above Sect. 3.3.2.

[173] On the object of SPVs, see art 1 (2) ibid. art 2, art 6 (2), Art 14 Loi du 22 mars 2004 relative à la titrisation (Lux).

[174] For example, under Luxembourg law, SPVs cannot assign to third parties or create security interests over their assets. See art 61 (1) – 61 (3) ibid.; under Italian law, see art 7.1 c 2 L 130/1999 (IT).

[175] See above Sect. 3.3.2.

[176] Bonavita (n. 16) 34.; Bavoso, 'Legal Issues' (n. 9). 14. See also Chambers and partners, 'Practical guide Securitisation – EU' (2022) Global Practice Guides', 16.

[177] Ibid., p. 16.

Consequently, the law as well as constitutional documents could serve a 'programmatic' function by clarifying that promoting sustainability goals constitutes SPVs' exclusive mission.[178]

3.4.3.2 Directors' Duties?

Regardless of the clauses articulating the scope of SPVs' activities, another crucial component of the governance of these entities is the role of directors. When SPVs are created as companies, the traditional regime applicable to directors' duties should, in principle, apply.[179] While the regime has its specificities in every EU country, it broadly involves a duty to promote the success of the company and a duty to act with care, skill and competence.[180] Similar standards, articulated at EU level, also apply to management companies of investment funds.[181]

It is, however, practically challenging to hold SPVs' directors accountable, given their allegedly limited discretion in the selection of the portfolio assets and in the management of the operation.[182] Some have also noted the challenges of transposing the traditional regime, since directors generally owe their duties to the corporate entity, which should indirectly capture those with a residual claim, i.e. shareholders. In contrast, the structure of SPVs generally lacks shareholders, as they predominantly or exclusively issue debt securities (for securitisation programmes) and also in the light of their limited 'object'. Unlike traditional companies, the 'residual claimants' may not be shareholders, but rather originators themselves, when they hold the last tranche of securities.[183]

Correspondingly, previous case law across EU and European countries shows seminal evidence of SPVs' liability. For instance, there has been one case in England where the directors were found to have breached their

[178] On the 'programmatic' function of the law see Chap. 2, Sect. 2.4. on the limitations of object clauses, especially in the UK, see Chap. 4, Sect. 4.4.3.

[179] Bavoso, 'Filling the Accountability Gap' (n. 23) 14.

[180] As highlighted by various comparative studies on directors' duties across EU countries, see, among others, Carsten Gerner-Beuerle, Philipp Paech and Edmund-Philipp Schuster, 'Study on Directors' Duties and Liability in the EU' (2013) LSE Enterprise, prepared for the EU Commission DG Markt, 64–172.

[181] Art 14 (1)(a)-(b) Regulation 2009/ 65/EC of the European Parliament and of the Council of 13 July 2009 on the coordination of laws, regulations and administrative provisions relating to undertakings for collective investment in transferable securities (2009) OJ L 302/ 32. Art 12(1)(a) Alternative Investment Funds Directive.

[182] Bavoso, 'Filling the Accountability Gap' (n. 23) 13–14.

[183] See above Sect. 1.1.4. Ramos Muñoz, *Transnational Securitisation* (n. 24).143–145.

duties towards investors in ABSs for carrying on business in a speculative manner on the verge of insolvency.[184] However, this SPV was also peculiar since it had a broader business activity than in most securitisations: it was not only used for these operations, but also for the purchase of land.[185] This specificity also gave leeway for the court to establish liability on its directors.

The restrictions to directors' liability found in previous case law are problematic for transactions promoting sustainability objectives. SPVs are those that effectively issue ABSs. The related directors are, in principle, in charge of managing the expenses for the operation. However, if they are considered not to be liable, there will be no legal avenues to establish SPVs' liability. As a result, investors may not have a direct remedy vis-à-vis SPVs in critical circumstances, such as when the funds of the operation are deviated and used for non-sustainability-related purposes ('greenwashing' practices).

At the same time, the above-mentioned UK case shows that SPV directors can, in principle, incur liability, suggesting that SPVs need not be viewed merely as 'empty shells'. Therefore, further legal initiatives could also clarify the extent to which SPVs' directors must and should promote additional sustainability-related objectives. Such provisions would 'enable' them to prioritise sustainability considerations when managing the operation as well as enhancing their accountability.[186]

3.4.4 Investors' Representation Mechanisms

As mentioned above, leaving some limited exceptions aside, ABSs generally take the form of debt securities.[187] These do not give investors any formal rights over a portion of the capital of the investee company, and thereby do not automatically grant any rights to be involved in SPVs' governance. In other words, unlike shareholders, ABS holders generally only have residual 'voice' rights.[188] For instance, they do not participate in SPVs' general meetings or appoint SPVs' directors.[189] Another illustration

[184] On the basis of the provisions concerning directors' liability prior to insolvency. See S 214 Insolvency Act (UK).

[185] *Roberts v Frohlich* [2011] EWHC 257 (Ch) (UK).

[186] On the 'enabling' role of law see Chap. 2, Sect. 2.4.

[187] Like the countries where SPVs can issue ABSs as shares, see above Sect. 3.3.3.

[188] See Chap. 2, Sect. 2.6.

[189] See also Chap. 6, Sect. 6.7.

of the limited scope of their rights is their exclusion from all the decisions regarding the identification of the portfolio assets when the portfolio is 'revolving'. The responsibility for such decisions generally lies with originators, servicers or 'collateral managers'.[190]

When SPVs are set up as trusts or as funds, investors also have limited 'voice' rights. In these circumstances, it is the very legal form of these entities which inevitably prevents investors' participation within internal decision-making dynamics and processes.[191]

Additional coordination costs and challenges further constrain the ability of ABS holders to raise their 'voice'. To a certain extent, these constraints are inherent in the characteristics of debt instruments and instruments issued by funds. These attract investors with different time horizons and objectives[192], and not necessarily interested in engaging with investee entities. Besides, securitisation issuances inevitably cater to a wide range of investors with different risk appetite and different return expectations, due to the variety of 'tranches' they can choose from.[193] The variety of their preferences may prevent reaching a consensus on potential courses of action in case of dissatisfaction with investee entities, including on sustainability matters.

For this reason, investors often appoint one actor or entity to represent their interests. Due to the limited influence of ABS holders *individually*, the powers of investors' 'representatives' could be a strategic entry point for sustainability-oriented legal initiatives.

EU and national legislations on securitisation often do not require, but rather implicitly hint at, the appointment of an intermediating entity,[194] which may overlap or diverge from the underwriter.[195] Depending on the

[190] Ramos Muñoz, *Transnational Securitisation* (n. 24) 40. See also Sect. 3.3.3.–3.4.1.

[191] Couret, Dupont (n. 113) 51.

[192] UNPRI, 'ESG engagement for fixed income investors: managing risks, enhancing returns' (2018) 27 https://www.unpri.org/download?ac=4449 accessed 8 December 2025.

[193] See above, Sect. 3.2.

[194] For example, at EU level, the Securitisation regulation contains references to 'mechanisms for 'the timely resolution of conflicts between different classes of investors''. At the same time, it is not necessary for originators to put those in place. Only in the event that they do so, they will have to be transparent about them. Art 21(10) Regulation 2017/2402. For examples of national laws across EU Member States, see artt 67–84 Loi du 22 mars 2004 relative à la titrisation (Luxembourg); Artt 2417–2418 cc (Italy). The appointment of a trustee does not seem to be necessary under applicable UK provisions, see 'if there is a trustee, or other representative, of the holders of the securities', Financial Conduct Authority (FCA) Handbook, Listing rule 17.3.12 (d)

[195] See above Sect.ection 1.3.3.

surrounding framework as well as on parties' choice, the representative could be a trustee or a functionally equivalent actor, especially across jurisdictions where trusts are not acknowledged, or problematic. As a result, the designated entity could also be, among others, an 'investment manager' or an investment company.[196]

There is a combination of legal sources articulating the powers of investors' representatives. As regards trustees, some have advocated for the full application of fiduciary principles, translating into a general duty of loyalty and care to beneficiaries.[197] It would also entail a duty to act in strict accordance with the terms of the deed for their appointment.[198] One may also argue that they owe a duty of impartiality to all beneficiaries: hence, they may need to balance the interests between investors in different tranches.[199]

By contrast, others have underlined the limited application of fiduciary concepts to commercial transactions, especially insofar as sophisticated parties are involved (as the case may be for institutional investors investing in ABSs[200]). Thus, they have concluded that the documentation for their appointment constitutes the exclusive source of their duties.[201]

When representatives are not trustees, the breadth of their powers is a contentious subject, especially given the absence of a comparably

[196]For example, 'representative fiduciaries' under Luxembourg law, see artt 67–84 Loi du 22 mars 2004 relative à la titrisation; Artt 2417-2418 cc (Italy). The Dutch legal framework contemplates the appointment of trustees and external investment companies, see Latham & Watkins (n. 42) 271. Both options are allowed under UK law, see Financial Conduct Authority (FCA) Handbook, Listing rule 17.3.12 (d). There seems to be a preference for non-trust structures due to the more favourable tax treatment, Wood, 'Securitisation Structures' in *Project Finance, Securitisations* (n. 91) 6–002.

[197]They need to demonstrate the 'ordinary prudence' of a business man, as described, among others, in the case *Learoyd v Whiteley* (1887) UKHL 1 (UK); on trustees' duties more generally, see Alastair Hudson, *Understanding Equity & Trusts* (6th edn, Routdledge 2017) 65–69.

[198]See *'As a general rule the law requires of a trustee no higher degree of diligence in the execution of his office than a man of ordinary prudence would exercise in the management of his own private affairs. (...) Business men of ordinary prudence may, and frequently do, select investments which are more or less of a speculative character; but it is the duty of a trustee to confine himself to the class of investments which are permitted by the trust' Learoyd v Whiteley*, ibid.

[199]Bavoso, 'Filling the Accountability Gap' (n. 23) 17.

[200]On investors in ABSs, see also Chap. 6, Sect. 5.7.

[201]For both positions, see Bavoso, 'Filling the Accountability Gap' (n. 23). 30–34;. Sheilah D Gibson, 'The Case for the Expanded Role of Trustees in Securitisations' (2004) 121 *Banking Law Journal* 387, 391; 394–395l; *Katz v Oak Industries Inc,* 508 A2d 873 (Del. Ch. 1986) (US).

developed doctrine and case law on fiduciary principles across civil law countries. The source of their duties is similar, comprising statutory provisions for financial intermediaries and the clauses of the documentation for their appointment.[202] In some countries like France and Italy, ad hoc legislation on representation mechanisms for bondholders may also apply.[203] However, the inherent limitation of fiduciary principles and of statutory provisions for financial intermediaries is their open-ended wording. Formulas like 'duty of loyalty' or 'duty of care' effectively operate as standards. In other words, despite their hard law nature, market actors have leeway to specify on a case-by-case basis, through their contractual agreements, what behaviours are consistent with the standard.[204]

All in all, the clauses of the documentation remain in practice the primary source to 'frame' the powers of investors' representatives since they delineate what they can and cannot do and to what extent they are liable vis-à-vis ABS holders. Such documentation generally consists of a trust deed or a broader 'indenture' (especially in the US)[205], when representatives are legally trustees, or documents sometimes referred to as 'investment management agreements'[206]. Furthermore, there may be 'intra-creditor provisions', which may further regulate such rights, within the issuance documentation. For instance, the prospectus may also contain further provisions on decision-making fora among ABS holders, like regular meetings.[207]

In practice, investors' representatives tend to only monitor the transaction on a formal and superficial level under normal circumstances.

[202] As argued by Ramos Muñoz, see Ramos Muñoz, 'In praise of small things' (n. 24) 161, 164. E.g. under the Luxembourg securitisation regime, the agreement for the appointment of 'representative fiduciaries' delineates their powers and it is binding on investors, see art 68 Loi du 22 mars 2004 relative à la titrisation.

[203] Artt L228-38 – L228-90 Code civil (FR), arts 2410ff c.c. (It).

[204] For a theoretical account of the notion, of the strengths and weaknesses of standards, which tend to be 'insufficiently directive', see Alan Schwartz, Simone Sepe, 'Economic Challenges for the Law of Contract' (2020) 8. *Yale Journal of Regulation,* 11-12.

[205] American Bar Association (hereinafterABA), 'Revised Model Simplified Indenture' (2000) 55 *the Business Lawyer* 3 1115, particularly 1139–1144 https://www.jstor.org/stable/40657063?seq=1 accessed 8 December 2025.

[206] For a model agreement, see Investment Association, 'Model Discretionary Investment Management Agreement' (2021) https://www.theia.org/sites/default/files/2021-11/Model%20IMA%202021.pdf accessed 9 December 2025.

[207] For some examples, see Chap. 3, Sect. 3.4.4.

For instance, they receive reports on payments and defaults related to the asset pool and update investors and rating agencies on that basis.[208] They may also oversee the entities holding securitisation funds in dedicated accounts.[209] The clauses of securitisation agreements (like the above-mentioned 'intra-creditor provisions') may also introduce additional decision-making fora and powers for investors, like internal general meetings.[210]

General market practice envisages a considerably more prominent role for investors' representatives after Events of Default. They have the *choice* to exercise their enforcement powers, making the principal and interest immediately due and payable, or to waive them.[211]

While these rights may in principle concur with investors', 'no action clauses' frequently make them exclusive, by preventing other ABS holders from bringing individual enforcement action.[212] At the same time, the agreements for their appointment may also introduce waivers and exemptions to their liability.[213]

Therefore, it is in principle possible to entrust such actors with remarkable powers to 'voice' investors' concerns and interests with originating companies and SPVs.[214] This possibility is particularly valuable for transactions which also advance sustainability goals. By way of example[215], law could determine the extent to which representatives need to monitor originating companies' performance on sustainability matters before and after an Event of Default. It could also 'enable' representatives to bring legal proceedings on behalf of all investors if originating companies engage in 'greenwashing' practices.[216]

[208] Ramos Muñoz, *Transnational Securitisation* (n. 16) 163–165. Ramos Muñoz, In praise of small things (n. 24). 384–385.

[209] See Sect. 2.2.3.

[210] See further on internal decision-making fora in Chap. 5, Sect. 5.7.3.

[211] E.g., among others, *Citibank NA and MBIA Assurance SA v QVT Financial LP* [2007] EWCA Civ 11.

[212] *Electrim SA v Vivendi Holdings 1 Corp* [2008] EWCA Civ 1178 (UK).

[213] Ramos Muñoz (n. 191) paras 387–388. For example, under UK law, *Kelly v Coopers* [1993] AC 205. For an in-depth analysis of the liability standard applicable to representatives, see Chap. 5, Sect. 5.7.2., and Chap. 6, Sect. 6.7.3.

[214] On 'voice' rights, see Chap. 2, Sect. 2.6.

[215] See further Chap. 5, Sect. 5.7.3. and Chap. 6, Sect. 6.7.4.

[216] On the 'enabling' function of the law, see Chap. 2, Sect. 2.4.

3.4.5 Monitoring Processes

Another crucial aspect of managing securitisation operations after the assignment of the portfolio and the issuance of ABSs involves monitoring the payments coming from the asset pool as well as the proceeds from investors, and how these are used.

The law intervenes by admitting and regulating the appointment of third parties for the day-to-day management of the operation, like servicers or additional management companies.[217] While EU law acknowledges the existence of arrangements with third parties, it is silent on the scope of their powers .[218] As such, the sources of their duties include the terms for their appointment ('servicing' or 'administration' agreements) and national legislation, albeit to a minor extent.[219] Despite the wide deal-to-deal variations, both 'servicers' and 'management companies' tend to have the powers to enforce the payments over the portfolio, just like originating entities, and to foreclose assets.[220] They may also act as 'cash managers' by ensuring that the repayment of investors complies with their priority rights and making appropriate advances upon liquidity shortages.[221]

By contrast, 'management companies' tend to have broader powers under the transaction documentation, since they are also generally responsible for the selection of portfolio assets (across 'revolving' transactions).[222] In common law jurisdictions, trustees, who may also hold the portfolio assets and the related cash flows on trust, generally perform such servicing functions.[223]

Furthermore, statutory and contractual provisions can also in principle establish mechanisms to store and 'ring-fence' the cash flows, so that they are exclusively available to the entities with decision-making powers and exclusively used for the purposes of the operation. Ring-fencing the cash

[217] See above Sect. 3.3.2.

[218] See further in Chap. 6, Sect. 6.8.1.

[219] See above Sect. 1.2.3. Jurisdictions which contemplate external management companies include France and Portugal, Code monétaire et financier arts L214-49-1, L214-49-2, L214-49-6 or L214-49-7 (FR), Decree-Law 453/99 arts 9(1), 15(1), 23 (Portugal). For a comparative analysis see Ramos Muñoz, *Transnational Securitisation* (n. 16) 4; ibid., 'In praise of small things' (n. 25) 383.

[220] Ramos Muñoz, *Transnational Securitisation* (n. 25) 39–42, 234.

[221] Ibid., 4, 234.

[222] Ramos Muñoz, 'In praise of small things' (n. 25), 234–237. For an exhaustive analysis of servicers' role, see also a US case, *Re Capstead Mortg. Corp. Securities Litigation (2003) 258 F.Supp.2d 533 ND Tex.*

[223] Ramos Muñoz, *Transnational Securitisation* (n. 16) 93.

flows for the operation is yet another facet of the 'shielding' characterising securitisations.[224] For instance, statutory provisions may provide for the 'shielding' of the cash flows of *each* securitisation operation when SPVs engage in repeated issuances, or they may 'shield' the proceeds and payments on the portfolio from any business activity that SPVs conduct other than securitisation.[225] For instance, statutory provisions across civil law countries often consider each offer as giving rise to a separate patrimony, so that the securitised assets are only available to the respective investors and not to any other creditors of the SPV or of the originating entity.[226] At the same time, Luxembourg law allows SPVs to introduce different compartments for individual issuances.[227] Through this structure investors only have a claim over the assets of that specific compartment.[228] This would also allow a separate liquidation of every compartment in case of insolvency.[229]A similar solution in common law countries is to create distinct security interests for each issuance in favour of SPVs (acting as beneficiary).[230]

It may also be possible to design contractual mechanisms to ring-fence funds on a deal-by-deal basis. Indeed, it is common practice to conclude a 'bank account agreement', which requires that all the cash flows of the operation are deposited in a dedicated bank account. Management companies or additional entities (sometimes named 'depositors') may be in charge of opening and overseeing such accounts. [231] Across various jurisdictions, it is also possible and common practice to create 'ring-fenced' accounts, where the balance can only be disposed on the basis of servicers' instructions and is unavailable to third parties.[232] For instance,

[224]See above Sect. 3.2., 3.3.2, Chap. 2, Sect. 2.6.

[225]On the impact of rules on the separation of patrimonies and on 'shielding' in organisational law in general, especially in civil law countries, see Hansmann, Kraakman, 'The law and rise' (n. 139) 1338.

[226]Art 3 (2) L 130/1999 (IT); Art L214-46-1 Code Monétaire et financier (FR).

[227]Articles of association must allow the management body to introduce different compartments, see art 5, 8 Loi titrisation.

[228]Art 62 ibid.

[229]Art 33 ibid. for a commentary see Couret, Doupont (n. 108) 51–53, 57–59 Baker MacKenzie, 'A global guide to legal issues in securitisations' (2021) 137–138 https://www.bakermckenzie.com/-/media/files/insight/guides/2021/securitisation-guide-2021_29092021.pdf. accessed 30 September 2023.

[230]Especially in England, Ramos Muñoz, *Transnational securitisation* (n. 16) 79.

[231]Ibid., 66, 67, 235.

[232]Ibid., 67.

the use of ring-fenced accounts, out of reach of any other creditors, is not just the result of a contractual arrangement, but rather a statutory requirement, in France. Specific rules restrict all withdrawals and expenses from this account, so that they may only be performed when in line with the instructions from the management companies and when promoting the interests of the operation.[233] A segregated account may also take the form of a trust (a so-called *Quistclose trust*[234]) across Anglo-Saxon jurisdictions. This means that 'depositors' act as trustees on behalf of all investors and have to account for the profits when using the funds.[235]

Alternatively, parties may introduce restrictions and conditions on the use of the cash flows of the operation through contractual clauses. For instance, it is possible in several civil law jurisdictions to create a security interest over a bank account,[236] granting control rights over the funds raised through the issuance of ABSs or over the payments from the underlying credit claims.

All in all, legislation and contractual provisions for the monitoring of securitisation cash flows have a considerable strategic potential. The legal framework on these aspects is even more crucial across 'sustainable' securitisations than across traditional ones. The law notably provides the basis for the appointment of third parties. It also allows for the design of mechanisms to oversee the allocation of securitisation funds, particularly the extent to which they are used for the sustainability goals of the issuance. Such statutory and contractual mechanisms could thus serve an 'enabling' and 'framing' function, authorising or restricting the use of securitisation proceeds for sustainability purposes. Such mechanisms could, in turn, deter or prevent market actors from misusing investors' funds by allocating them to other non-sustainability-related activities.

[233] Ramos Muñoz (n. 16) 65–67. Under French law, see Article D214-228 Code monétaire et financier (FR).

[234] i.e. a trust where the trustee is bound to achieving a specific purpose, as defined in *Barclays Bank Ltd v Quistclose Investments Ltd* [1970] AC 567.

[235] Ramos Muñoz (n. 16) 66.

[236] See e.g. in France Article D 214-103 ibid. it is also common in Germany, Ireland, Italy and Spain, see Latham & Watkins (n. 206).152, 195, 216. On Spanish law, see also questions 5.3–5.9 Hector Bros, Arnau Pastor, *Securitisation – Spain* (27 May 2022 ICLG) https://iclg.com/practice-areas/securitisation-laws-and-regulations/spain accessed 8 December 2025.

3.5 'Strategic' Remarks on Transaction Costs

When reflecting on the existing, and on the desirable, legal framework for advancing sustainability goals in securitisations, it is also important to consider the distribution of 'transaction costs'. Within the law and economics scholarship, these encapsulate the costs arising when market actors engage in market exchanges or seek to establish their property rights.[237]

Transaction costs associated with securitisations are inherently high. For originators, these could include identifying suitable homogeneous assets[238] for the creation of portfolios, setting up SPVs (particularly if they do not regularly engage in securitisations), drafting the documentation, negotiating terms with investors, as well as complying with pre- and post-issuance regulatory requirements (such as periodic reporting[239]). Potential investors face significant costs in assessing the profile and characteristics of ABSs, which are often described as a deterrent.[240] After investment, considerable costs are also involved in monitoring and enforcing the securities. These, coupled with the challenges of coordinating with other investors and with the limited 'voice' rights, could lead to superficial monitoring, especially in non-default scenarios. Coordination and enforcement challenges are also the main drivers of the recurrent appointment of a representative for all ABS holders.[241]

When market actors use securitisation as a vehicle to achieve sustainability goals, all these costs are amplified. Especially smaller-sized entities may lack a critical amount of assets that can independently make a positive contribution to sustainability goals (e.g. 'green' loans, such as loans supporting energy-efficiency improvements[242]) for the creation of a portfolio. They could also find it too costly to obtain such assets from third parties just for securitisation purposes. 'Drafting costs' and 'negotiation costs'[243] are higher than those for 'traditional' securitisations, especially when parties have no prior expertise in similar transactions, partly due to the limited availability of standard documentation.[244] Structuring these transactions

[237] See Chap. 2, Sects. 2.2. and 2.3.

[238] On the relevance of the homogeneity of assets, see Chap. 6, Sect. 6.4.1.

[239] On disclosure requirements, see Chap. 5, Sect. 5.4.1., 5.5.1, 5.7.2.

[240] See Chap. 5, Sect. 5.7.1.

[241] See above Sect. 2.3.4.

[242] On 'green' loans see further Chap. 5, Sect. 5.4.2., and Chap. 6, Sect. 6.4.2.

[243] For theoretical accounts on negotiation costs before the conclusion of contracts see Ayres, Gertner, 'Incomplete contracts', 92–93.

[244] Chap. 5, especially sections 5.4, 5.5. and 5.7. will further elaborate on this point.

also involves high 'outsourcing costs' for external professionals, such as analysts, lawyers,[245] but also 'gatekeepers', attesting to the credibility of the sustainability profile through reviews or ratings.[246]

As regards potential investors, they bear increased 'information costs' in assessing the effective contribution of the transaction to sustainability goals, especially when they lack specific expertise on ESG matters.[247] Investors with stronger sustainability preferences, or those interested in buying large stakes of the issuance[248], are more likely to push for the inclusion of specific sustainability-related terms.

After the issuance of ABSs, considerable 'bonding costs' are likely to arise for investors willing to monitor originators' progress towards the achievement of sustainability goals, especially in instances where post-issuance reporting and monitoring processes are not standardised.[249] Compared to 'traditional' ABS holders, investors in 'sustainable' products may be more inclined to monitor the transaction and expect that originators meet the announced sustainability goals. 'Sustainable' issuances are also likely to exacerbate already high 'coordination costs' among investors, who are likely to display a wider range of sustainability preferences. Lastly, investors with stronger preferences in this regard may also be more interested in raising their 'voice'[250] by enforcing their securities when sustainability-related promises are not fulfilled.[251] Such higher bonding and coordination costs highlight the broader role of legal mechanisms to amplify investors' 'voice', such as the appointment of a representative for investors.

[245] On legal fees as a source of transaction costs, see Allen, 'Transaction Costs' in Boudewijn Bouckaert, Gerrit De Geest et al (eds), *Encyclopedia of Law and Economics* (Elgar 2000) 906.
[246] On the notion of 'gatekeepers' throughout the present work, see Chap. 2, Sect. 2.6.
[247] On 'Information costs', see Allen (n. 245) 906; Michael C Jensen, William H Meckling, 'Theory Of The Firm: Managerial Behaviour, Agency Costs and Ownership Structure' (1976) 3 *Journal of Financial Economics* 305, 338–342.
[248] On the role of investors' preferences, see also Chap. 5, Sect. 5.7.2.
[249] See above Sect. 2.3.5. and Chap. 5, Sect. 5.8. Jensen, Meckling (n. 248) 338–342; Allen (n. 245) 899.
[250] On the notion of 'voice' throughout the present work, see Chap. 2, Sect. 2.6.
[251] On coordination and enforcement costs, see Jensen, Meckling (n. 248) 338–342. Allen (n. 246) 899

3.6 Conclusions

The chapter has provided an overview of foundational aspects of securitisation law, which represent the backbone for the remaining analysis. It has also illustrated various promising entry points for legal initiatives aiming to embed sustainability considerations within securitisation deals.

First, national and EU legislation ('public law'), as well as contractual provisions ('private law') can restrict the types of assets which can be securitised ('framing' function), also on the basis of the environmental and social impact and risks of the underlying credit claims. In a similar fashion, legislation and central banks' frameworks can strongly disincentivise or incentivise the use of certain assets. By requiring originators to be more transparent about the types of assets they pool together, they may also boost or curb demand for ABSs across investors.

Second, legislation and the contractual documentation can allow investors, and specifically those underwriting the issuance, to monitor the environmental and social risks of the transaction through 'due diligence' processes and to require issuers to meet certain conditions before marketing the ABSs. Even the documentation for the offer of ABSs to the public, such as prospectuses, can provide further information on the environmental and social impact of the operation. It can require issuers to meet certain sustainability-related targets and outline corresponding remedies for non-compliance.

Third, statutory provisions and the constitutional documentation for the incorporation or creation of SPVs can also in principle shape their 'governance' with regard to sustainability goals. In particular, they could further clarify the scope and purpose of SPVs' activities ('programmatic' function). They may also, in turn, require SPVs' directors to take into account environmental, social, and governance objectives in their decisions ('framing' function).

Fourth, mechanisms for the representation of investors' interests, such as the appointment of a third party or the creation of additional decision-making fora, are crucial to introduce 'voice' powers, through which ABS holders can monitor and take action vis-à-vis originating entities and SPVs regarding their performance on sustainability matters.

Fifth, contractual and statutory provisions can also require mechanisms to monitor the allocation of the funds, as well as the appointment of actors in charge of this function. This could ensure that proceeds are exclusively used to meet the sustainability objectives of the operation.

The analysis of the 'focal points', particularly those related to the purpose of SPVs, prospectus clauses on the final impact of ABSs, or the final allocation of the cash flows, illustrates the distinctive narrative underpinning 'sustainable' securitisations. Traditional scholarly accounts and policy instruments for securitisations have focused on the management of risks and the 'refinancing' of *existing* credit claims. By contrast, a 'strategic' discourse on securitisation and the design of related policy initiatives, requires attention to the broader contribution of these transactions to *further* sustainability goals.

The analysis of the law of 'sustainable securitisations' in the following chapters centres on these five focal points. It will examine how market actors and legislators have intervened in these strategic areas to shape environmental, social, and objectives of securitisation operations. It will also culminate in normative recommendations on how law could unlock the full potential of these focal points.

References

ACPR, Banque de France, 'Le marché de la titrisation en Europe: caractéristiques et perspectives' (2014) 31 *Analyses et syntheses* https://acpr.banque-france.fr/sites/default/files/medias/documents/20140701-marche-titrisation-en-europe-caracteristiques-et-perspectives2.pdf

Albrecht TW and Smith SJ, 'Corporate Loan Securitisation: Selected Legal and Regulatory Issues' (1998) 8 *Duke Journal of comparative & international law* 411

Allen DW, 'Transaction Costs' in Bouckaert B and De Geest G (eds), *Encyclopedia of Law and Economics*, vol I (Edward Elgar2000)

American Bar Association, 'Revised Model Simplified Indenture' (2000) 55 *The Business Lawyer* 3 1115 https://www.jstor.org/stable/40657063?seq=1

Ayres I and Gertner R, 'Filling Gaps in Incomplete Contracts: An Economic Theory of Default Rules' (1989) 99 *The Yale Law Journal* 87

BakerMacKenzie, 'A global guide to legal issues in securitisations' (2021) https://www.bakermckenzie.com/-/media/files/insight/guides/2021/securitisation-guide-2021_29092021.pdf

Basel Committee on Banking Supervision, Board of the International Organisation of Securities Commissions, 'Criteria for Identifying Simple, Transparent and Comparable Securitisations' (2015) 5 https://www.iosco.org/library/pubdocs/pdf/IOSCOPD494.pdf

BCBS, 'General Provisions, Basel Framework' (2023) https://www.bis.org/basel_framework/chapter/CRE/40?inforce=20230101&published=20201126

Bavoso V, 'Filling the Accountability Gap in Structured Finance Transactions: The Case for a Broader Fiduciary Obligation' (2016) 23 *Columbia Journal of European Law* 5

Bavoso V, 'Good Securitisation, Bad Securitisation and the Quest for Sustainable EU Capital Markets' (2015) 4 *Butterworth Journal of International Banking and Financial Law* 221

Bavoso V, 'Legal Issues on Financial Innovation and Structured Finance' (2012) 34 *The Company Lawyer* 1

Beale H, Bridge M and others, *The Law of Security and Title-Based Financing* (3rd edn, Oxford University Press 2018)

Bonavita J, 'Asset Securitisation in Germany: Risk Transfer or Legal Transformation?' (2016) 17 *European Business Organisation Law Review* 15

Bontempi P, *Diritto bancario e finanziario* (Giuffrè, 6th edn 2019)

Busch D and others, *Prospectus Regulation and Prospectus Liability* (Oxford University Press 2020)

Calabresi G and Melamed AD, 'Property Rules, Liability Rules, and Inalienability: One view of the Cathedral' (1972) 85 *Harvard Law Review* 6 1089, 1090, 1092

Castro DI, 'Structured Investment Vehicles (SIVs)' in Strumeyer G and Swammy S (eds), *Capital Markets: Evolution of the Financial Ecosystem* (John Wiley 2017)

Chambers and Partners, 'Securitisation 2023 – UK law and practice' (2023) https://www.lw.com/admin/upload/SiteAttachments/Chambers%20Securitisation%202023%20UK.pdf

Chartered Governance Institute UK and Ireland (CGI), *What is corporate governance?* (2026) https://www.cgi.org.uk/resources/factsheets/factsheets/what-is-governance/

Climate Bonds Initiative, 'Green Securitisation: Unlocking Finance for Small-Scale Low Carbon Projects' (2018) https://www.climatebonds.net/data-insights/publications/green-securitisation-unlocking-finance-small-scale-low-carbon-projects

Commission de Surveillance du Secteur Financier (hereinafter: CCSF), 'Frequently Asked Questions – Securitisation' (2013) https://www.cssf.lu/en/Document/faq-securitisation/

Couret A, Stechen A, Prüm A and others, *Titrisation* (Anthemis 2008)

Damerow F, KidneyS and Clenaghan S, 'How Covered Bond Markets Can Be Adapted for Renewable Energy Finance and How This Could Catalyse Innovation in Low-Carbon Capital Markets – Unlocking Bank Lending in an Era of Capital Constraint and Limited Public Budgets' (2021) Climate Bonds Initiative Discussion paper http://ecosystemservices.co.uk/_docs/Climate%20Bonds%20RE%20covered%20bonds%2022May2012%20%28published%29.pdf

De Gioia Carabellese P and Della Giustina C, *The law of securitisations: from crisis to techno-sustainability* (Routledge2023)

Deakin S and others, 'Legal Institutionalism: Capitalism and the Constitutive Role of Law' (2017) 45 *Journal of Comparative Economics* 188 http://dx.doi.org/10.1016/j.jce.2016.04.005

Delmas MA and Burbano VC, 'The Drivers of Greenwashing' (2011) 54 *California Management Review* 64

Di Filippo D, 'Green Securitisation: Developments and Challenges' (2021) Fitch Ratings special report, https://c2e2.unepccc.org/wp-content/uploads/sites/3/2021/07/fitch-green-Securitisation-developments-and-challenges-15-april-2021.pdf

ECB and Bank of England, 'The Case for a Better Functioning Securitisation Market in the European Union' (2014) https://www.ecb.europa.eu/pub/pdf/other/ecb-boe_case_better_functioning_securitisation_marketen.pdf accessed 20 December 2025

EBA, 'Developing a Framework for Sustainable Securitisation' (2022) https://www.eba.europa.eu/sites/default/files/document_library/Publications/Reports/2022/1027593/EBA%20report%20on%20sustainable%20securitisation.pdf

EBA, 'Report on synthetic securitisation' (2015) EBA/Op/2015/26 https://www.eba.europa.eu/sites/default/files/documents/10180/983359/03ed077f-cdf9-4f6c-901b-fb7fda0e44a3/EBA-Op-2015-26%20EBA%20report%20on%20synthetic%20securitisation.pdf

ESMA, 'Consultation Paper – Consultation Paper on the revision of the disclosure framework for private securitisation under Article 7 of the Securitisation Regulation' (2025) ESMA12-2121844265-4462 https://www.esma.europa.eu/press-news/esma-news/esma-consults-revised-disclosure-requirements-private-securitisations

ESMA, 'TRV risk analysis – EU securitisation markets overview' (2023) https://www.esma.europa.eu/sites/default/files/2023-09/ESMA50-524821-2908_TRV_risk_analysis_-_EU_securitisation_markets_overview.pdf

EU Commission, 'Report from the Commission to the European Parliament and the Council On the functioning of the Securitisation Regulation' (2022) https://eur-lex.europa.eu/legal-content/EN/TXT/PDF/?uri=CELEX:52022DC0517

Federal Reserve Bulletin, 'Asset Securitization: A supervisory perspective' (1989) https://fraser.stlouisfed.org/files/docs/publications/FRB/pages/1985-1989/32502_1985-1989.pdf

Fitch Ratings, 'Representations, Warranties and Enforcement Mechanisms in Global Structured Finance Transactions' (2016) https://www.fitchratings.com/research/corporate-finance/representations-warranties-enforcement-mechanisms-in-global-structured-finance-transactions-31-05-2016

Fuller TGF, *The law and practice of international capital markets* (LexisNexis Butterworths, 3rd edn 2012)

Gerner-Beuerle C, Paech P and Schuster E-P, 'Study on Directors' Duties and Liability in the EU' (2013) LSE Enterprise, prepared for the EU Commission DG Markt
Gibson SD, 'The Case for the Expanded Role of Trustees in Securitisations' (2004) 121 *Banking Law Journal* 387
Gorton GB and Metrick A, 'Securitised Banking and the Run on Repo' (2012) 104 *Journal of Financial Economics* 3 425, 430–432
Graf G, Hinteregger M, Weissenbacher M and others, 'Case 11: Securitisation' in Smith L, Graziadei M and Mattei U (eds), *Commercial Trusts in European Private Law* (Cambridge University Press2005)
Green Storm2021, 'Prospectus dated 23 March 2021'https://pcsmarket.org/wp-content/uploads/Green-STORM-2021-Prospectus43433997.13.pdf
Hansmann H and Kraakman R, 'The Essential Role of Organisational Law' (2000) 110 *The Yale Law Journal* 387
Hansmann H, Kraakman R and Squire R, 'Law and the rise of the firm' (2006) 119 *Harvard Law Review* 5 1333
Hudson A, *The Law of Finance* (Sweet & Maxwell ed, 2nd edn, 2009)
Hudson A, *Understanding Equity & Trusts* (6th edn, Routledge2017)
ICMA, 'Sustainable Securitisation – Related questions' (2022) https://www.icmagroup.org/assets/documents/Sustainable-finance/2022-updates/Sustainable-Securitisation-QAs_CLEAN-and-FINAL_2022-06-24-280622.pdf
Jensen MC and Meckling WH, 'Theory Of The Firm: Managerial Behaviour, Agency Costs And Ownership Structure' (1976) 3 *Journal of Financial Economics* 305
Jouffray JB and others, 'Leverage Points in the Financial Sector for Seafood Sustainability' (2019) 5 *Science Advances* 1
Kim J, 'Property-Ization: The Process by Which Financial Corporate Power Has Risen' (2018) 3 *Review of Capital as Power* 1 58
Kraakman R, Armour J and Davies P, *The Anatomy of Corporate Law: A comparative and functional approach* (3rd edn, Oxford University Press2017)
Kullig S, Bertalot L and Johnson J, 'European Covered Bond Fact Book 2024' (2024) https://hypo.org/app/uploads/sites/3/2024/08/FactBook-2024_web.pdf
Latham & Watkins, *The International Comparative Legal Guide to: Securitisation 2013* (2013) Global Legal Group
Lipson JC, 'Defining Securitisation' (2012) 85 *Southern California Law Review* 1229
Mayer C, *Prosperity: Better Business Makes the Greater Good* (Oxford University Press ed, 2019)
Natwest, Mortgage Sale Agreement (2010) https://investors.natwestgroup.com/~/media/Files/R/RBS-IR-V2/securitisation/reseidential-mortgages/pdf/

7-arran-residential-mortgages-2010-1-plc/principal-transaction-documents/010 mortgage sale agreement-2010-a1.pdf
Nyborg KG, 'Central Bank Collateral Frameworks' (2017) 76 *Journal of Banking and Finance*198, 204
OECD, 'Due Diligence for Responsible Corporate Lending and Securities Underwriting – Key considerations for Banks implementing the OECD Guidelines for Multinational enterprises' (2019) https://mneguidelines.oecd.org/due-diligence-for-responsible-corporate-lending-and-securities-underwriting.pdf
Pistor K, 'Law in Finance' (2013) 41 *Journal of Comparative Economics*311, 312
Pistor K, 'A Legal Theory of Finance' (2013) 41 *Journal of Comparative Economics* 2 315
Pistor K, *The Code of Capital: How the Law Creates Wealth and Inequality* (Princeton University Press, 2019)
Plank TE, 'Regulation and Reform of the Mortgage Market and the Nature of Mortgage Loans: Lessons from Fannie Mae and Freddie Mac' (2009) 60 *South Carolina Law Review* 3 9
Quiquerez A, *La titrisation des actifs intellectuels* (Larcier2013)
Ramos Muñoz D, 'Bankruptcy-Remote Transactions and Bankruptcy Law – a Comparative Approach (Part 1): Changing the Focus on Vehicle Shielding' (2015) 10 *Capital Markets Law Journal*239
Ramos Muñoz D, 'In Praise of Small Things: Securitisation and Governance Structure' (2010) 5 *Capital Markets Law Journal*363, 365–366
Ramos Muñoz D and Ingram S, *The Law of Transnational Securitisation* (Oxford University Press, 2010)
Schwarcz SL, 'Commercial Trusts as Business Organisations: Unraveling the Mystery' (2003) 58 *Business Lawyer*559
Schwarcz SL, 'What Is Securitisation? And for What Purpose?' (2012) 85 *Southern California Law Review*1283
Schwartz A and Sepe S, 'Economic Challenges for the Law of Contract' (2020) 8 *Yale Journal of Regulation*11–12
Scopelliti A, 'Securitisation, Bank Capital and Financial Regulation: Evidence from European Banks' (2016) ECB Forum https://www.ecb.europa.eu/pub/conferences/ecbforum/shared/pdf/2016/scopelliti_paper.pdf
UNPRI, 'ESG engagement for fixed income investors: managing risks, enhancing returns' (2018) 27 https://www.unpri.org/download?ac=4449
Wagner H and Heinzmann A, 'Structured finance in Luxembourg: a new dawn?' in Bourin P, Thieltgen N (eds.) *Droit bancaire et financier au Luxembourg* (Larcier 2014) vol IV.
Wood P, *International Loans, Bonds, Guarantees, Legal Opinions* vol 5 (3rd edn, Sweet & Maxwell, 2019)
Wood P, *Project finance, securitisations and subordinated debt* vol 7 (3rd edn, Sweet & Maxwell2019)

CHAPTER 4

Market Actors as Rule-Makers of 'Sustainable Securitisations': The (Lack of) 'Private Hard Law'

4.1 Introduction

Drawing from the conceptualisation of the role of law in supporting sustainability objectives and from the 'anatomy' of *traditional* securitisation, it is essential to move to an in-depth analysis of the intersection of the two spheres.

In line with the broader momentum for sustainability considerations across financial transactions and investments,[1] a growing number of market actors, including financial institutions and companies, have started to use securitisation as a vehicle to further environmental, social, and governance objectives.[2]

While the emergence of 'sustainable securitisations' is a global trend, the following analysis will focus on the EU market. Empirical data confirms the growth of these operations within the EU market since 2016,[3]

[1] See also Introduction, 'Background and aims'.

[2] See the report by Climate Bonds Initiative, 'Green Securitisation: Unlocking Finance for Small-Scale Low Carbon Projects' (2018) Briefing Paper https://www.climatebonds.net/data-insights/publications/green-securitisation-unlocking-finance-small-scale-low-carbon-projects accessed 3 June 2026. This trend continues to date, see, among others, Banque de France, 'Green securitisation: a lever for financing the transition' (2025) https://www.banque-france.fr/en/publications-and-statistics/publications/green-securitisation-lever-financing-transition accessed 9 December 2025.

[3] AFME, 'Q3 2023 – ESG Finance Report – European Sustainable Finance' (2024) 18 https://www.afme.eu/Portals/0/DispatchFeaturedImages/AFME%20Sustainable%20Finance%20Report%20Q3%202023.pdf accessed 9 December 2025. Others have reported

F. Agostini, *The Law of Sustainable Securitisations*, EBI Studies in Banking and Capital Markets Law,
https://doi.org/10.1007/978-3-032-25046-9_4

but the use of this financing technique remains niche and sporadic.[4] As the market for other 'sustainable' financial products, such as bonds and loans, has experienced a real boom, ABSs (as the final product of securitisations) constitute 2% of all 'sustainable' debt securities within the EU.[5] In 2021, these operations account for only around 6% of overall EU securitisations.[6] The European Banking Authority (hereinafter: EBA) has tracked only 13 'sustainable securitisations' in the EU market up to 2022 and the involvement of very few entities as issuers.[7]

The OECD has estimated the further rise of these transactions globally, especially in the renewable energy sector (with an expected supply of $280–380 billion by 2035).[8] A broader use of this financing technique would also be welcome within the EU context, where securitisation issuances have stagnated after an all-time peak in 2008.[9] Institutions and

a growing use of exposures to zero- or low-emissions vehicles as assets within securitisation pools in the EU between 2016 and 2023. Carmelo Latino, Loriana Pelizzon, Max Riedel, How to green the European auto ABS market? A literature survey' (2025) 31 European Financial Management 627.

[4] As also compared with the volume of other types of 'green' debt instruments like bonds and loans, AFME, ibid., 14–19 (only in relation to 2023); European Central Bank, 'Eurosystem Reply to the European Commission's Public Consultations on the Renewed Sustainable Finance Strategy and the Revision of the Non- Financial Reporting Directive' (2020) 14. https://www.ecb.europa.eu/pub/pdf/other/ecb.eurosystemreplyeuropeancommission publiconsultations_20200608~cf01a984aa.en.pdf accessed 9 December 2025.

[5] AFME, 'ESG Finance Report – –Q2 2021' https://www.afme.eu/Portals/0/DispatchFeaturedImages/AFME%20Sustainable%20Finance%20Report%20-%20Q2%202021-%20v2.pdf accessed 9 December 2025; EBA, 'Developing a Framework for Sustainable Securitisation' (2022) 6, 20 https://www.eba.europa.eu/sites/default/files/document_library/Publications/Reports/2022/1027593/EBA%20report%20on%20sustainable%20securitisation.pdf accessed 3 June 2026.

[6] AFME, 'Q2 2021 Securitisation Report – –European Structured Finance' (2021) 20 https://www.afme.eu/Portals/0/DispatchFeaturedImages/AFME Q2th 2021 Securitisation Report1.pdf accessed 9 December 2025.

[7] EBA (n. 5) 20–21.

[8] OECD, 'Mobilising Bond Markets for a Low-Carbon Transition' (2017) Green Finance and Investment, 86 https://doi.org/10.1787/9789264272323-en accessed 9 December 2025.

[9] 'True sale' securitisations reached EUR 711 bn in 2008. In the decade after the crisis, issuance dropped and remained stable at around EUR 200 bn per year. Deutsche Bank, 'European securitisation market – ready for a comeback?' (2024) EU Monitor – Global Financial Markets, 3, https://www.dbresearch.com/PROD/RPS_EN-PROD/European_securitisation_market_%E2%80%93_ready_for_a_comeb/RPS_EN_DOC_VIEW.calias?rwnode=PROD0000000000435631&ProdCollection=PROD0000000000535114 accessed 9 December 2025.

stakeholders have identified the revitalisation of the securitisation market as an imperative for the strengthening of the Capital Markets Union.[10]

At least until 2025, when the first statutory definition of 'sustainable securitisation' has came into force[11], it has been exclusively a market-driven phenomenon, in which market actors have independently named certain transactions 'green', 'sustainable', or attached similar labels to them.438 They have used various contractual tools to articulate how the portfolio and the funds raised through the issuance of ABSs contribute to environmental, social, and governance objectives. External financial service providers have then reviewed and assessed the sustainability characteristics of the operation, also drawing on external principles (such as the ICMA Green Bond Principles[12]).

From a legal standpoint, the predominant role of market actors in articulating (sustainability-related) commitments reflects a broader trend in the securitisation market. Indeed, Chap. 3 has explored the interrelatedness of proprietary and contractual rights and statutory rules, which jointly provide the building blocks for 'traditional' securitisations.[13]

In the following chapter, the focus will be on market actors' rule-making powers, which manifest in the clauses of the securitisation documentation. These clauses can create binding legal obligations between the parties, which I will describe as 'private hard law'. In the present chapter, I will analyse, also based on examples of previous securitisation transactions, how sustainability commitments were articulated in practice. The analysis will, however, highlight the very limited use of documentation so far, with previous clauses largely failing to establish credible and enforceable commitments on sustainability objectives.

In particular, Sect. 2 will conceptualise the meaning of 'private hard law' and contextualise the theoretical background of the analysis. Sect. 3 will examine how sustainability considerations have permeated different types of securitisation operations. Sect. 4 will map the various types

[10] In 2020, it was expected that new RMBS issuances in the following 5–10 years would lead to an overall volume of around 800bn €, Final Report of the High Level Forum on the Capital Markets Union', Capital Markets – –A New Vision For Europe's capital markets' (2020) https://finance.ec.europa.eu/system/files/2020-06/200610-cmu-high-level-forum-final-report_en.pdf accessed 9 December 2025.

[11] With the entry into force of the European Green Bond Standard see also Chap. 6, 6.4.2–6.5.2.

[12] On the Principles, see Sect. 4.3. and Chap. 5, Sect. 5.4.2.

[13] See also Chap. 3, Sect. 3.2.

of deals, classifying them according to what makes them 'sustainable'. In Sect. 5, the analysis will delve into how market participants have introduced 'private hard law' on sustainability matters on a deal-by-deal basis, with a focus on the 'focal points' of sustainable securitisations identified in Chap. 3.

Sect. 6 will summarise the key challenges that the current market-designed, deal-specific legal framework for 'sustainable securitisations' poses. It will anticipate the need for further regulatory coordination mechanisms, in line with the legal strategies discussed in Chap. 2.[14]

4.2 Conceptualising 'Private Hard Law'

As 'sustainable securitisations' have emerged as a market-driven phenomenon, the parties have independently designed the elements that render them 'sustainable' on a deal-by-deal basis. To achieve this, they have utilised various agreements and transactional documentation. In essence, market actors have set their own rules in the absence of binding state-designed laws, aligning with a recurrent trend in the financial context.[15] *Pistor* has extensively discussed this phenomenon, highlighting how private parties 'code the capital' by negotiating their deal-by-deal contractual commitments, along with related enforcement mechanisms.[16] This power is exercised *ex ante,* before regulators engage with innovation.[17] The binding legal obligations between securitisation parties will be referred to as instances of 'private hard law'.

Some commentators have depicted cases of market actors *independently* formulating rules for their own transactions as instances of 'self-regulation', [18] yet the term could be misleading for the purposes

[14] Sect. 2.6.

[15] See, *inter alia*, Awrey's analysis on the growth of OTC derivatives, Dan Awrey, 'Towards a Supply-Side Theory of Financial Innovation' (2013) 41 Journal of Comparative Economics 401, 402, 406, 413. On the interplay between innovation and regulation in financial markets, Cristie Ford, *Innovation and the State: Finance, Regulation, and Justice* (Cambridge University Press 2017) 26–44.

[16] For an example in relation to securitisation, see Katharina Pistor, *The Code of Capital: How the Law Creates Wealth and Inequality* (Princeton, 2019) 80–83. for the theoretical framework, see ibid., 'A Legal Theory of Finance' (2013) 41 *Journal of Comparative Economics* 2 315, 320.

[17] Pistor, 'A Legal Theory of Finance', ibid., 315, 321, 324.

[18] Cary Coglianese, Evan Mendelson, 'Meta-Regulation and Self-Regulation' in *The Oxford Handbook of Regulation* (Oxford University Press 2010); Baldwin R, Cave M, Lodge M, 'Self-Regulation, Meta-Regulation, and Regulatory Networks'in *Understanding*

of the present analysis. Throughout other scholarly work, this notion has also been used to capture initiatives with a *multilateral* effect on market actors, like standard clauses.[19] Chap. 5 will examine the latter.[20] Instead, in this chapter I will refer to 'private hard law' as a legal exercise resulting in the allocation of proprietary rights across different market actors and the negotiation of transactional clauses on a deal-by-deal basis.

The present work also draws from the insights of contract theorists in economics, who have acknowledged the role of contracts as sources of binding legal obligations. They are functional for the long-term planning between parties and for designing mechanisms protecting both parties against risks.[21] Contract theory also constitutes the background of the present analysis since it acknowledges the inherent incompleteness and unavoidable costs of private bargaining.[22]

Another noteworthy trend, also highlighted by *Pistor*, is the tendency among market actors to ensure that they act 'inside, rather than outside, the law'.[23] When designing and communicating what makes their transactions 'sustainable', parties tend to make extensive reference to various legal sources, ranging from state-designed binding measures to market-designed non-binding initiatives[24]. This strong reliance on various forms of law indirectly legitimises the labelling of transactions as 'sustainable'.

Regulation: Theory, Strategy, and Practice in (Vol. 15, 2011). For a review of all the various meanings of the term 'self-regulation', see J Black, 'Decentreing Regulation: Understanding The Role Of Regulation And Self-Regulation in a 'Post-Regulatory' World' (2001) 54 *Current Legal Problems* 103, 113–121.

[19] For example, Doreen Mcbarnet, Aurora Voiculescu and Tom Campbell (eds), *The New Corporate Accountability: Corporate Social Responsibility and the Law* (Cambridge University Press 2007) 120–121.

[20] See Chap. 5, Sect. 5.2.3.

[21] A Mitchell Polinsky, Steven Shavell, 'Economic Analysis of Law' in Matias Vernengo, Esteban Perez Caldentey, Jayati Ghosh (eds.) *The New Palgrave Dictionary of Economics* (Palgrave MacMillan 2008) 14; Iain R MacNeil, 'Contracts: adjustment of long-term economic relations under classical, neoclassical, and relational contract law' (1978) 72 *Northwestern University Law Review* 854; Charles J Goetz, Robert E Scott, 'Principles of Relational Contracts' (1981) 67 *Virginia Law Review* 1089, 1090.

[22] Ian Ayres, Robert Gertner, 'Filling Gaps in Incomplete Contracts: An Economic Theory of Default Rules' (1989) 99 The Yale Law Journal 87; Michael Jensen, William Meckling, 'Theory of the firm: Managerial Behaviour, Agency Costs and Ownership Structure' (1976) 72 Journal of Financial Economics 305, 308–323.

[23] Pistor, 'A legal theory of finance' (n. 16) 313.

[24] Like the ICMA Green Bond Principles, See below Sect. 4.3.

In turn, it also has the potential to encourage and stimulate the growth of 'sustainable' deals.[25] Therefore, all these external legal sources serve an 'enabling' function in that they lend support to market actors labelling securitisations as 'sustainable'.[26]

The main advantages of market actors' involvement in the design of the 'private law' for these deals include their proximity to transactions and their power to innovate. On the one hand, they are better suited than policy-makers to agree upon the arrangements for their own contractual relationships,[27] and they possess more information for this purpose.[28] On the other hand, they are more likely to design up-to-date solutions, while regulators may inevitably fall short of responding to innovation.[29] Both these aspects are pivotal for transactions supporting sustainability objectives, which tend to vary significantly from deal to deal and to evolve in line with the global agenda.

The other point of reference for the present analysis is the scholarly analysis of the boundaries to private bargaining. Chicago School and neo-institutionalist scholars have pointed out the complementary role of (state-designed) law when private transactions lead to higher costs or when market actors fail to independently achieve efficient outcomes ('market failures').[30] With this in mind, the present chapter will also highlight the various shortcomings of 'private hard law', which call for the complementary intervention of other regulatory coordination techniques.

[25] Awrey has drawn similar conclusions in relation to Over-the-counter (OTC) derivatives, Awrey, 'Financial Innovation' (n. 15) 412.; see also more generally Pistor (n. 16), 315, 321.

[26] In line with Javier Solana, 'The Roles of Law in Finance' (2022) Working Paper, on file with the author, 4. On the 'enabling' function, see also Chap. 2, Sect. 2.4.

[27] Mary M Shirley, Claude Ménard, *Handbook of New Institutional Economics* (Springer 2008) 43–44. Arrangements between private parties have also been referred to as instances of 'Contractual governance', see, *inter alia*, Peter Zumbansen, Osgoode Hall, 'The Law of Society: Governance Through Contract' (2007) 14 Indiana Journal of Global Legal Studies.

[28] See Black, 'Decentreing Regulation' (n. 18) 122.

[29] As extensively argued and demonstrated by Ford (n. 15) 26–44.

[30] Ronald H Coase, 'The Problem of Social Cost' (1960) 3 Journal of Law and Economics 1, 17–18; Shirley, Ménard, (n. 25) 43, 44, 209.

4.3 The Intersection Between Sustainable Finance and Securitisation

The notion of 'sustainable securitisation' is so vague that tracing back to its first uses is not a straightforward exercise. In the absence of a statutory definition, qualifying these transactions as 'sustainable' is primarily the result of the parties' decision. This means that there have been some cases of deals which furthered sustainability objectives but were not named or advertised as such. For example, *Credit Suisse* pooled together carbon offset credits in the early 2000s and issued securities to support forestry projects, paving the way for a series of similar deals.[31]Even *Crédit Agricole* has used securitisation to free up part of its capital, with the aim to extend its funding to tackle poverty across emerging markets.[32]

One case where parties have specifically named a securitisation transaction 'sustainable' involved the company *Hannon Armstrong,* which used a portfolio of leases for solar, wind, and energy-efficiency installations as collateral. The company announced the issuance as one of 'Sustainable Yield' bonds.[33] However, the Climate Bonds Initiative Database only traces the first example of green securitisation transactions back to 2013.[34] It also

[31] Hedge Week, 'Credit Suisse and SFM launch USD 200m carbon-backed forestry finance facility' (2007)

https://www.hedgeweek.com/credit-suisse-and-sfm-launch-usd-200m-carbon-backed-forestry-finance-facility/ accessed 9 December 2025; on the application of carbon credits to the environmental sustainability cause, see Felipe Alviar-Baquero JP Sweny, Mark Nicolaides, 'The Applications Of Structured Finance Techniques To The Cleantech Industry' (2012) Latham & Watkins, 7. https://www.americanbar.org/content/dam/aba/publications/nr_newsletters/eemf/201108_eemf.authcheckdam.pdf accessed 21 May 2023; in the specific context of securitisation, see John Hay, Carbon permits – Securitisation finds a hidden seam, (*Global Capital,* 27 September 2021) https://www.globalcapital.com/article/29425pg7njwqa46q9a9kw/Securitisation/abs/carbon-permits-Securitisation-finds-a-hidden-seam accessed 9 December 2025.

[32] Stephen Foley, 'A synthetic path to green lending that might just work' (FT, 6 March 2017) https://www.ft.com/content/7042b704-fad1-11e6-bd4e-68d53499ed71 accessed 9 December 2025; Tracy Alloway, 'World Bank Backs Crédit Agricole Loans' (FT, 6 June 2014) https://www.ft.com/content/4cbee3c0-ece7-11e3-8963-00144feabdc0 accessed 9 December 2025.

[33] Hannon Armstrong Sustainable Infrastructure, 'Hannon Armstrong (HASI) Completes $100,000,000 Asset-Backed 'securitisation of 2.79% Sustainable Yield Bonds' https://www.prnewswire.com/news-releases/hannon-armstrong-hasi-completes-100000000-asset-backed-'securitisation-of-279-sustainable-yield-bonds-236999811.html accessed 9 December 2025.

[34] Climate Bonds Initiative, 'Green Securitisation' (n. 2) 1.

reported the wide circulation of the model globally across various jurisdictions in Latin America,[35] as well as Australia[36] and China,[37] while identifying some barriers in other jurisdictions like India.[38]

Yet, specialist press has referred to the first case of a 'green securitisation' deal in the EU as the one launched by *Obvion* in 2016. The company inaugurated a series of deals named 'Green Storm'.[39] Since then, the EBA has reported 13 cases of securitisation deals with 'green' or other 'sustainability-related' features[40] within the EU and UK market between 2016 and 2022.[41]

Sustainability considerations have indeed pervaded securitisation deals with very disparate proprietary structures.[42] Alongside traditional

[35] Andrew Whiley, 'First round of UN Green Climate Fund projects kick-starts green securitisation markets in Latin America' (10 November 2015) https://www.climatebonds.net/2015/11/first-round-un-green-climate-fund-projects-kick-starts-green-securitisation-markets-latin accessed 2 October 2023. Climate Bonds Initiative, 'Brazil Sustainable Securitisation – State of the market Q3 2022' (2022) https://www.climatebonds.net/data-insights/publications/brazil-sustainable-securitisation-state-market-q3-2022 accessed 9 December 2025.

[36] Andrew Whiley, 'Sunny side up! FlexiGroup issues first Australian green ABS. Proceeds for solar & 5bps pricing benefit for green label! And.... It's Climate Bonds Certified!' (*Climate Bonds Initiative,* 21 April 2016) https://www.climatebonds.net/news-events/blog/sunny-side-flexigroup-issues-first-australian-green-abs-proceeds-solar-5bps-pricing-benefit-green-label-climate-bonds-certified accessed 9 December 2025.

[37] Climate Bonds Initiative, 'China green securitisation report State of the market' (2020) 9 https://www.climatebonds.net/data-insights/publications/china-green-Securitisation-state-market-2020-report accessed 9 December 2025.

[38] Climate Bonds Initiative, 'Securitisation as an enabler of green asset finance in India' (2020) 18–19 https://www.climatebonds.net/data-insights/publications/securitisation-enabler-green-asset-finance-india accessed 9 December 2025.

[39] See For example, Ionova M, 'Obvion plans Europe's first Green ABS bond' *(Reuters,* 24th May 2016) www.reuters.com/article/green-abs/obvion-plans-europes-first-green-abs-bond-idUSL5N18K4NJ accessed 9 December 2025; see also Sustainalytics, 'Green Storm 2016 Obvion: Framework Overview and Second Opinion by Sustainalytics' (2016) https://www.sustainalytics.com/corporate-solutions/sustainable-finance-and-lending/published-projects/project/obvion-n-v/obvion-n.v.-green-storm-2016-cbi-pre-issuance-verification-letter-(2015)/obvion-n.v.-green-storm-2016-cbi-pre-issuance-verification-letter-(2015) accessed 9 December 2025.

[40] For the distinction, see below Sect. 4.4.

[41] EBA (n. 5) 20–21.

[42] For the differences between traditional securitisation deals and all the other proprietary structures mentioned below, see Chap. 3, Sect. 3.1.–3.2. others have noted how green securitisation developed at market level, and regulation was not necessary to enable its development. Pierre De Gioia Carabellese, Camilla Della Giustina, *The law of securitisations: from crisis to techno-sustainability* (2023) Routledge, 168–169.

securitisation deals, leading to the issuance of asset-backed securities (ABSs), there have also been some examples of asset-backed commercial paper (ABCP) transactions and programmes which furthered sustainability objectives.[43] The EU legislation defines ABCP programmes as a bundle of securitisation deals comprising securities with a maturity of up to one year.[44]

Moreover, some originators have started to use on-balance-sheet structures to promote sustainability objectives.[45] For example, *Crédit Agricole* transferred the risk, but retained the title, over a bundle of loans to *Marine Investment.* The loans backed bonds issued to develop environmentally friendly projects. [46] Similarly, Multilateral Development Banks (MDBs) engaged in these kinds of deals with a view to supporting the UN Sustainable Development Goals: for instance, the *Room2run* deal allowed the African Development Bank to transfer some risk to investors and to use the released capital to support further lending towards renewable energy, employment and other similar objectives.[47] The European Investment Bank has been actively involved in this market, providing credit protection for securitisation deals furthering sustainability goals, e.g. in the context of the InvestEU Programme.[48]

[43] For example, Snam SA, 'Euro-Commercial Paper Programme For The Issuance Of Notes And Esg Notes – –Information memorandum update 2020' (2020) https://www.snam.it/content/dam/snam/pages-attachments/it/investor-relations/documents/euro-commercial-paper/Snam_ECP_-2020_Update-Information_Memorandum.pdf accessed 9 December 2025; see also 'Lafayette Asset securitisation 2020' as described in the external review, Vigeo Eiris, 'Second-party opinion on the Green ABCP Programs Annex to Credit Agricole's Green Bond Framework' (2020) https://www.ca-cib.com/sites/default/files/2020-10/20200803_VigeoEiris_SPO_Credit-Agricole.PDF, accessed 9 December 2025.

[44] Article 2 (7) Securitisation Regulation.

[45] In the EU it has occurred in 2017, 2020 and 2021, see AFME, 'Q3 2023 – –ESG Finance Report ' (n. 3) 18.

[46] Climate Bonds Initiative, 'Deep-dive: Green synthetic securitisation deal: Frees USD 2bn of capital for green investments – –Crédit Agricole and Mariner Investment Group take the plunge' (2017) https://www.climatebonds.net/news-events/blog/deep-dive-green-synthetic-securitisation-deal-frees-usd-2bn-capital-green-investments-cr%C3%A9dit-agricole-mariner-investment-group-take-plunge accessed 9 December 2025.

[47] for the details of the deal, see UNPRI, *PRI Awards 2019 Case Study: Room2Run* (2019) https://www.unpri.org/showcasing-leadership/pri-awards-2019-case-study-room2run/4848.article accessed 20 July 2025.

[48] See, inter alia, European Investment Bank, 'INBANK climate action for private individuals' (2024) https://www.eib.org/de/projects/pipelines/all/20240110 accessed 9 December 2025; ibid., 'DNB enhanced support for climate action' (2024) https://www.eib.org/en/projects/pipelines/all/20210715, accessed 9 December 2025.

Across all these deals, the issuance of securities occurred both within publicly regulated markets and through 'private placements'.[49] The portfolios have generally been 'static', but there has also been a surge of 'sustainable CLOs', where managers or management companies have actively been in charge of the selection of the underlying assets.[50]

In parallel to the evolution of sustainable securitisations, companies have also issued similar financial instruments to further sustainability objectives, like covered bonds. These are securities also backed by a pool of assets, such as loans, but they differ in not requiring the involvement of SPVs.[51] One example oriented to sustainability goals was the 'green covered bond' issued by *BerlinHyp* to support the green building sector. There were also a series of sustainability-related requirements for inclusion of assets in the portfolio.[52] Both 'Green' covered bonds and 'sustainable securitisations' have experienced a very slow growth in the EU market.[53]

Moreover, certain market actors have started to issue 'structured' notes combining a traditional bond and a derivative linked to the performance of an index. For instance, *Credit Suisse* and *Société Générale* have linked the repayment of the coupon to an index tracking environmental and other sustainability-related data.[54]

[49] For example, The above-mentioned securitisation deal involving Credit Agricole and Mariner Investment and the ABCP programme, Climate Bonds Initiative, 'Deep-dive' (n. 46), ibid. SNAM SA (n. 43).

[50] The first example was launched by the Australian NIBC bank, see S&P Global Ratings, 'North Westerly CLO VI B.V' (2019) 4 https://www.spglobal.com/_assets/documents/ratings/research/2342405.pdf accessed 30 September 2023. On the definition of CLOs, see Chap. 3, Sect. 3.2.

[51] Steven L Schwarcz, 'The Conundrum of Covered Bonds' (2011) 66 *Business Lawyer* 561, 566 – –567. For the differences between traditional securitisation and covered bonds, see Chap. 3, Sect. 3.2.

[52] ISS Oekom, 'Second-party opinion – –Re-Verification of the Sustainability Quality of the Issuer and Green Bond Asset Pool' (2019) https://spowidget.iss-corporate.com/icsspo app/file/documents/spo/spo-berlinhyp-29apr2019.pdf accessed 9 December 2025. On eligibility criteria across sustainable securitisations, see below 4.5.1.1.

[53] Fitch ratings, 'Covered Bonds Still a Limited Share of Sustainable Bond Market' (9 December 2021) https://www.fitchratings.com/research/structured-finance/covered-bonds/covered-bonds-still-limited-share-of-sustainable-bond-market-09-12-2021 accessed 9 December 2025.

[54] Sweny, Nicolaides (n. 31) 7. In Société Générale's deal, the performance of Lyxor Dynamic Environment Fund had an impact on the coupon, see Raffaele Della Croce, Christopher Kaminker, Fiona Stewart, 'The Role of Pension Funds in Financing Green Growth Initiatives' (2011) OECD Publishing, 34, 44–45 https://www.oecd.org/en/publications/the-role-of-pension-funds-in-financing-green-growth-initiatives_5kg58j1lwdjd-en.html accessed 9 December 2025; Credit Suisse, *Credit Suisse launches inaugural structured notes with green use of issuance proceeds* (8 December 2020) https://www.credit-suisse.com/about-us-news/en/articles/media-releases/credit-suisse-launches-inaugural-structured-notes-with-green-use-202012.html?t=485_0.0737152877794387 accessed 20 September 2023.

4.4 Towards a Typology of 'Sustainable Securitisations'

The lack of consensus around the meaning of 'sustainability' and 'green' has strongly affected the development of the sustainable securitisation sector, creating many inconsistencies across the transactions that have been defined as such.

Throughout the present work, I will use 'sustainable securitisations' as an umbrella term capturing all the deals furthering environmental, social, and governance objectives, following the UN Sustainable Development Goals.[55] Thus, it covers, but is not restricted to, 'green' transactions, enabling originating companies to pursue environmental objectives. These were and still remain the largest share within the EU market.[56] In parallel, the notion of 'sustainable securitisations' can also comprise deals focusing on other aspects of the sustainability agenda. The COVID-19 pandemic has been the main driver behind the structuring of an increasing number of transactions advancing social objectives ('social securitisations'), like social housing, health, or the protection of income, especially in the UK market,[57] with a couple of examples in the EU market.[58] Other denominations that have had growing use include 'sustainable',[59] 'sustainability-linked' or 'ESG-linked' securitisation.[60]

[55] As set out in the 2015 UN resolution, Transforming our world: the 2030 Agenda for Sustainable Development (2015) A/RES/70/1.

[56] For a break-down of all types of 'sustainable' bonds and loans issued in EU, AFME (n. 3) 7.

[57] See Kensington's 'social' MBS, Alex Maddington, 'Kensington completes inaugural £ 472 Million Social Bond STS Securitisation' (*Kensington Mortgages*, 2021) https://www.kensingtonmortgages.co.uk/blog/article/kensington-completes-inaugural-472-million-social-bond-sts-securitisationaccessed 3 June 2026; see also the 'Gemgarto' issuance in 2020 and the 'Brass' issuances of 'social' RMBS, as reported by Isabel Tinsley, Tim Conduit, 'Implications of Sustainability and ESG for the Securitisation Market' (Allen & Overy 2021) https://www.jdsupra.com/post/fileServer.aspx?fName=698231e7-4302-4500-a87e-f44307e60e3e.pdf accessed 9 December 2025.

[58] For example, PMI100 and Fortuna 2021, issued in 2021, according to AFME, European Green Securitisation -
Regulatory State of Play (2022) 9 https://www.afme.eu/Portals/0/DispatchFeaturedImages/AFME_ESGSecuritisation_2022_07_final-2.pdf accessed 9 December 2025.

[59] For example, Bayfront Infrastructure's securitisation had one 'sustainable' tranche, Bayfront Infrastructure, 'Bayfront Prices US $ 402 . 7 Million Infrastructure Asset-Backed Securities' (Clifford Capital 2022). https://api.cliffordcapital.sg/dev/file/download?id=1862065557811945474 accessed 9 December 2025.
According to ICMA, 'Sustainable' bonds would be those that simultaneously contribute to environmental and to social objectives. ICMA, 'Sustainability Bond Guidelines' (2021) https://www.icmagroup.org/assets/documents/Sustainable-finance/2021-updates/Sustainability-Bond-Guidelines-June-2021-140621.pdf accessed 9 December 2025. I will, however, refer to 'sustainable' deals as an overarching term which can also capture those featuring exclusively green or exclusively social aspects.

[60] On the difference between 'green' and 'sustainability-linked' deals, see below.

Market actors have adopted several different arrangements when designing such transactions, as summarised at the end of this section. The first broad demarcation lies between the deals which are 'sustainable' because of the final allocation of the proceeds or because of the underlying portfolio. These qualities may refer to the transaction as a whole or to specific tranches of it.[61]

Under the first scenario, the originating entity uses securitisation to obtain liquidity in order to implement projects or to pursue specific targets in the environmental, social, and governance remits ('sustainable proceeds securitisation'[62]). For instance, *Toyota* was one of the first companies to announce in 2014 that it would raise finance to develop hybrid and electrical vehicles by issuing 'green ABSs'.[63] While operations of this kind still ultimately contribute to environmental, social, and governance objectives, they are not always backed by a pool of assets making an equally positive contribution to the sustainability agenda. In other words, the originating entities sometimes include in the portfolio credit claims which have a neutral or even potentially adverse environmental and social impact. For example, *Toyota*'s transaction was backed by general consumer car loans and leases.[64]

The second scenario comprises the cases where the operation has 'sustainable' qualities only at portfolio level, or at both the proceeds and portfolio level ('sustainable portfolio securitisations'[65]). For instance, the SPV *Hannon Armstrong* issued 'Sustainable Yield' bonds, backed by leases for solar, wind and energy-efficiency

[61] See For example, Bayfront Infrastructure's transaction, which had one tranche of 'sustainable infrastructure asset-backed securities', Bayfront Infrastructure (n. 59). For the role of tranching in securitisation deals, see Chap. 3, Sect. 3.2.–3.3.

[62] The terminology builds on the one proposed by James and Parker, who only referred to 'green proceeds' and 'green collateral' securitisations. Kathryn James, Isabel Parker, 'New Growth in the Green Securitisation Market' (2019) Simmons & Simmons, 1. www.simmons-simmons.com/-/media/files/articles/2019/capital%20markets/new%20growth%20in%20the%20green%20securitisation%20market%20january%202019.pdf accessed 5 December 2025.

[63] Climate Bonds Initiative, 'Green securitisation' (n. 2), 4.

[64] Ibid.

[65] James, Parker (n. 62).

installations.[66] Similarly, the Spanish financial institution CETELEM will pool together a portfolio of loans to individuals and households for renovation of residential properties, small-scale renewable energy projects and for the purchase of equipment to enhance residential energy efficiency.[67]

Drawing from this broad distinction, I will further distinguish the transactions based on the meaning that market actors have given to the 'sustainable' label. Arguably, this is either based on the activities, or the targets, linked to the issuance.[68]

First, the 'sustainable' denomination has sometimes reflected the nature of the *activities* linked to the transaction, i.e. the projects which the operation finances or the assets which are part of the portfolio. This link between the 'sustainable' label and specific activities resonates with the idea behind 'taxonomies'. These emerging initiatives by policy makers and market actors, like the one introduced in the EU[69], provide a classification system to identify what assets, projects and sectors are 'sustainable' or harmful to sustainability objectives.[70]

'Sustainable proceeds securitisations' of this kind have comprised operations supporting specific activities, like electric vehicles or renewable energy installations, like the deal by *Toyota* illustrated above. The focus on the 'use of proceeds' makes these deals functionally similar to green

[66] OECD, Mobilising Bond Markets (n. 8) 34; Hannon Armstrong Sustainable Infrastructure Capital Inc, 'Hannon Armstrong (HASI) Completes $100,000,000 Asset-Backed 'securitisation of 2.79% Sustainable Yield Bonds™' (PR Newswire 23rd December 2013) https://investors.hasi.com/news/press-releases/detail/214/hannon-armstrong-hasi-completes-100000000-asset-backed-Securitisation-of-2-79-sustainable-yield-bonds accessed 9 December 2025.

[67] European Investment Bank, 'CETELEM enhanced ABS for climate action' (2025) https://www.eib.org/en/projects/all/20230971 accessed 9 December 2025.

[68] Correspondingly, some have distinguished between 'Input-based' sustainable finance standards, which only provide a catalogue of eligible projects, and 'output-based', which also set thresholds and outcomes, Christoph Nedopil, Truzaar Dordi and Olaf Weber, 'The Nature of Global Green Finance Standards – Evolution, Differences, and Three Models' (2021) 13 Sustainability 7 3723, 2, 15, 16 https://www.mdpi.com/2071-1050/13/7/3723. Accessed 9 December 2025.

[69] Under Regulation 2020/852 (EU). For its application to securitisation, see Chap. 6, Sect. 6.4.1.–6.5.1.

[70] NGFS, 'Enhancing Market Transparency in Green and Transition Finance' (2022) 11 https://www.ngfs.net/en/publications-and-statistics/publications/enhancing-market-transparency-green-and-transition-finance accessed 9 December 2025.

bonds, i.e. debt instruments issued in order to develop environmentally friendly projects.[71] Among these transactions, there has also been a series of deals where originating entities have placed the emphasis on the consequences of the assignment of the portfolio, like the 'Premium Green' deal by *Crédit Agricole.* [72] By removing certain assets from originators' balance sheets, these deals have left more space to expand or implement sustainability-related activities or targets, like lending to environmentally-friendly sectors.[73]

By contrast, 'sustainable portfolio securitisations' with an activity-oriented focus have been those where originating entities have undertaken to exclusively assign to SPVs credit claims in sectors and activities that contribute to sustainability objectives, like loans towards energy-efficiency installations and renewable energy.[74] An example is the above-mentioned CETELEM case.[75] In very few cases, the collateral comprised project finance loans towards environmental and social projects.[76] Conversely,

[71] See the definition of green bonds as 'any type of bond instrument where the proceeds or an equivalent amount will be exclusively applied to finance or refinance, in part or in full, new and/or existing eligible Green Projects' under the Green Bond Principles. One of the core principles is the 'Use of Proceeds'. ICMA, Green Bond Principles, 3–4. Commentators have described 'sustainable' financial instruments like these as 'activity-based', Rohit Goel, Deepali Gautam, Fabio Natalucci, 'Sustainable Finance in Emerging Markets: Evolution, Challenges, and Policy Priorities' (2022) IMF Working Papers WP/22/182, 10–11. https://www.imf.org/-/media/Files/Publications/WP/2022/English/wpiea2022182-print-pdf.ashx accessed 9 December 2025.

[72] arranged by Credit Agricole, through which it committed to extend the volume of its Green Loans, see Mariner Investment Group, *Landmark $3 Billion Socially Responsible Synthetic 'securitisation* (PR Newswire, 6 March 2017) https://www.prnewswire.com/news-releases/landmark-3-billion-socially-responsible-synthetic-Securitisation-300418309.html accessed 9 December 2025.

[73] Some commentators have referred to these transactions as 'green capital securitisations', see James, Parker (n. 62). Other commentators have rightly pointed out that they would in practice constitute a sub-species of other 'green proceeds securitisations', see Christy Petit, Pierre Schlosser, 'Rationale, Potential and Pitfalls of Green Securitisation' (2020) RSCAS Working Paper 35, 2 https://cadmus.eui.eu/bitstream/handle/1814/67018/RSCAS%20 2020_35.pdf?sequence=1 accessed 9 December 2025. For this reason, these deals will not be treated as a different category throughout the present analysis.

[74] Like Obvion described above, see Sustainalytics, 'Green Storm 2016 Obvion' (n. 39) 1.

[75] On the requirements that the underlying loans must fulfil, see also European Investment Bank, 'Environmental and Social Data Sheet' (2025) https://www.eib.org/attachments/registers/241965713.pdf accessed 9 December 2025.

[76] SCI, 'Specialist-squared' solution? (2022) https://www.structuredcreditinvestor.com/Article.asp?SID=79770 accessed 9 December 2025.

other market actors have defined their deals as 'sustainable' due to the kinds of assets *not* eligible for the portfolio, due to their adverse environmental and social impact. This approach has been especially common across CLOs, such as the one managed by ***Bardin Hill Advisors.*** It has translated into a series of exclusion criteria preventing the inclusion of certain types of loans within the portfolio based on the activities that they financed, such as palm oil or speculative oil and gas extraction. [77] In these cases, CLO managers have also at times indicated how material their business activity in 'sustainable' sectors would be[78] or provided thresholds for ineligibility.[79]

Second, the alternative approach is target-oriented rather than activity-oriented. Market actors have increasingly named their transactions as 'sustainable' after setting certain quantitative environmental, social, and governance targets and linking them to the use of the 'sustainability' label or to the repayment conditions of the securities. This different focus builds on a recently and rapidly growing trend in the sustainable finance landscape, initiated by ***Enel.*** The Italian energy company tied the issuance of bonds to specific environmental targets. Failure to attain such objectives would translate into an increase in the interest rate.[80] Such an example paved the way for a series of similar transactions and for the design of market standards for Sustainability-Linked Bond Principles (hereinafter: SLBs) and Sustainability-Linked Loan Principles (hereinafter: SLLPs). As I will discuss in Chap. 4, they recommend choosing relevant sustainability performance targets (SPTs) and related Key Performance Indicators (KPIs)

[77] The documentation for the issuance of CLOs managed by Bardin Hill Loan Advisors lists a series of ineligible activities. Bardin Hill Loan Advisors European Funding 2019–1 designated activity company, 'Listing Particulars' (2019) 120 https://ise-prodnr-eu-west-1-data-integration.s3-eu-west-1.amazonaws.com/legacy/ListingParticulars_b4733386-9ceb-4590-a12e-2b613746ab3c.pdf accessed 9 December 2025. The first example was launched by the Australian NIBC bank, see S&P Global Ratings (n. 50).

[78] For a general analysis, see Fitch Ratings, 'More US CLO Managers considering ESG in investment decisions' (May 2022) https://www.fitchratings.com/research/structured-finance/more-us-clo-managers-considering-esg-in-investment-decisions-27-05-2022 accessed 9 December 2025.

[79] Deutsche Bank, 'ESG and Securitisation: Early Days' (2020) 14–15 https://structuredfinance.org/wp-content/uploads/2020/02/Deutsche-Bank-Markets-Research-ESG-and-Securitisation-February-2020.pdf accessed 9 December 2025.

[80] Enel, 'Sustainability-linked bonds', and specifically 'Enel's General Purpose SDG Linked Bond – –Context and Principles' (2019) https://www.enel.com/investors/investing/sustainable-finance/sustainability-linked-finance/sustainability-linked-bonds accessed 9 December 2025.

to measure the progress against them. Not reaching the SPTs should affect the financial conditions of these instruments.[81] The key feature of bonds like *Enel*'s and of these market standards is the focus on the behaviour of the entities issuing financial instruments, rather than on the types of activities they finance.[82] While the *Enel* transaction and the aforementioned market standards relate to bonds and loans[83], market actors have recently replicated similar mechanisms within their securitisations.

Across these seminal examples, market actors have predominantly used the 'sustainability' label across 'sustainable proceeds securitisations'. For instance, *Jupiter* launched a 'SDG-linked' transaction, whereby it promised a reduction of the coupon if it successfully used the liquidity from the issuance to increase its lending to projects with positive impact.[84] The sponsor of the 2021 *Project Boqueron* deal, *Santander*, also undertook to extend further loans to renewable energy projects in exchange for a step-down in the coupon.[85] *SNAM* set up a commercial paper programme where the securities had to achieve certain ESG rating scores or they would have to stop using the 'ESG' denomination.[86]

At the same time, there have been cases of 'sustainable portfolio securitisations' where issuers undertook to adjust the repayment conditions of the ABSs based on whether the underlying portfolio assets met or failed to meet certain targets. For instance, the 2021 *Project Boqueron* deal was also designed with a series of targets regarding the environmental impact of the portfolio. As it was a revolving transaction,[87] the issuer achieved a

[81] ICMA, 'Sustainability-Linked Bond Principles' (2020) 3. See further Chap 5, Sect. 5.3.–5.4.–5.5.

[82] Torsten Ehlers, Diwen Nicole Gao, Frank Packer, 'A Taxonomy of Sustainable Finance Taxonomies' (2021) 8–9 https://www.bis.org/publ/bppdf/bispap118.pdf accessed 9 December 2025. These financial instruments have also been described as 'behaviour-based', Goel, Gautam, Natalucci (n. 71) 10.

[83] EBA confirms the applicability of SLBs to securitisation, see EBA (n. 5) 16.

[84] Societe Generale, 'Societe Generale and Mariner Investment Group Complete a $ 3.4b Impact Investment Risk Transfer Transaction' (2019) 24 https://www.societegenerale.com/en/news/newsroom/societe-generale-and-mariner-investment-group-complete-34b-impact-investment-risk accessed 9 December 2025.

[85] Newmarket Capital, 'Sustainable investing – –solutions, impact, Innovation' 7, 16 https://www.newmarketcapital.com/wp-content/uploads/2021/11/ESG-Analysis_v9_2021-11-16_ARM_spreads.pdf; Global Capital, 'Private securitisation Deal of the Year – Project Boqueron' (2 June 2021) https://www.globalcapital.com/article/28wq98676i4ve6c5xrsxc/polls-and-awards/private-securitization-deal-of-the-year-project-boqueron accessed 3 June 2026.

[86] SNAM SA (n. 43) 5.

[87] See Chap. 3, Sect. 3.2.

further step-down in the coupon if it managed to replenish the portfolio by increasing the share of assets in the renewable energy sector.[88]

Across all these examples of 'sustainable securitisations', originators' commitments to sustainability objectives have varied in intensity. For instance, when they have announced the activities linked to the operation, they have sometimes provided high-level lists of eligible, or ineligible sectors and projects.[89] In other instances, they have also indicated specific thresholds or outcomes that the eligible activities should reach in order to be awarded or to maintain the 'sustainable' denomination.[90]

Another common practice across market actors has been to undertake additional commitments on top of the selected activities and targets before and after the issuance. For instance, following the ICMA Green Bond Principles (hereinafterGBPs) and Sustainability-Linked Bond Principles (hereinafterSLBPs), originators have also committed to tracking expenses, reporting on their sustainability impact or obtaining periodic external reviews.[91] Self-evidently, these additional practices ensure compliance with external standards, which 'enable' the use of the 'sustainable' label and confirm its integrity and reliability.The following table (Table 4.1)[92] summarises the various meanings of 'sustainability' within securitisation practice, linking them to the discussed examples.

Overall, this analysis shows that market actors have qualified securitisations as 'sustainable' in disparate contexts. Following the previously mentioned

[88] Global Capital (n. 85); Newmarket Capital (n. 85) 16.

[89] See, among others, Bardin Hill Loan Advisors (n. 77).

[90] For a more elaborate articulation of quantitative eligibility criteria for the inclusion of Loans, see also the US government-sponsored agency Fannie Mae and its framework for green securitisations, Fannie Mae, 'Multifamily Green Bond Framework' (2024) 11–13 https://capitalmarkets.fanniemae.com/media/20616/display.

[91] Updated several times, with the last update tracing back to 2025 for green bonds and to 2024 for sustainability-linked bonds. See ICMA, 'Green Bond Principles – –Voluntary Process Guidelines for Issuing Green Bonds' (2025) 5–7. https://www.icmagroup.org/sustainable-finance/the-principles-guidelines-and-handbooks/green-bond-principles-gbp/ accessed 8 December 2025 (*hereinafter*GPBs); ICMA, 'Sustainability-Linked Bond Principles' (2024), 5 https://www.icmagroup.org/sustainable-finance/the-principles-guideli nes-and-handbooks/sustainability-linked-bond-principles-slbp/ accessed 8 December 2025 (*hereinafter*SLBPs). An analysis of recurrent and best practices across sustainable securitisation deals will also follow in s. 4.5.

[92] Which builds on, expands and updates the table in Federica Agostini, 'One Step Forward, One Step Back': 'Sustainable Securitisations' Before and After the EU Green Bond Standard' in Mario La Torre, Sabrina Leo (eds.), *Contemporary Issues in Sustainable Finance* (Palgrave Studies in Impact Finance 2025), 227.

Table 4.1 Definitions of sustainable securitisation

Definitions of 'sustainable' securitisation	ACTIVITY-BASED (under market practice 'green' or 'social')	TARGET-BASED (under market practice 'sustainability-linked')
Proceeds ('sustainable proceeds securitisations')	1) The proceeds of the issuance finance projects considered instrumental to sustainability objectives. (e.g. Toyota) or 2) The transaction frees up capital, which originators use to expand or develop sustainability-related activities (e.g. Premium Green).	Originating entities undertake to reach sustainability performance targets, which have an impact on the repayment conditions of the securities. (e.g. Jupiter, Santander)
Portfolio ('sustainable portfolio securitisations')	1) Portfolio assets include loans financing sectors and activities considered instrumental to sustainability objectives. (e.g. Hannon Armstrong, CETELEM) and/or 2) Portfolio assets include loans financing sectors and activities considered harmful to sustainability objectives. (e.g. Bardin Hill Advisors)	Portfolio assets have to meet certain sustainability performance targets, which have an impact on repayment conditions. (e.g. Project Boqueron, Marcegaglia)

trend, they have designed sustainability characteristics entirely independently or drawing from external standards and legislation.

4.5 The 'Private Hard Law' of Sustainable Securitisations, Between Law and Practice

Drawing upon the typology of the distinctive attributes of 'sustainable' securitisations, it is essential to delve into how market actors have articulated sustainability goals. As elaborated above, the transactional documentation should be the key conduit for introducing binding legal obligations among the parties ('private hard law'). The following subsections will particularly examine how parties have embedded sustainability considerations into the five 'focal points' of these transactions, as identified in Chap. 3. The analysis will aim to assess whether parties' commitments have prompted meaningful practices aligned with sustainability goals.

4.6 The (Frequently Unclear) Composition of the Portfolio

As previously mentioned, the most evident consequence of the lack of a legal definition of 'sustainable securitisation' is the wide divergence of assets which can be included in the portfolio. Empirical evidence points to the use of a wide variety of credit claims as collateral, including, among other things, loans financing electric vehicles and energy-efficiency upgrades in residential and commercial buildings, and infrastructure or technologies for agriculture. While the trend is not so widespread in the EU market yet, loans for solar, wind, and other renewable energy projects have also been frequently pooled together in other jurisdictions, such as the US and Australia.[93] At the same time, originators of other transactions pooled together credit claims not making any positive contribution to sustainability objectives ('grey' or 'brown').[94] Regardless of the type of underlying assets, the parties involved in the design of these transactions are likely to set the environmental, social, and governance criteria that credit claims have to meet throughout securitisation agreements or additional documentation.

Some constraints on the originators' side may also impair their ability to fully disclose the composition of portfolios. For instance, companies engaging in previous 'sustainable securitisations' have highlighted the lack of available information on the environmental and social characteristics of portfolios.[95] The risk may be higher especially when they act as 'sponsors',

[93] Petit, Schlosser (n. 31) 3–7, 9–11; Climate Bonds Initiative, 'Green securitisation' (n. 2) 1. For an analysis of four deals revolving around residential mortgages, see also Pascale Cornut St-Pierre, Jasmine Gareau-Lindsay, 'Green Securitisation, A Legal Structure Currently Unfit for Ecological Transition' (2026) 16 *Accounting, Economics and the Law* 2, 6 ff.

[94] See s. 4.3.

[95] UNPRI, 'ESG engagement for fixed income investors: managing risks, enhancing returns' (2018) 35–36 https://public.unpri.org/download?ac=4449 accessed 7 December 2025. On the need for further transparency around the composition of the portfolio see also a survey across investors in 'sustainable' securitised products in US, Structured Finance Association, 'ESG Market Sentiment Survey' (2023) 15 https://structuredfinance.org/wp-content/uploads/2023/02/SFA-ESG-2022.pdf accessed 7 December 2025.

buying assets from third parties.[96] There are also limited tools to measure certain sustainability factors. For instance, I will discuss in Chap. 4 the preliminary work by certain financial industry associations towards the design of metrics to measure the environmental impact of some financial assets.[97] However, they remain generally underdeveloped, especially when it comes to the quantification of social and governance factors. It may also be more challenging to quantify the impact of non-uniform portfolios, consisting of assets like commercial mortgages with different characteristics.

The following paragraphs will critically discuss how parties articulate sustainability objectives for the portfolio across the agreements governing the assignment of the assets to the SPV and across additional documents specifically focusing on sustainability issues. They will also analyse how external review processes may affect sustainability-related terms of the documentation concerning the portfolio.

4.6.1 Within Traditional Portfolio-Related Documentation

In line with a recurrent trend across the financial sector[98], securitisation agreements generally tend to exhibit similar structures and clauses.52 The terms describing sustainability-related elements of portfolios are, however, not comparable. The analysis of previous examples of documentation reveals how parties tend to distil such terms across four different types of clauses and agreements.

First, they can incorporate clauses introducing sustainability-related 'eligibility criteria' for the inclusion of assets within the portfolio. Receivables would then need to meet 'eligibility criteria' either at the closing of the deal or upon assignment to SPVs (if the deal is 'revolving'[99])[100]. By way of example, such clauses may specify that the asset pool may not contain

[96] For the difference between originators and sponsors, see Chap. 3, Sect. 3.2.

[97] Among others, ICMA, 'Illustrative Examples for the Selection of Key Performance Indicators (KPIs) for Sustainability-Linked Bond Issuers, Underwriters and Investors'. (2023) https://www.icmagroup.org/assets/documents/Sustainable-finance/2022-updates/Registry-SLB-KPIs_Final_2022-06-24-280622.xlsx accessed 20 July 2025.

[98] Julia Black, David Rouch, 'The Development of the Global Markets as Rule-Makers: Engagement and Legitimacy' (2008) 2 Law and Financial Markets Review 218, 226.; see further Chap. 5, Sect. 5.2.2.–5.2.3.

[99] See Chap. 3, Sect. 3.2.1.

[100] Daniela Di Filippo, 'Green Securitisation: Developments and Challenges' (2021) 3 https://c2e2.unepccc.org/wp-content/uploads/sites/3/2021/07/fitch-green-Securitisation-developments-and-challenges-15-april-2021.pdf accessed 9 December 2025.

loans to high-emitting companies. They may be part of the 'loan purchase agreement' or of other similar agreements for the assignment of rights over the assets to SPVs, like trust deeds, agreements for the creation of security interests or other contracts, depending on the proprietary arrangements between the parties.[101] The shortcoming of this positioning is the confidentiality of these agreements, which would prevent end investors from reading 'eligibility criteria' before purchasing sustainable ABSs.

Second, parties may add 'sustainability eligibility criteria' to prospectuses, which issuers use to trade ABSs in the EU market, since EU law only prescribes the minimum requirements thereof. There may also be similar documents, sometimes defined as 'offering circulars', for the deals falling outside the EU regime, such as securitisation programmes exclusively addressed to institutional investors, private placements and synthetic deals. [102] Chap. 2 has discussed how prospectuses trigger a broader liability regime. False or misleading statements within prospectuses, as well as omissions of material information, could result in fines by financial authorities as well as entitling investors to take action vis-à-vis issuers.[103] Nevertheless, parties' autonomy in drafting such documentation also leaves room for provisions with the opposite effect. Prospectuses can contain disclaimers, where originating entities can, in principle, waive liability for not meeting sustainability requirements.[104] Indeed, previous examples suggest that this could be a recurrent pattern: some publicly available prospectuses of

[101] For the arrangements governing the assignment of rights over the asset pool, see Chap. 3, Sect. 3.3.1. For a sample agreement for the assignment of the proprietary rights over residential mortgages, see Natwest, Mortgage Sale Agreement (2010) https://investors.natwestgroup.com/~/media/Files/R/RBS-IR-V2/securitisation/reseidential-mortgages/pdf/7-arran-residential-mortgages-2010-1-plc/principal-transaction-documents/010--mortgage-sale-agreement-2010-a1.pdf accessed 9 December 2025.

[102] For prospectus requirements and exemptions, see Chap. 3, Sect. 3.3.3., and Chap. 6, Sect. 6.5.1.

[103] On the requirements for Prospectus liability, see Chap 6, Sect. 6.5.1.

[104] See For example, Green Storm's 2021 Prospectus, where the issuer excludes any obligation to ensure that the assets maintain the Energy Performance certifications, and excludes any duty to repurchase the receivables if the assets no longer meet the ' Green Eligibility Criteria ', Green Storm 2021, 'Prospectus dated 23 March 2021' 15–16 https://pcsmarket.org/wp-content/uploads/Green-STORM-2021-Prospectus43433997.13.pdf accessed 9 December 2025. A study of four green securitisation transactions has also highlighted the recurrent use of disclaimers. Cornut St-Pierre, Gareau-Lindsay (n. 93) 18–19.

previous 'sustainable' ABSs contain similar wording,[105] as is also recurrent in the broader green bond universe, as I have discussed elsewhere.[106]

Third, originating entities may make certain 'representations' and 'warranties' on the qualities of the underlying assets within the agreements for the assignment of the portfolio or within an additional contract ('warranty and indemnity agreement').[107] Such clauses set the conditions for the assignment of the assets at the root of the contract: for instance, originating entities may, among others, represent or warrant the validity of the documentation or that the underlying assets comply with regulations. [108] In the context of sustainable securitisation, one could imagine that originating entities could represent, among others, that the portfolio contains assets which will contribute to an environmental goal or comply with environmental regulations (e.g. loans towards energy-efficiency buildings). The

[105] eg, Société Générale excludes in its Prospectus any liability for the assessment of the eligibility criteria, see Société Générale SFH, 'First Supplement to the Base Prospectus Dated 3 June 2020' (2020) 32. https://www.societegenerale.com/sites/default/files/documents/Notations%20Financi%C3%A8res/societe-generale-sfh-emtn-base-prospectus-dated-3-june-2020.pdf accessed 9 December 2025. Similarly, Cornut St-Pierre, Gareau-Lindsay, ibid., 18–19.

[106] Federica Agostini, 'From 'Green Bond Principles' to 'Green Bond Clauses': Mitigating Greenwashing Through Contract Law' in Mads Andenas, Maren Heidemann (eds), *Quo vadis Commercial Contract? Reflections on Sustainability, Ethics and Technology in the Emerging Law and Practice of Global Commerce* (Springer 2023)

[107] See e.g. the reference to a standalone 'Warranty and Indemnity Agreement' within the Prospectus by Profamily SPV, Profamily SPV SRL, 'Prospectus Asset Backed Fixed Rate Notes due December 2040' (2021) https://www.securitisation-services.com/it/reports/scarica_prospetto.php?f=files/Profamily%20SPV/Profamily%20SPV%20Prospectus.pdf accessed 9 December 2025; see also sample Representations and Warranties, Fannie Mae, 'Contractual representations and warranties – overview' (2023) A2/0/21, 9, 24 ff https://selling-guide.fanniemae.com/Selling-Guide/Doing-Business-with-Fannie-Mae/Subpart-A2-Lender-Contract/Chapter-A2-2-Contractual-Representations-and-Warranties/1645977541/A2-2-01-Representations-and-Warranties-Overview-02-05-2020.htm#Overview. accessed 9 December 2025; Freddie Mac, 'Selling Representation and Warranty Framework' https://sf.freddiemac.com/working-with-us/selling-delivery/delivery-options-pricing/selling-representation-and-warranty-framework accessed 9 December 2025.

[108] For the definition of representations and warranties under common law see *Oscar Chess v Williams* [1957] 1 WLR 370; *Hong Kong Fir Shipping Ltd v Kisen Kaisha Ltd* (1962) EWCA Civ 7. For an analysis of the most common representations and warranties in securitisation documentation, see Patricia Ann McCoy and Susan M Wachter, 'Representations and Warranties: Why They Did Not Stop the Crisis' in LA Fennell, BJ Keys (eds.) *Evidence and Innovation in Housing Law and Policy* (Cambridge University Press 2017) 8–10; Phillip Wood, *Project Finance, Securitisations, Subordinated Debt* (Sweet & Maxwell, 3rd edn 2019), 7-003.

inclusion of sustainability-related representations and warranties within 'loan purchase agreements' or similar agreements would have important implications for investors' enforcement powers. While the main counterparty to the assignment of the assets is technically the SPV, these agreements also include some remedies for investors or their representative[109] in case of non-compliance with representations and warranties. For instance, a widely used clause allows investors to give prompt notice to originators when discovering any breaches of warranties, and requires originators to repurchase the loans under these circumstances.[110] This also means that if the examples above were reflected in the documentation, originators would assume all the risks related to the portfolio, such that they may have to repurchase the securities if these did not meet the requirements. This would represent a considerable deterrent to misstatements.

At the same time, it is hard to ascertain to what extent sustainability-related representations and warranties are used in market practice across 'loan purchase agreements' or 'warranty and indemnity agreements' due to the confidential nature of these agreements. Only in certain limited circumstances have large issuers made publicly available large standardised templates of all the representations and warranties governing all their securitisation operations.[111]

As a result, even prospective investors may not be able to *directly* read these provisions. Rather, they are more likely to become aware of the 'eligibility criteria' and of representations and warranties through the 'pre-issuance reports' rating agencies tend to publish to outline the key aspects of securitisation deals.[112] However, as the aim of these documents is to justify the awarded rating, they may not always cover sustainability information, or they may only include what is relevant for credit risk.[113] As a result,

[109] On the role of trustees and investors' representatives see Chap. 3, Sect. 3.4.5.; see also below Sect. 4.9.

[110] Miller reports an illustrative example of standard clause introducing a repurchase obligation, Robert T. Miller, 'The RMBS Put-Back Litigations and The Efficient Allocation Of Endogenous Risk Over Time' (2014) 34 Review of Banking and Financial Law 225, 274. For some examples across the prospectuses analysed in the present paper, see below, s. 4.7.

[111] For example, Fannie Mae's Framework, also indicating general 'terms and conditions', see Fannie Mae 'A2-2, Contractual Representations and Warranties' in 'Selling Guide' (n. 107) 9, 24 ff; Freddie Mac (n. 106).

[112] See For example, the pre-issuance report ahead of the 'social' Gemgarto deal (cited in footnote 55), Fitch Ratings, 'Gemgarto 2021-1 PLC' https://www.fitchratings.com/research/structured-finance/gemgarto-2021-1-plc-24-02-2021 accessed 9 December 2025.

[113] For example, the Fitch rating report only covers 'credit-relevant' ESG information, see ibid., 26.

it is possible for investors to only have a partial and indirect knowledge of the characteristics of portfolios.

Fourth, agreements governing the assignment of the portfolio or the issue of securities may contain 'conditions precedent' that originators must fulfil in order for the transactions to be effective. These may require the delivery of certain documents or the guarantee that there are no defaults on the portfolio.[114] Therefore, it is reasonable to assume that conditions precedent may, among other things, require originating entities to provide documentation on environmental or social aspects of the deal, such as energy efficiency performance certificates for securitised loans over commercial and residential buildings. An example is the clause in the Prospectus of the *2021 Green Storm* deal, which lays down specific energy performance requirements for the portfolio assets and requires the submission of energy-efficiency certificates.[115] Not complying with 'conditions precedent' may prevent the assignment from taking effect.[116]

Fifth, a further mechanism may be a document which originating entities may voluntarily circulate among investors before the issuance, outlining the composition of the portfolio ('portfolio file').[117] However, due to the lack of publicly available examples of these documents, there is insufficient evidence of how they reflect and describe the environmental and social qualities of securitisation portfolios. On the contrary, a UNPRI survey has evidenced the lack of sustainability-related information therein and investors' frequent unmet expectations.[118]

Overall, while there are in principle various options for originators to incorporate sustainability-related terms within their documentation, there are arguably few incentives for originators to do so. Most of the abovementioned provisions and agreements, like 'eligibility criteria' or 'representations and warranties' are not accessible to prospective investors. The comparison of previous examples of portfolio-specific securitisation

[114]On the definition of conditions precedent for loans, see Phillip Wood, *International Loans, Bonds, Guarantees, Legal Opinions* (3rd edn, Sweet & Maxwell, 2019) Vol. 5, 5-011, 5-013; 28–092.

[115]Green storm, 'Prospectus dated 23 march 2021' (n. 104) 265; on the series of Green Storm deals by Obvion, see Sect. 4.3.

[116]Wood (n. 114) 5-009-5-011, 14-001, 25-040.

[117]According to UNPRI, this would be standard market practice, especially before publicly traded deals. See 'ESG incorporation in securitised products: The Challenges Ahead' (2022) 5 https://gsfo.org/resource/esg-incorporation-securitised-products-challenges-ahead accessed 9 December 2025.

[118]Ibid., 25.

documentation also does not provide any evidence around the wide use of sustainability-related terms. As a result, investors are unlikely to possess sufficient information about the environmental and social impact and risks of the portfolio assets, as previous surveys have also anecdotally confirmed.[119] More importantly, the flexibility in the drafting of such agreements leaves room for disclaimers, which considerably restrict investors' enforcement avenues. In certain cases, such disclaimers may even prevent investors from bringing proceedings against originating companies if the portfolio does not comply, or ceases to comply, with sustainability-related eligibility criteria. This could represent a genuine threat to the credibility of these transactions, which also *de facto* raises the question of the distinction between traditional and 'sustainable' securitisations.

4.6.2 Within 'Sustainability-Related' Documentation

Originating companies may also use other or additional formats to communicate how securitisation portfolios contribute to sustainability objectives to the public.

In certain cases, they would only make public statements about their recent 'sustainable' securitisation operation (e.g. in the form of press releases). While these may provide some information about the final terms of the issuance, they often remain vague or silent about the characteristics of the portfolio.[120] In these cases, investors and other market actors may only acquire such information through the statements and reports issued by external reviewers.[121]

[119] On the lack of information about portfolios across 'traditional' securitisations, Structured Finance Association, 'RMBS 3.0' (2017) 315, 320, 330–331, 340 https://structuredfinance.org/wp-content/uploads/2019/05/RMBS-3.0-Sixth-Edition-Final-1109.pdf accessed 9 December 2025.

[120] see For example, Obvion's 2016 Green storm transaction: the main sources of information are the press release and the external review, KFW, 'Obvion NV Issues EUR 500m Green STORM 2016 RMBS to finance energy-efficient housing' (2016) https://www.kfw.de/nachhaltigkeit/PDF/Nachhaltiges-Investment/Press-Release-Green-Storm.pdf accessed 30 September 2023; Sustainalytics (n. 39); See also Marcegaglia's sustainability-linked securitisation in the steel sector, where the sustainability performance target has not been disclosed, Ahren Lester, 'Steel firm Marcegaglia signs sustainability-linked securitisation' (26 May 2022, Environmental Finance) https://www.environmental-finance.com/content/news/steel-firm-marcegaglia-signs-sustainability-linked-securitisation.html accessed 9 December 2025.

[121] For example, see the SPO over Obvion's transaction, see Sustainalytics (n. 39); see also the SPO about RIVER Green Finance securitisation, Sustainalytics, 'River Green Finance

More frequently, issuers have drafted ad hoc documentation to outline their commitments to environmental, social, and governance objectives (often referred to as 'sustainability frameworks'). Unlike CSR policies and codes of conduct, which tend to have a wide-ranging scope553, these 'sustainability frameworks' illustrate how they undertake to utilise their 'sustainable' financing techniques such as bonds, loans, and securitisations to advance sustainability goals.[122] The most authoritative market-designed principles for 'sustainable securitisations', i.e. the ICMA GBPs[123] recommend issuers to publish a 'framework' or other legal documentation providing further information around the characteristics of the securities and linking them to the overall issuer's sustainability strategy.[124] A 'framework' may also be compulsory to obtain third-party reviews, like the 'climate bond' certification by the Climate Bonds Initiative.[125] In some cases, it may be part of a broader report which issuers would publish in order to comply with the legal requirements to disclose sustainability-related information.[126]

Some market actors have included environmental and social 'eligibility criteria' or the expected impact of their asset pools within their frameworks.[127] When disclosing such information, they have often relied on external market-designed principles and metrics as well as legislation,

Green Securitised Bond Framework' (2019) https://mstar-sustops-cdn-mainwebsite-s3.s3.amazonaws.com/docs/default-source/spos/river-green-finance-green-bond-second-party-opinion.pdf?sfvrsn=e00cf8ac_3 accessed 9 December 2025. On the function and types of external reviews, see below 4.6.4.–4.7.3.

[122] See For example, the 'Green financing framework' referring to both bonds and loans, see Hitachi Capital, 'Green Financing Framework' (2021) https://www.mitsubishihccapital.co.uk/media/pvkl4srm/green-financing-framework-2021.pdf accessed 9 December 2025.

[123] See further Chap. 4, Sect. 4.5.

[124] ICMA, 'GBPs' (n. 91) 4, 7, 9.

[125] See Climate Bonds standard – –version 4.0 https://www.climatebonds.net/files/documents/Climate-Bonds_Climate-Bonds-Standard_V4-0_Apr-2024.pdf accessed 9 December 2025; at the time of writing, the latest amendment is the 4.3 Version, introduced in August 2025. 'Climate Bonds Standard – version 4.3' (2025) CBI_Standard_V4.3_FINAL_2025-08-20-102147_gbqn.pdf accessed 15 December 2025. On the functioning of certification mechanisms, see Sect. 4.6.4.

[126] For example, the disclosure requirements for certain types of 'sustainable' funds under SFDR, see Regulation 2019/2088, as discussed in Chap. 6, Sect. 6.7.

[127] See For example, the eligibility criteria, in terms of savings in energy and water consumption, within Fannie Mae's framework, see Fannie Mae, 'Multifamily Green Bond Framework' (n. 90) 5–6; Sustainalytics reveals that Obvion has an internal Green Bond framework, also covering asset pool eligibility criteria, see Sustainalytics (n. 39) 4.

such as taxonomies, as also recommended by the GBPs.[128] For instance, *Obvion*'s framework for *Green Storm* deals[129] links the energy efficiency requirements and the types of energy performance certificates eligible for inclusion within the portfolio to Dutch legislation.[130]

However, the format of these documents and the reference to external standards is not always informative on the characteristics of the asset pool. External principles serve an 'enabling' function, since they attest to the reliability of 'sustainable' qualities and make the transaction more predictable. Therefore, market actors will have an interest in demonstrating their alignment. This also means that they are very likely to draft their 'sustainability frameworks' around the 'use of proceeds', which constitutes the core aspect of the GBPs and of the Climate Bond Certification.[131] Conversely, such frameworks do not always contain detailed information on the characteristics of the securitisation portfolio and, when they do, they are also likely to present such material inconsistently.

4.6.3 *Implications: (Limited) Legal Consequences*

The publication of 'sustainability frameworks' or similar documentation also triggers certain, albeit limited, legal consequences. In certain cases, market actors have cross-referred to these frameworks across other transactional documents, such as prospectuses.[132] It results in the 'incorporation' of the frameworks, which then acquire the same value and prescriptiveness as the other prospectus provisions, including the civil liability and sanction regime.[133] Relatedly, after recognising the 'greenwashing' risk deriving from untruthful statements on environmental matters within prospectuses, the European Securities and Markets Authority (*hereinafter*ESMA) has undertaken to develop common guidelines to support authorities in the enforcement.[134]

[128] ICMA, 'GBPs' (n. 91), 5.
[129] See above Sect. 4.3.
[130] As indicated by Sustainalytics (n. 39) 3–4.
[131] As examined later in Sect. 4.6.3. others have also highlighted that sustainability frameworks at times 'ignore certain fundamental issues', Cornut St-Pierre, Gareau-Lindsay (n. 93) 11.
[132] For example, Société Générale, when issuing 'sustainable positive impact notes' as part of its programme, see SFH (n. 105) 5-6–7.
[133] See Chap. 6, Sect. 6.5.1.
[134] ESMA 'Sustainable finance roadmap 22–24 (2022) 12–13 https://www.esma.europa.eu/sites/default/files/library/esma30-379-1051_sustainable_finance_roadmap.pdf. Accessed 9 December 2025.

At the same time, the reference to sustainability frameworks across prospectuses is not a universal trend.[135] Certain prospectuses contain specifications that either exclude the incorporation of sustainability frameworks[136] or allow single securitisation issuances within a programme to deviate from these criteria. [137] Thus, while the 'incorporation' may in principle enable investors to monitor the compliance with the framework, other provisions within prospectuses could effectively weaken their function.

The publication of sustainability frameworks may also have legal implications. To assess the potential contours of issuers' liability, corporate 'codes of conduct' can serve as references.[138] Like sustainability frameworks, these codes also list a series of business practices and targets to which certain actors commit.[139] Academic commentary has identified two possible legal avenues to challenge their content under consumer and competition law.

On the one hand, false and misleading claims on the portfolio may constitute examples of 'false advertising'. Notably, statements to the public (which may, in the context of the present analysis, include sustainability frameworks or other press releases) may give rise to a 'promise to the public' and create a binding expectation that the products comply with the description. [140] This may give rise to a contractual remedy as long as

[135] See the Obvion 2021 Prospectus, see Green storm (n. 104).

[136] See For example, the example of green bond prospectus by the Saudi Arabia sovereign wealth fund, GACI First Investment Company, 'Public Investment Fund Euro Medium Term Note Programme Offering Circular' (2022)vii. GACI First Investment Company, 'Public Investment Fund Euro Medium Term Note Programme Offering Circular' (2022). https://docs.londonstockexchange.com/sites/default/files/documents/PIF%20Euro%20Medium%20Term%20Note%20Programme%20Offering%20Circular.pdf accessed 9 December 2025.

While the document refers to a green bond, it could be reasonably assumed that similar clauses may be found across the issuances of 'sustainable' ABSs, whose prospectuses follow a similar structure, see Chap 2, Sect. 2.2.3.

[137] In the example above, Société Générale clarifies that individual prospectuses under the programme may or may not comply with the framework, see SFH (n. 105) 5-6-7.

[138] See For example, Sean D Murphy, 'Taking Multinational Corporate Codes of Conduct to the next Level' (2005) 43 Columbia Journal of Transnational Law 389, 431; Haley Revak, 'Corporate Codes of Conduct: Binding Contract or Ideal Publicity?' (2011) 63 Hastings Law Journal 1645, 1667–1668.

[139] See the definition of 'codes of conduct' under Art 2 (f) Consolidated text directive 2005/29 concerning unfair business-to-consumer commercial practices in the internal market.

[140] As argued by Tom Campbell Doreen McBarnet, Aurora Voiculescu, *The New Corporate Accountability* (Cambridge, 2007) 124–125.

the statement directly influenced the decision to purchase a product.[141] At the same time, a more general EU provision characterises the violation of 'codes of conduct' as 'misleading commercial practices' 'as long as they cause or are likely to cause the average consumer to take a transactional decision that they would not have taken otherwise'. [142] In different instances, courts have scrutinised 'greenwashing' practices and held them as incompatible with consumer law.[143] Therefore, originating companies not substantiating the claims regarding the portfolio (or regarding other aspects of the transaction) may face different claims under consumer law. However, one potential constraint on the applicability of these remedies is their scope, which is restricted to the transactions between businesses and consumers.[144] EU law provides several restrictions to the sale of securitised products to retail investors,[145] meaning that sustainable securitised products are generally purchased by institutional investors. The latter would not be able to rely on such remedies.

On the other hand, untruthful claims regarding the portfolio may also give an 'unfair competitive advantage' to originating companies, which may run counter to competition law principles. Although there is a recent Italian case, where misleading information on environmental matters was

[141] 'The seller shall not be bound by public statements (...) if the seller shows that: (c) the decision to buy the goods could not have been influenced by the public statement'. Art 7(2)(c) Directive (EU) 2019/771 of the European Parliament and of the Council of 20 May 2019 on certain aspects concerning contracts for the sale of goods, amending Regulation (EU) 2017/2394 and Directive 2009/22/EC, and repealing Directive 1999/44/EC (2019) OJ 136/28.

[142] Article 6 (2)(b) consolidated text directive 2005/29 concerning unfair business-to-consumer commercial practices in the internal market.

[143] There is considerable case law in Germany, see the case law cited by McBarnet (n. 141) 127; for a case in Italy, see Autorità garante della concorrenza e del mercato, Delibera ACGM 26173/2016 https://www.agcm.it/dotcmsDOC/allegati-news/PS10211_chiusura.pdf.accessed 9 December 2025.

[144] See the reference to the 'average consumer' In art 6(2)(b) ibid.; see also Art 2(1) directive 2019/771.

[145] I.e. they need to pass a 'suitability test' and to provide additional information to the originating company. They can only invest on securitised products for an aggregate amount of less than 10% of their portfolios, art 3 (1)-(2) Regulation (Eu) 2017/2402 of the European Parliament and of the Council of 12 December 2017 laying down a general framework for securitisation and creating a specific framework for simple, transparent and standardised securitisation, and amending Directives 2009/65/EC, 2009/138/EC and 2011/61/EU and Regulations (EC) No 1060/2009 and (EU) No 648/2012 (2017) OJ L 347/35. I will elaborate on these aspects of securitisation legislation in Chap. 5, Sect. 5.7.1.

held to be incompatible with competition law rules;[146] case law on this issue appears to be limited.

Therefore, parties have clear incentives to curtail enforcement avenues linked to securitisation portfolios through disclaimers in their prospectuses. Limited and predominantly underexplored remedies arise under consumer and competition law. Therefore, sustainability frameworks could be the appropriate context to broaden investors' remedies, encompassing misleading and false statements on sustainability matters that culminate in 'greenwashing'.

4.6.4 *The Role of External Reviews: Reviewing What and to What End?*

A common practice is for originating entities to entrust third parties, such as rating agencies, audit companies, or other independent service providers, to review sustainability characteristics ahead of, or after, the issuance of 'sustainable' financial instruments, such as bonds and ABSs. For instance, it is possible for several providers to certify the environmental characteristics of specific types of financial assets, such as the environmental standards for buildings, such as the Leadership in Energy and Environmental Design (LEED) by the US Green Building Council, or the Building Research Establishment Environmental Assessment Method (BREEAM) by BRE Global. These could be relevant to assess the portfolios of RMBSs and CMBSs.[147] Originators can then publish or refer to the assessment of these providers throughout their securitisation documentation to further prove the reliability of the 'sustainable' label.

At the same time, a burgeoning sector is the one that brings together the providers which specifically assess 'sustainable' financial products and issue reports on the reviewed products. The GBPSs, which recommend pre- and post-issuance reviews over 'sustainable' ABSs[148] classify such

[146] See Tribunale di Gorizia 2021/712, https://valori.it/wp-content/uploads/2022/01/greenwashing-Alcantara-vs-Miko-ordinanza-del-25-novembre-2021-del-Tribunale-di-Gorizia-R.G.-2021-712.pdf; for a summary, see Clifford Chance, 'Italy's first greenwashing case between corporates' (2022) https://www.cliffordchance.com/insights/resources/blogs/business-and-human-rights-insights/2022/01/italys-first-greenwashing-case-between-corporates.html accessed 9 December 2025.

[147] International Finance Corporation, 'Green Buildings – –A Finance And Policy Blueprint For Emerging Markets' (2019), 60 https://www.ifc.org/content/dam/ifc/doc/mgrt/59988-ifc-greenbuildings-report-final-1-30-20.pdf accessed 9 December 2025. As also noted by Petit, Schlosser (n. 31) 5; see ibid. for further examples.

[148] ICMA, GBPs (n. 91), 9.

processes as second party opinions (SPOs), second party ratings, verifications, and certifications. These processes are not mutually exclusive and sometimes overlap in practice. [149] Second Party Opinions (SPOs), generally conducted before the operation[150] often revolve around the assessment of issuers' documentation, such as sustainability frameworks. 'Verifications' also entail the use of a benchmark to evaluate the accuracy and integrity of issuers' claims against their own standards (like their own frameworks) or against external sustainability standards, like the GBPSs. They are often performed by audit firms. Accredited service providers can also award a 'Certification' to testify the full alignment with an external sustainability standard or label, like the Climate Bonds Standard.[151] Lastly, sustainability bond ratings involve the use of ad hoc methodologies to provide a sustainability score, frequently before a programme or the issuance of securities.[152] They would be distinct from standard credit risk ratings, even in circumstances where the latter also takes environmental, social, and governance aspects into account.[153]

The expertise and reputation of external reviewers should effectively make them 'Gatekeepers', who ensure the reliability of the 'sustainability' qualities of the operation.[154] However, review processes also come with several shortcomings.

The voluntary nature of external reviews has inevitable consequences. In the context of voluntary certification systems in the environmental sphere, commentators have highlighted the potential 'selection bias' deriving from the free choice to align with the requirements. This makes the certification only attractive to those who are already committed to

[149] ICMA, 'Guidelines for Green, Social and Sustainability Bonds External Reviews' (2020), 3 https://www.icmagroup.org/assets/documents/Regulatory/Green-Bonds/June-2020/External-Review-GuidelinesJune-2020-090620.pdf. Accessed 9 December 2025.

[150] See ibid., 5. As also noted by Asian Bonds online, Green Bond factsheet 7 – Engaging the External Reviewer (2022) https://asianbondsonline.adb.org/green-bonds/gbfactsheet7.html accessed 9 December 2025.

[151] See above Sect. 4.6.2.

[152] see ICMA, 'Guidelines on external reviews' (n. 149) 3, 5; EU Commission, 'Study on the Potential of Green Bond Finance for Resource-Efficient Investments' (2016) 26 https://op.europa.eu/en/publication-detail/-/publication/5a95e1b0-dc82-11e6-ad7c-01aa75ed71a1/language-en accessed 3 June 2026; Asian Bonds online (n. 150).

[153] ICMA (ibid) 5. See For example, the scores provided by Standard Ethics, Standard Ethics, 'Security Standard Ethics Rating – Overview' (2020) 1. https://standardethicsrating.eu/SE_Bond_Rating_Overview_2020-1.pdf Accessed 9 December 2025

[154] As defined in Chap. 1, Sect. 1.5. On the definition see also John C Coffee, 'Introduction', *Gatekeepers: The Professions and Corporate Governance* (Oxford University Press 2006).

environmental objectives, and not to all.[155] However, it must be noted that this risk may not be as pressing in the context of the sustainable securitisation market, since engaging in these transactions is arguably already a sign of the commitment to these objectives. Data on green bonds also demonstrates a widespread application of review processes.[156]

Instead, a more pressing issue inherent in the voluntary external review system is the risk of conflicts of interest. Preliminary data on ESG ratings seems to confirm such a conclusion.[157] Therefore, review statements may not always be fully transparent and provide a too positive assessment of the operation. The risk may be higher for SPOs due to their narrower focus on the issuer's documentation. In the past, legislation has intervened to mitigate conflicts of interest for other categories of 'gatekeepers', like rating agencies.[158] Thus, it is clear that more effective regulatory strategies to address these issues include legislation on external reviewers or on the methodology for the review process, as I will discuss in Chap. 6.[159]

Furthermore, other challenges of external reviews, especially for the assessment of the portfolio, relate to their unclear and limited scope. Most

[155] Cary Coglianese, 'Environmental Soft Law As a Governance Strategy' (2020) 61 *Jurimetrics* 19, 12 https://ezp.lib.cam.ac.uk/login?url=https://www.proquest.com/scholarly-journals/environmental-soft-law-as-governance-strategy/docview/2501511700/se-2?accountid=9851%0Ahttps://libkey.io/libraries/603/openurl?genre=article&au=Coglianese%2C+Cary&aulast=Cogli accessed 9 December 2025.

[156] Climate Bonds Initiative reported that 86% of green bonds added to their database (which also covers asset-backed securities) in 2021 had a form of external review, see Climate Bonds Initiative, 'Sustainable debt: Global State Of The Market 2021' (2022) 13 https://www.climatebonds.net/files/documents/publications/Sustainable-Debt-Global-State-of-the-Market-2021.pdf accessed 9 December 2025.

[157] see Ruoke Yang, 'What Do We Learn From Ratings About Corporate Social Responsibility (CSR)?' (2021) Columbia Business School Research Paper No. 18–37, 22 https://ssrn.com/abstract=3165783 accessed 9 December 2025; Cadwalader, Wickersham & Taft LLP. 'ESG Ratings: A Call for Greater Transparency and Precision' (2022) https://www.cadwalader.com/uploads/cfmemos/5d29869eec33512f55a05b5d195e4bea.pdf accessed 9 December 2025. More specifically on potential conflicts of interests within sustainable reviews, Cornut St-Pierre, Gareau-Lindsay (n. 93) 13. The authors also highlight that reviewers ultimately have the same business model as credit rating agencies, ibid.

[158] For example, by introducing a maximum term for contractual relationships and by allowing investors to bring proceedings vis-à-vis rating agencies for intentional or negligent behaviours, see Art 6b, Art 35a Regulation (EU) No 462/2013 of the European Parliament and of the Council of 21 May 2013 amending Regulation (EC) No 1060/2009 on credit rating agencies (2013) OJ L 146/1. On the conflicts of interest risks for 'gatekeepers', see Coffee, 'Introduction' (n. 154) 323–324;

[159] See Sect. 6.4.–6.5.

of the reviews aim to assess the compliance of the securitisation programme or of the issuance with a set of external principles, like the GBPs and CBI. As anticipated above, these standards, originally designed for bonds, predominantly focus on the final allocation of the issuance.[160] This means that reviews of documents such as sustainability frameworks may be inaccurate on the qualities of the asset pool. For instance, *Sustainalytics* confirmed in its SPO over the 2017 Green Storm deal the compliance with the GBPS, while only drawing very general conclusions on the environmental qualities of the portfolio.[161] Review statements following the Climate Bonds Standard's certification have equally been superficial or silent. [162] Even for the reports that do include such information, the lack of common standards on how to review securitisation portfolios may exacerbate the confusion and divergencies across review statements.

Besides, external review processes, especially SPOs and ratings, strongly focus on the pre-issuance commitments[163] rather than on the subsequent progress. It is true that the GBPs also recommend obtaining post-issuance reviews, based on all types of processes mentioned above, on a regular basis or at least after the full allocation of proceeds.[164] If the operation has 'sustainability-linked' targets, post-issuance verification also becomes a requirement. [165] However, the reviews which are most likely to take place include verifications and certifications. In the latter case, the process may also result in the certification being revoked if the transaction no longer fulfils the requirements[166] However, this power may not suffice to deter

[160]On the 'use of proceeds', see ICMA GBPs (n. 91), Principle 1, similarly, see Principle 1, Climate Bonds Standard – version 4.0' (n. 125) 12; 'Part A', Climate Bonds Standard v 4.0' (n. 125) pp. 17–22.

[161]'Given the established process to select eligible mortgages, deriving energy performance information from credible third parties, Sustainalytics is confident that the refinanced loans will adhere to the eligibility criteria' Sustainalytics (n. 26) 9.

[162]See For example, the review statement for the certification of Ecoagro Securitizadora's deal, not commenting on the portfolio, Bureau Veritas Certification, Assurance Statement (2020) https://www.climatebonds.net/files/files/Ecoagro-Rizoma-pre-issuance-statement-PUBLIC.pdf, https://ecoagro.agr.br/public/storage/Arquivo/1662064095Relat%C3%B3rio%20de%20Verifica%C3%A7%C3%A3o%20P%C3%B3s%20Emiss%C3%A3o%20-%20Bureau%20Veritas%20-%20Ingles%20-%20Ago-2022_compressed.pdfaccessed 9 December 2025.

[163]As noted by ICMA, 'Guidelines for external reviews' (n. 149) and Asian Bonds Online (n. 150).

[164]ICMA, ibid.

[165]At least once per year or at any relevant period for the assessment of performance targets, ICMA, SBPs (n. 91) 5; .

[166]Climate Bonds Initiative, 'Climate Bonds Standard 4.3' (n. 125) 1, 29.

investors from disregarding their commitments about the portfolio. As with the pre-issuance review, service providers will not necessarily focus on the impact of the securitised receivables, but rather on the compliance with external principles, and thereby on aspects like the allocation of the securities or the implementation of the sustainability framework.[167]

All in all, the focus of external review processes on external principles like ICMA GBPs may not ensure a sufficient critical insight into the 'sustainability' profile of the securitised assets and their progress. It may result in a limited, and misleading, assessment, which may incentivise the originating company to omit relevant information on the real environmental and social impact of the portfolio. Such partial disclosure can, however, leave room for 'greenwashing' practices and may send a misleading price signal to investors.

4.7 The (Clearer but Unenforceable?) Terms of Issue of ABSs

Market actors engaging in previous transactions, while not always using credit claims that make a positive contribution to environmental, social, and governance factors[168], have in most cases named their securitisations 'sustainable' based on how they planned to allocate the proceeds raised from the operation. As mentioned above, this has translated into a series of commitments to implement specific sustainability-related activities (across 'green' and 'social' deals) or to reach certain targets, which also impact the repayment conditions of the securities (across 'sustainability-linked' deals). The following sections will discuss how and to what extent the transactional documents underpinning the issuance of ABSs, the additional documentation exclusively drafted for sustainability purposes, and the external review statements capture the commitment to sustainability objectives in binding terms.

4.7.1 Within Traditional Issuance Documentation

A closer analysis of the documentation circulated ahead of the issuance highlights fragmentation in market practices for communicating

[167] See also Asian Bonds Online (n. 150).
[168] See above Sect. 4.6.1.–4.6.2.

the sustainability purpose behind the operation to investors and to the broader market.

Most securitisation transactions follow the approval and publication of prospectuses, which, among other things, outline the 'reasons for the offer' and the 'use of proceeds'.[169] When companies are not required to publish a prospectus, these elements are often included in similar but more succinct issuance documentation, such as 'offering circulars' or in more general 'final terms and conditions', summarising the terms of the issuance. Issuers are liable for all the information included in prospectuses, notably if they add false and misleading information or if they omit material information.[170]

However, 'reasons for the offer' and 'use of proceeds' clauses have not always referred to sustainability objectives. In certain cases, market actors have only reported that the funds raised from the operation will be used to refinance the underlying loans.[171] While this may not be in principle problematic, such clauses may still give rise to transparency issues as long as loan purchase agreements remain confidential and sustainability frameworks are vague on portfolios.[172] In other cases, green bond prospectuses have also contemplated using the proceeds for non-sustainability-related projects, like general corporate purposes.[173] Given the similarities between green bonds and sustainable securitisations[174], it is reasonable to assume the recurrence of similar clauses across ABS prospectuses too. Lastly, press coverage has reported that the documentation for one of the above-mentioned deals only required the issuer to use its 'best

[169] Annex I Regulation (Eu) 2017/1129 of the European Parliament and of The Council of 14 June 2017 on the prospectus to be published when securities are offered to the public or admitted to trading on a regulated market, and repealing Directive 2003/71/EC (hereinafterProspectus Regulation). See also Chap. 6, Sect. 6.4.

[170] On the exemptions to prospectus requirements and liability regime, see Chap. 3, Sect. 3.3.3., and Chap. 6, Sect. 6.4.

[171] For example, in the Green storm's prospectus 'The Issuer will apply the net proceeds from the issue of the Notes (other than the Class E Notes) on the Closing Date towards (i) payment of part of the Initial Purchase Price for the Mortgage Receivables purchased by the Issuer pursuant to the provisions of the Mortgage Receivables Purchase Agreement and (ii) payment of any initial swap payment to the Swap Counterparty in connection with the entering into the Swap Agreement'. Green storm (n. 104) 52; SFH (n. 104) 10.

[172] See above Sect. 4.3.–4.4.

[173] GACI First Investment Company, 'Public Investment Fund Euro Medium Term Note Programme Offering Circular' (n. 136) 69.

[174] Especially for securitisations that are 'sustainable' for the use of proceeds, see above Sect. 4.4.

efforts' to deploy the funds for environmental projects.[175] These clauses, which do not require an exclusive use of the proceeds for sustainability-related goals, would *de facto* blur the distinction between 'sustainable' and traditional securities.

At the same time, when it comes to 'sustainability-linked' ABSs, there is also current empirical evidence that market documentation for 'sustainability-linked' bonds tends not to set robust sustainability-related targets, leaving issuers large freedom to deviate from their commitments.[176]

More importantly, there are other clauses of the issuance documentation that tend to fail to fully capture sustainability objectives. In particular, common terms of prospectuses and similar documents include 'representations and warranties' or 'conditions precedent'.[177] Other widely spread clauses are 'covenants', where issuers set certain commitments, indirectly allowing investors to exercise their 'voice' over their business activities. [178] These clauses are generally drafted as affirmative promises to adopt specific behaviours (e.g. to produce regular reports) or as restrictions to certain managerial decisions (e.g. a ban from paying dividends).[179] They can also list 'Events of Default', i.e. a series of circumstances allowing some or all investors to accelerate the repayment. These may also result in the issuer 'cross-defaulting' in all other obligations vis-à-vis third parties. [180] Due to the highly deterrent power of the acceleration threat, parties can also in

[175]i.e. one of the first transactions by Credit Agricole, see Foley (n. 32).

[176]For example, the penalties of sustainability-linked bonds are often small compared to the coupon rate, and uncorrelated. ESMA, 'TRV Risk report – –Report on Trends, Risks and Vulnerabilities' (2023) 1, 36 https://www.esma.europa.eu/sites/default/files/library/ESMA50-165-2438_trv_1-23_risk_monitor.pdf accessed 9 December 2025.

[177]See above Sect. 4.6.1.

[178]William W Bratton, Bond and loan covenants, theory and practice (2016) 11 Capital Markets Law Journal 4 461, 461–462; Lars Hornuf, Markus Reps, Stefan Schäferling, 'Covenants in European Investment-Grade Corporate Bonds' (2015) 10 *Capital Markets Law Journal* 345; see Wood (n. 114), para. 10-001.

[179]Bratton (2016) 462

[180]Wood (n. 114), para. 10-004, 14-001, 16-001. At times, the documentation only allows investors of the higher tranche to trigger Events of default, Wall, Chambers et al, *Collateralised loan obligations (CLOs): overview* Practical law https://uk.practicallaw.thomsonreuters.com/9-385-8080?originationContext=knowHow&transitionType=KnowHowItem&contextData=(sc.Search)&comp=pluk&navId=A69723AC3AE5089BE4072F83F0FF8512&OWSessionId=a54e585f08e24e878731cec06d2aa050&skipAnonymous=true&firstPage=true#co_pageContainer

principle negotiate that non-compliance to Warranties and to covenant constitutes an 'Event of Default'.[181]

However, the analysis of previous examples of documentation has revealed that the sustainability-related activities linked to the issuance, or the targets, are generally absent from the list of Representations, Warranties and Events of Default.[182] Instead, prospectuses often include disclaimers which seem to exclude sustainability-related representations: for instance, certain provisions warn investors that the securities may not meet investors' expectations, guidelines or rules on sustainability-related matters.[183] Alternatively, there have also been examples of disclaimers explicitly ruling out that not meeting sustainability targets would trigger an Event of Default.[184] The UK Financial Conduct Authority (hereinafterFCA) and the EBA have confirmed similar trends.[185]

[181] Wood (n. 114), paras 7-021, 7-023. Although it is in practice rare to call on default for the breach of warranties, ibid., para 9-010.

[182] For a similar trend regarding green bond prospectuses, see From 'Green Bond Principles' to 'Green Bond Clauses' (n. 106); For example, the circumstances for 'Mandatory repurchase of Mortgage Receivables' and the 'Events of Default', Green Storm 2021 prospectus, Green storm (n. 104) 63, 96–98.

[183] See For example, '*no assurance is given by the Issuer that an investment in the Notes will satisfy, whether in whole or in part, any present or future investor expectations or requirements as regards any investment criteria or* guidelines with which investors or their investments are required to comply in relation to so-called 'green' or 'sustainable' investments. Furthermore, it should be noted that there is currently no clearly defined definition (legal, regulatory or otherwise) of, nor market consensus as to what precise attributes are required for Notes to be defined as 'green' or 'sustainable' or such other equivalent label nor can any assurance be *given that such a clear definition or consensus will develop over time. Accordingly, no assurance is or can be given to investors that the investment in the Notes will meet any or all investor expectations regarding such 'green', 'sustainable' or other equivalently labelled performance objectives'*. Green storm (n. 104) 16. The 'supplement' to the base prospectus of the Euro medium term note Programme by Societe Generale also uses a similar wording, SFH (n. 104) 6–7.

[184] This wording is common across all prospectuses for green debt securities, including traditional green bonds, as I have argued in Federica Agostini, 'From 'Green Bond Principles' to 'Green Bond Clauses' (n. 106). A recent comparative analysis of prospectuses has reached the same conclusion, Mark C Weidemaier, Quinn Curtis, 'Green Bonds, Empty Promises' (2023) Public Law and Legal Theory Research Paper Series 2023-14, 18, 20–24.

[185] FCA, 'Enhancing Climate-Related Disclosures by Standard Listed Companies and Seeking Views on ESG Topics in Capital Markets' (2021) CP21/18, 30 https://www.fca.org.uk/publication/consultation/cp21-18.pdf accessed 9 December 2025; EBA, 'Sustainable securitisation' (n. 5) 35–37. in relation to Events of Default, EBA, 'Report on the monitoring of TLAC-/MREL – Eligible Liabilities Instruments of European Union Institutions' (2020) 35–37.

The missing link between the wording of the clauses and sustainability goals has remarkable implications for the enforcement of these securities. It prevents investors from being able to demand compensation if issuers' claims regarding their projects and targets are intentionally wrong or do not substantiate. Even when issuers undertake in press releases or other statements to conduct external reviews before or after the issuance, the lack of a dedicated covenant may prevent this promise from being prescriptive. The same would happen if they promise to report regularly on the use of proceeds, but they fail to do so or they include wrong or misleading information within their reports.

Lastly, not linking sustainability-related objectives to Events of Default also deprives investors of the crucial possibility of demanding immediate repayment of their bonds should issuers not attain these goals. It is true that investors may not necessarily *want* to trigger this remedy, due to the additional costs and to the resulting end of the contractual relationship with issuers. Nevertheless, the mere threat of defaulting and of cross-defaulting across other obligations should incentivise issuers to comply with the provisions. In all these circumstances, investors may only rely on limited remedies under consumer and competition law, with the caveats discussed above.[186]

All in all, several shortcomings arise from the examination of issuance documentation, where clauses do not effectively bind issuers to allocate the proceeds towards sustainability-related activities, to meet certain targets or to provide external reviews or reports. Their formulation effectively leaves room for 'green-' or 'ESG washing' practices, where originators could mislead investors and not abide by their commitments.[187] In the absence of dedicated clauses on sustainability matters, investors would have the same remedies that would arise from investments in *ordinary* securitisation operations. This raises the question of the effective distinctive feature of these deals.

[186] See Sect. 4.6.3.

[187] See the definition of 'Greenwashing' by Magali A Delmas, Vanessa Cuerel Burbano, 'The Drivers of Greenwashing' (2011) 54 California Management Review 64, 65. While the literature first focussed on 'greenwashing', a broader debate on 'social' and 'ESG-washing' risks has also come to the fore. Michela Scatigna, Dora Xia et al, 'Achievements and challenges in ESG markets' (2021) *BIS Quarterly Review*, 94 https://www.bis.org/publ/qtrpdf/r_qt2112f.pdf accessed 9 December 2025.

4.7.2 Within 'Sustainability-Related' Documentation

As an alternative to prospectuses, market actors may outline their commitments to sustainability-related activities or targets in ad hoc documentation. In practice, however, the analysis of previous examples of these documents also reveals shortcomings.

For instance, they may issue public statements. This may be more common across private placements, not subject to the same prospectus disclosure requirements.[188] Such statements would not, however, suffice to give rise to a binding legal obligation for issuers to comply with what they have promised.

Oftentimes, market actors also describe how their issuance contributes to sustainability goals across the above-mentioned 'sustainability frameworks'. Indeed, because these documents often directly replicate the structure of ICMA's GBPs and SBPs[189], they tend to be more explicit regarding the allocation of proceeds than about the portfolio. For instance, they may provide quantitative metrics and provide further information on how the issuance aligns with external standards. Issuers may also make further pre-issuance and post-issuance commitments: for example, they may commit to an articulated process for the selection of the eligible sustainability-related projects, to track expenses or to report on the allocation of their securities, their sustainability impact or on progress against their targets.[190]

As discussed above, referring to sustainability frameworks across prospectuses may expose the issuer to the prospectus civil liability regime. This means that they could be liable for false and misleading information, as well as for omitting material information.[191] One may think of the case where issuers refer to these sustainability frameworks throughout their 'use of proceeds' clauses[192]. In principle, if these latter frameworks do not include material information on the potential adverse impacts of the operation, issuers could be liable.

[188] On private placements, see Chap. 3, Sect. 3.2., 3.3.3. Examples discussed above include Jupiter's and Project Boqueron's deals, Société Générale (n. 84); Global Capital (n. 85).

[189] As discussed above, Sect. 4.2.–4.6.2.

[190] See above Sect. 4.6.2.; see For example, the information about Obvion's green bond framework contained within Sustainalytics SPO, Sustainalytics (n. 26).

[191] Under EU law, prospectuses 'shall contain the necessary information which is material to an investor for making an informed assessment', Art 6(1) Prospectus Regulation. See also Chap. 6, Sect. 6.5.

[192] See Annex I, ibid., for the mandatory requirements of prospectuses.

However, previous transactions show that the references to sustainability-related documentation have not been consistent. While certain market actors have mentioned such frameworks,[193] there have been instances where green bond prospectuses have explicitly clarified that green bond frameworks and periodic reports are not 'incorporated by reference'.[194] The wording of these clauses effectively prevents transposing the prospectus regime to the terms of the framework. Thus, it deprives investors of the related enforcement remedies, further adding to the other enforcement challenges of prospectuses discussed above. When investors cannot rely on the frameworks' terms, they may only be able to take action under consumer law or competition law, with the same caveats discussed above.[195]

4.7.3 External Reviews

As long as there is no statutory obligation or contractual covenant requiring external reviews over 'sustainable' ABS, the involvement of third parties is a voluntary decision for issuers. However, as the majority of the issuers have, in fact, resorted to external service providers,[196] the most pressing issue may not be *whether* issuers will have an external review, but rather *to what extent* this review will be effective. An interrelated question is whether the costs and burdens for external reviews may disincentivise the issuance of sustainable ABSs altogether. For example, the Climate Bonds' Initiative database has reported that there were only 19 issuers globally in 2021.[197] Industry associations such as the ICMA have identified external review procedures as a barrier, preventing market actors from accessing this market.[198]

[193] Like Société Générale, when issuing 'sustainable positive impact notes' as part of its programme, see SFH (n. 104) 5-6–7.

[194] See For example, the clause in the green bond prospectus 'None of the GBPs, the Second Party Opinion or the Framework or any of the above reports, verification assessments or the contents of any of the above websites are incorporated in, or form part of, this Offering Circular', GACI First Investment Company (n. 136) 70.

[195] See above Sect. 4.6.2.

[196] See above Sect. 4.6.4.

[197] Climate Bonds Initiative (n. 156), 11.

[198] As highlighted by ICMA when commenting about the introduction of the EU Green Bond Standard, ICMA, 'Analysis of the amendments to the EuGB Regulation proposed by the Rapporteur of the EU Parliament' (2022) 3 https://www.icmagroup.org/assets/ICMA-update-to-its-analysis-of-the-EuGB-Regulation-05012022.pdf accessed 9 December 2025.

As explained above, the frequently used review methodologies focus on compliance with external standards like ICMA GBPs: in so doing, the emphasis tends to be on the final allocation of the securities (the 'use of proceeds' or the sustainability performance targets) rather than on the portfolio.[199]

At the same time, the scope of the external reviews may still not suffice to effectively ensure 'gatekeeping' of the qualities of the securities. It appears from the ICMA Guidelines on external reviews and from review statements that the primary goal is to assess whether and to what extent *the documents provided by the issuer*, like the sustainability frameworks, abide by an overarching set of principles.[200] However, the formal compliance of the documentation with external criteria does not necessarily signal a positive environmental, social, and governance performance. For instance, while an energy efficiency certificate may testify the maximum levels of energy efficiency, it does not certify how a building performs. [201]

For those who voluntarily resort to external SPOs, certifications, verifications or ratings, the degree and frequency of the assessment inevitably varies. As with securitisation portfolios, issuers may undertake to only have a pre-issuance review, which would leave no room for third parties to evaluate the actual progress on the implementation of the sustainability-related projects or against the performance targets. One example could

[199] See For example, the 2022 Saudi Arabian green bond prospectus, clarifying that any external reviews would *'not be deemed to be an expression of an opinion as to the Issuer and/or the Guarantor or any aspect of any Eligible Green Projects (including but not limited to the financial viability of any Eligible Green Projects) other than with respect to conformance with the Climate Bonds Standard'*, GACI First Investment Company (n. 136) 92.

[200] See For example, the climate bonds standard certification scheme 4.0, para 4.1; see also CICERO, mentioning a 'desk-review' over the frameworks, see also the old certification scheme by CICERO (the so-called 'Shades of Green'), CICERO, 'Green Best Practices' (2022) 12 https://www.nasdaq.com/docs/2021/06/08/Methodology_Description_CICERO_Shades_of_Green.pdf. Accessed 3 June 2026; following the acquisition of CICERO by S&P ratings, the latter continues to deploy a similar procedure to assess the degree of alignment of instruments and transactions with external principles (in the form of a 'second-party opinion). The company confirms that the assessment is 'provides a point-in-time opinion, *reflecting the information provided to us* at the time the SPO was created and published', see For example, S&P Ratings, 'Second Party Opinion: Sparebank 1 Ringerike Hadeland Green Bond Framework' (2025) https://www.spglobal.com/ratings/en/regulatory/delegate/getPDF?articleId=3496992&type=COMMENTS&defaultFormat=PDF accessed 20 December 2025.

[201] As highlighted in relation to energy-efficiency certificates for buildings by Coglianese (n. 155) 12.

be the *2021 Green Storm* Programme, where the issuer has only undertaken to request a second-party opinion on the *prospectus,*[202] but not over any additional documents at the pre-issuance stage or on progress. Only allowing a documentary and *ex ante* assessment of the securities may give investors a false and misleading impression about the extent to which third parties have effectively assessed and certified the reliability of the operation. Misleading statements regarding external labels and certifications have been described by commentators as sources of 'greenwashing'.[203]

A comparison between different reviewers and reviews also reveals wide inconsistencies across the methodologies and scoring systems applied. Academic commentators analysing 'sustainability ratings' have highlighted wide discrepancies at three levels. First, they diverge in scope as they all cover different sustainability aspects (such as GHG emissions, carbon intensity, compliance with human rights standards). Second, even ratings with a similar scope use different indicators to measure the same attributes. Third, they give them a different relative importance, so that one factor may have a more evident impact on the rating than others. Overall, these differences complicate the interpretation of the ratings, creating 'aggregate confusion'.[204] Similarly, commentators have already noted the divergences across the methodologies to review green bonds.[205] While these analyses focus on ratings and on green bonds, there is merit in transposing them to all types of external review for sustainable securitisations. Reports range from those providing quantitative information[206] and scores (especially in the case of sustainability ratings[207]) to those focusing on the

[202] 'the Seller has requested Sustainalytics, a provider of environmental, social, and governance (ESG) research and analysis, to evaluate Obvion's green residential mortgage-backed securities transaction set forth in this Prospectus and the alignment thereof with the Green Bond Principles', Green storm (n. 104) 247.

[203] 'the sin of worshipping false tables', see TerraChoice Group, Inc., 'The Sins of Greenwashing', (2009) https://www.twosides.info/wp-content/uploads/2018/05/Terrachoice_The_Sins_of_Greenwashing_-_Home_and_Family_Edition_2010.pdfaccessed 3 June 2026; Delmas, Burbano (n. 187) 67.

[204] Florian Berg, Julian Kölbel, Roberto Rigobon, 'Aggregate Confusion: The Divergence of ESG Ratings' (2019) 26 Review of Finance 6 1315. see also Cadwalder (n. 157).

[205] Torsten Ehlers, Frank Packer, 'Green Bond Finance and Certification' (2017) BIS Quarterly Review 89, 94.

[206] Like the report by Sustainalytics over Obvion. Sustainalytics (n. 26).

[207] Like the 'sustainability rating reports by Standard ethics, see For example, the outlook of a 2022 programme, Standard Ethics, 'Eco-SISMA bonus securitisation programme' (2022) https://standardethicsrating.eu/component/finances/?view=items&category_id=59 accessed 9 December 2025.

qualitative and formal compliance with procedural steps and best practices.[208] Some providers also exclusively conduct pre-issuance reviews.[209] These divergencies may foster confusion and represent a considerable obstacle to the assessment of the qualities of the securities for investors and other market participants.

More importantly, the competition between providers gives rise to distortions in the review process altogether. The published review reports so far have almost exclusively expressed positive judgments on sustainable securitisations. This suggests that issuers have an incentive to choose providers using the most favourable methodology[210] and that they are more likely to turn to another provider rather than to wait for a negative review.

All in all, external reviews focus heavily on the allocation of proceeds. However, review processes are inherently partial in that they are centred on the documentation that originators provide and on the situation at the moment of assessment (generally before the issuance). They are also fragmented due to the divergences in methodologies. The narrow scope and fragmentation of these assessments may add further complexity to investors' due diligence exercises[211] and mislead them about the qualities of the securities.

4.8 The (Narrowly Defined) Governance of SPVs

Chap. 2 has also identified the law concerning the purpose and management of SPVs (its 'governance') as instrumental to sustainability goals. However, the analysis of transactional documentation reveals very limited attention to the role of these entities when describing sustainable securitisations to the public.

[208] Like the Bureau Veritas statement analysed above (n. 62).

[209] Like the 'Shades of Green' reviews previously performed by Cicero. One example may be the assessment of the compliance with the EU Taxonomy, where the provider 'consider(s) the likely alignment of eligible project categories in the issuer's framework with the mitigation or adaptation criteria and the Do No Significant Harm (DNSH) criteria contained in the technical screening criteria (TSC)'. See CICERO, 'Best Practices' (2022) 9, 11–12 https://www.spglobal.com/_assets/documents/ratings/green_best_practices_2022.pdf accessed 9 December 2025.

[210] As discussed above, Sect. 4.6.4.

[211] On investors' duty to conduct 'due diligence' before investing in ABSs, see Chap. 3, Sect. 3.4.4.

Both securitisation agreements and ad hoc sustainability-related documentation tend not to disclose about SPVs' purpose and activities, or about the link between these entities and sustainability objectives.

It is at times possible to infer some information on SPVs from public announcements, like press releases,[212] or from publicly available documentation such as prospectuses, which must necessarily disclose the identity of the 'issuer' under EU law.[213] However, the mere identity does not reveal how sustainability objectives inform SPVs' activities.

Instead, the documents for the constitution of SPVs, such as the articles of association when they are companies[214] should describe their core objective or mandate.[215] This means that they should, in principle, also outline the link between sustainability goals and SPVs' strategy. The prospectus itself may also cross-refer to such documents. However, while the documents for the incorporation of companies may be subject to some publicity requirements[216], the articles of association are not always directly circulated to potential investors.

Reviewers' reports could shed light on the governance of SPVs. Based on a survey of a series of market participants, the PRI has indicated that the assessment of SPVs' governance structure is one of the factors that informs sustainability ratings, especially across CLO transactions.[217] The survey suggests that at least rating reports should provide some information about the findings of this assessment. Indeed, a couple of examples of review statements have contextualised the programme or the securities

[212] For example, the related press release provides limited information about the capitalisation and mandate of the SPV within the Bayfront infrastructure transaction. See Bayfront Infrastructure (n. 59).

[213] Annex II, II, Prospectus Regulation. For instance, green storm's 2021 prospectus provides some information on Obvion, including the corporate purpose, Green storm (n. 102) 68–69. The prospectus for the CLO managed by Bardin Hill identifies securitisation as the only activity where the SPV engages, but does not mention how sustainability objectives fit within this purpose, Bardin Hill Loan Advisors (n. 77) 234. .

[214] For the different proprietary arrangements for the assignment of the securitised assets to the SPV and for the segregation thereof, see Chap. 3, Sect. 3.3.1.–3.4.1.

[215] On the relevance of the documentation for the creation of the SPV to determine the SPV's mandate and the administration rules, see D Ramos Muñoz, 'In Praise of Small Things: 'securitisation and Governance Structure' (2010) 5 Capital Markets Law Journal 363, 381. See For example, the Luxembourg legislation, requiring the SPV's statute to have an 'Object clause, art 10 (1) Loi du 22 mars 2004 relative à la titrisation (Luxembourg).

[216] For example, under Art 2332 cc (Italy)

[217] UNPRI (n. 117) 18. On the difference between CLOs and other types of securitisation operations, see Chap. 3, Sect. 3.2.

within issuers' overarching strategy on environmental, social, and governance matters.[218] However, various examples of previous SPOs and verifications have neglected the assessment of these entities.[219] This is consistent with the previous observation that the methodologies heavily focus on the use of proceeds or on sustainability performance targets.[220]

Admittedly, some information regarding SPVs may not directly affect the 'sustainable' profile of the operation. However, it may be useful for investors to gain insights into SPVs' purpose and business activity. Understanding whether these entities exclusively engage in sustainable securitisation or securitisation in general may be of interest to potential investors, especially when performing compulsory due diligence exercises.[221] This has a financial implication, since other business operations may diminish the SPVs' patrimony and frustrate creditors' protection. More importantly, contextualising the issuance within the remainder of the SPVs' activities may also support the assessment of the overarching commitment and strategy of these entities on sustainability matters.

4.9 (Steps Towards) Investors' Representation Mechanisms

Another strategic component of the documentation is the set of provisions laying out investors' rights and the respective representation mechanisms in response to the coordination costs and challenges. The regulation of these aspects will not only be contractual but may also depend on a series of statutory provisions in certain jurisdictions.[222] There are various options for parties to agree on such aspects: Sometimes they condense a series of

[218] For example, the SPO over Berlin Hyp's 'covered bonds', see ISS ESG, Berlin Hyp Sustainability-Linked Bond
Framework (2021) 10 https://www.icmagroup.org/assets/Uploads/berlin-hyp-sustainability-linked-bond-framework-second-party-opinion-2021.pdf accessed 9 December 2025.

[219] For example, the SPO over River Green Finance, focusing on ICMA GBPs and not mentioning River's strategy, Sustainalytics, 'River Green Finance Green Securitised Bond Framework' (n. 121); see also the SPO over 2017 Green Storm, which only has one paragraph on the sustainability strategy of the company, see Sustainalytics (n. 26) 4.

[220] See above 4.6.4.; 4.7.3.

[221] See Chap. 3, Sect. 3.4.4.

[222] See Chap. 6, s. 6.8. David Ramos Muñoz, *The Law of Transnational 'securitisation* (Oxford Uni, 2010) 154; Sheilah D Gibson, 'The Case for the Expanded Role of Trustees in 'securitisations' (2004) 121 Banking Law Journal 387.

'intra-creditor' provisions within prospectuses.[223] There is also often an additional agreement, which may take the form of a 'trust deed' or a more general contract, appointing one entity as trustee or representative for all investors.[224] In most cases, this appointment translates into the delegation of powers and deprives individual noteholders from enforcing their individual rights under the securities.[225] 'Intra-creditor' clauses or such supplementary agreements may also contain further provisions on the right to consult with all noteholders within 'noteholders' meetings' about issues like the modification of the terms and conditions.[226]

For the purposes of the present analysis, it is essential to investigate whether these agreements also define the duties of these representatives on sustainability matters. Due to the coordination costs and challenges, it may be reasonable to rely on these actors to raise investors' voice on sustainability matters.[227] For instance, they could be in principle, be required to co-design the sustainability clauses of the transaction with issuers and to conduct due diligence assessments on sustainability aspects ahead of the operation. Thereafter, they could be entrusted with the duty to monitor the performance and progress against these commitments as well as enforce the securities in cases such as a missed target or 'greenwashing' in issuers' statements.

A point of reference for the role of these representatives could be the emerging market practice of 'sustainability coordinators'. The leading LMA Principles for sustainability-linked Loans[228] recommend the appointment of these intermediaries, who should be in charge of leading the dialogue

[223] See the condition 14. 'Meetings of Noteholders; Modification; Consents; Waiver; Removal Director', Green storm (n. 102) 99.

[224] For a sample agreement, see Natwest group, Arran Residential Mortgages Funding 2010-1 PLC – Trust Deed (2010) https://investors.natwestgroup.com/~/media/Files/R/RBS-IR-V2/securitisation/reseidential-mortgages/pdf/7-arran-residential-mortgages-2010-1-plc/principal-transaction-documents/015--trust-deed-2010-a1.pdf accessed 9 December 2025.

[225] See Sect. 11 ('Enforcement'), Green storm (n. 102).

[226] See Sect. 14 ('Meetings of Noteholders; Modification; Consents; Waiver; Removal Director') within the 'terms and conditions', ibid.

[227] See Chap. 2, Sect. 2.6., Chap. 3, Sect. 3.4.4.

[228] As recommended by the LMA Principles, which, like the GBPs, have also been updated, with the last update tracing back to 2025. LSTA LMA, APLMA, 'Sustainability-Linked Loan Principles' (2025) https://www.lsta.org/content/sustainability-linked-loan-principles-sllp/#; see also LMA, 'Guidance on sustainability-linked loan Principles' (2025) 6 https://www.lma.eu.com/application/files/9416/4873/5604/Sustainability-Linked_Loan_Principles_Guidance_31_March_2022.pdf.

with borrowers and of negotiating the terms of the documentation (such as the sustainability performance targets or the identification of the projects to be financed). The same industry association has also recommended throughout specific guidance, the involvement of coordinators to gather information on sustainability matters in the post-loan phase.[229]

However, while press coverage has at times mentioned the involvement of sustainability coordinators across bond and securitisation deals[230], and early literature appears to confirm the emergence of this practice,[231] publicly available documentation has so far failed to clarify how regularly they participate. Nor is it yet possible to draw general conclusions about the extent of their involvement.

This suggests that the degree of investors' involvement in the negotiation of the sustainability-related provisions of the documentation is very deal-specific. However, the lack of a coordination mechanism would place the burden of assessing the sustainability characteristics of the operations on each individual investor. It may also disincentivise both traditional investors and those with a preference for 'sustainability' objectives, who may opt for other less complex products. Correspondingly, the above-mentioned surveyed institutional investors have identified the too high due diligence costs as one of the most pressing challenges of sustainable ABSs.[232]

As regards the post-issuance phase, it is not clear to what extent certain investors are in charge of monitoring the issuers' progress on sustainability matters and of taking action when the claims are not substantiated. On the contrary, there have been cases in publicly available documentation

[229] LMA, 'An Introduction to the Sustainability Coordinator Role' (2022) https://www.lma.eu.com/download?p=5942446308-6796 accessed 9 December 2025.

[230] For example, within Marcegaglia's sustainability-linked securitisation, Legal Community, 'GOP e Nctm nel finanziamento ESG-linked da 1 miliardo per Marcegaglia' (2022) https://legalcommunity.it/gop-e-nctm-nel-finanziamento-esg-linked-da-1-miliardo-per-marcegaglia/ accessed 9 December 2025; see also Environmental Finance, 'Bond Awards 2022 – – Lead manager of the year, social bonds – –financial institution; Lead manager of the year, sustainability-linked bonds; and Loan structurer/arranger/coordinator of the year: BNP Paribas' (2022) https://www.environmental-finance.com/content/awards/environmental-finances-bond-awards-2022/winners/lead-manager-of-the-year-social-bonds-financial-institution;-lead-manager-of-the-year-sustainability-linked-bonds;-and-loan-structurer/arranger/coordinator-of-the-year-bnp-paribas.html

[231] This statement is also consistent with the findings of interviews with the industry. Agnieszka Smoleńska, 'The Constitutive Role of Law in Sustainable Finance' (2025) *Industrial Law Journal* 20, 29.

[232] UNPRI (n. 117) 24.

where clauses explicitly ruled out the duty of investors' representatives to assess the compliance of the transaction with external sustainability standards[233] or to monitor the use of proceeds.[234] If such waivers were standard practice, investors' representatives would have limited powers and incentives to monitor the originating company's progress on sustainability matters. The lack of Events of Default on sustainability matters mentioned above[235] also further limits the powers of these representatives to act in case of 'greenwashing' practices. Due to the limited enforcement duties and avenues, they are also unlikely to check whether issuers have substantiated their commitments on sustainability matters under the sustainability framework.[236]

It is not clear whether 'sustainability coordinators' are often involved across 'sustainable CLO' deals either. These transactions see the involvement of a manager with the power to purchase the underlying credit claims on behalf of all the investors.[237] However, the survey has also confirmed the lack of transparency on the powers and discretion of these managers. In certain cases, they were exclusively in charge of, or had significant discretion over, the assessment of the compliance of the operation with ESG criteria and regulations. At the same time, the vast majority of investors have reported that their assessment of the operation only consisted in the application of 'negative screening' criteria, i.e. they excluded investments in certain controversial sectors, but not in the positive integration of ESG scores.[238] However, exclusionary strategies, like divestments, have shown to have scant real world effects in that they only facilitate progress towards sustainability and climate goals to a limited extent .[239]

[233]For example, see the clause in the Gemgarto Prospectus (also mentioned in fn 55), 'The Arranger, Joint Lead Managers and ESG Structuring Banks have not undertaken, nor are responsible for, any assessment of the Social Bond Framework and any verification of whether the notes or the Social Bond Framework achieve any of ICMA's Social Bond Principles', Gemgarto 2021-plc, Prospectus (2021) 52 https://ise-prodnr-eu-west-1-data-integration.s3-eu-west-1.amazonaws.com/legacy/Prospectus+-+Standalone_1fe8519a-a55e-4826-8e3d-74f87136927e.PDF accessed 9 December 2025.

[234]See 'None of the Arranger, Joint Lead Managers or ESG Structuring Banks will verify or monitor the proposed use of proceeds of the Notes issued'., ibid., 52.

[235]See s. 4.7.1.–4.7.2.

[236]See Chap. 3, Sect. 3.4.4.

[237]On the difference between traditional and CLO securitisations, see Chap. 3, Sect. 3.2.

[238]UNPRI, (n. 117) 21.

[239]Tom Gosling, Iain Macneil, 'Can Investors Save the Planet? – NZAMI and Fiduciary Duty' (2023) 18 *Capital Markets Law Journal* 2 172, 182–184.

All in all, the lack of binding clauses within publicly available documentation, as well as qualitative data emerging from surveys, suggest that most sustainable securitisation deals do not have one or more actors raising the 'voice' of all investors on sustainability matters. If investors have insufficient incentives to engage in these aspects before and after the issuance, originating companies may have leeway to only provide a suboptimal performance, i.e. to engage in 'moral hazard' behaviours.[240] In other words, there may be room to make exaggerated promises on their sustainability commitments and not to comply with them.

As I shall elaborate in the remaining chapters, the peculiarities and challenges of sustainable securitisations require strong regulatory mechanisms to ensure the adequate representation of their interests.

4.10 (Fragmented) Monitoring Processes

Chap. 3 also shed light on the transformational impact that law and contractual provisions have when they design mechanisms to monitor how securitisation funds are spent. Ad hoc legislation or clauses may ensure that originating entities exclusively devote the cash flows of the operation, such as the proceeds and the payments from the portfolio, to the implementation of sustainability-related projects (such as extending further 'sustainable' loans) or to sustainability-related targets.

In securitisation practice, arrangements on how expenses are monitored tend to be highly deal-specific. One possibility is to appoint third parties, such as 'servicers' (or 'cash managers'), [241] to process or even enforce the payments of the underlying portfolio receivables and arrange the repayments to investors. 'Trustees', who also represent investors, may also perform this role.[242]

[240] Theoretical accounts have highlighted the higher likelihood of moral hazard behaviours in situations in which a contracting party wishes to 'shade' on his performance, see Oliver Hart, John Moore, 'Contracts as Reference Points' (2008) 123 *Quarterly Journal of Economics* 1, 2–9; Alan Schwartz, Simone Sepe, 'Economic Challenges for the Law of Contract' (2020) 8.Yale Journal of Regulation, 14, 18–19.

[241] In certain jurisdictions, like Spain, France, Luxembourg, statutory provisions explicitly require this model, see David Ramos Muñoz, Kevin Ingram, *The Law of Transnational 'securitisation*, 64–65.

[242] Especially in US, see Sheilah D Gibson, 'The Case for the Expanded Role of Trustees in 'securitisations' (2004) 121 *Banking Law Journal* 387. On the role of trustees, see Sect. 4.5.4.

Alternatively or on top of these appointments, parties may negotiate additional mechanisms to 'ringfence' the funds, such as a segregated bank account for the payments on the asset portfolio and for the repayment to investors. They may also simply undertake to 'earmark' the proceeds, i.e. to track the expenses and to regularly report about them to end investors. [243]

These arrangements may be part of the servicing and 'cash management agreements'[244] or of dedicated 'bank account agreements'.[245] When trustees are involved, the deed for their appointment may also be relevant.[246] However, as such agreements tend not to be publicly available, it is hard to ascertain what mechanisms are effectively deployed in practice, unless they are also indicated in the prospectuses and final terms and conditions.[247]

When it comes to previous examples of 'sustainable securitisations', prospectuses have not always provided information about these arrangements. In some cases, some provisions even allowed originating companies to use the proceeds for general corporate purposes, suggesting that some issuers may not have any tracking procedures in place.[248]

Only in some instances, internal sustainability frameworks and review statements have included information on monitoring processes over the cash flows. For instance, they were covered in the section on the 'management of proceeds', which also constitutes one of the core pillars of the ICMA GBPs.[249] However, as I shall discuss in Chap. 5, soft law principles for green bonds, also applicable to securitisations, should represent the reference point for market actors. These recommend the implementation of more robust monitoring processes, such as the appointment of third parties, the use of a sub-account or sub-portfolio, 'formal internal processes'

[243] As also recommended by the ICMA's Green Bond Principles, see ICMA, 'GBPs' (n. 91) 5.

[244] For a sample cash management agreement, see Natwest Group, Administration Agreement and a Cash Management Agreement dated 29 September 2010 (2010) https://investors.natwestgroup.com/~/media/Files/R/RBS-IR-V2/securitisation/reseidential-mortgages/pdf/7-arran-residential-mortgages-2010-1-plc/principal-transaction-documents/2010-1-cash-management-and-administration-agreements.pdf accessed 9 December 2025.

[245] For a sample agreement, see Natwest Group, 'Arran Residential Mortgages Funding 2010-1 PLC – –Account Bank Agreement' (2010) https://investors.natwestgroup.com/~/media/Files/R/RBS-IR-V2/securitisation/reseidential-mortgages/pdf/7-arran-residential-mortgages-2010-1-plc/principal-transaction-documents/014-account-bank-agreement-2010-a1.pdf accessed 9 December 2025.

[246] Wall, Chambers et al (n. 180).

[247] For example, Green storm 2021 (n. 104) 74, 164.

[248] See above 4.6.1.–4.7.1.

[249] ICMA, 'GBPs' (n. 91), 8.

to attest the tracking of expenses, or external auditing mechanisms to confirm the effective allocation of funds towards sustainability objectives.[250]

The comparison between various examples of sustainability frameworks and external review reports highlights wide discrepancies across market practices and an inconsistent application of these soft law principles. Some originators have put in place segregated accounts exclusively for the expenses of the operations.[251] Some have hired third parties[252] or created internal committees to track the balance of the funds. [253] By contrast, other reports have also highlighted that issuers have at times only generally committed to certain sustainability objectives, while not implementing any tracking mechanisms.[254]

Furthermore, due to the divergencies across the methodologies of external service providers,[255] not all external reviews and ratings attach the same value to monitoring processes, or they may even neglect them altogether. Thus, it comes as no surprise that, despite all the deal-to-deal differences, all the above-mentioned transactions have received positive assessments and scores by reviewers.[256]

A certain level of variety across transactions is to some extent inevitable in securitisations, whose design heavily depends on market actors' innovative powers.[257] However, cases where originators have not reported the existence of *any* contractual arrangements, governance systems or

[250] For example, ibid., 6–7. See more extensively Chap. 5, Sect. 5.7.2.

[251] https://www.fao.org/3/ap294e/ap294e.pdf 31

[252] Like Green storm 2021, see Sustainalytics (n. 26), 3–4.

[253] For a few examples emerging from sustainability frameworks, see Natwest, 'GSS Bond framework' (2020), 22 https://investors.natwestgroup.com/~/media/Files/R/RBS-IR-V2/green-social-and-sustainability-bonds/gss-bond-framework-oct-2020.pdf accessed 9 December 2025; for a SPO, see Vigeo Eiris over Lafayette asset securitisation LLC, Vigeo Eiris, Second party opinion on the green ABCP programs – –annex to Credit Agricole's Green Bond Framework (2020) 19 https://www.ca-cib.com/sites/default/files/2020-10/20200803_VigeoEiris_SPO_Credit-Agricole.PDF accessed 9 December 2025; Fannie Mae has put in place a Green Financing team, entrusted with post-financing monitoring powers. Fannie Mae, 'Multifamily Green Bond Framework' (n. 90) 11.

[254] Like the River Green Finance transaction, as reviewed by Sustainalytics, 'River Green Finance' (n. 120) 10. Relatedly, others have also criticised the lack of powers for servicers to proactively inquire about sustainability-related data or enforce the progress on sustainability-related commitments. Cornut St-Pierre, Gareau-Lindsay (n. 93) 22.

[255] See the analysis on the 'aggregate confusion' (n. 204).

[256] Like the SPO over River Green finance, where the reviewer concludes that the transaction is 'In line with the market practice', see Sustainalytics, 'River Green Finance' (n. 120) 3.

[257] See above Sects. 4.1.–4.2.

verification processes over expenses are inadequate, leaving room for 'greenwashing' behaviours. In other words, they cannot reassure investors about the efficient allocation of their funds. Reviewers' reports not highlighting the weaknesses of such practices, or the non-compliance with authoritative market standards do not provide a full picture of how sustainable the operations are. In so doing, it raises the question of the extent to which reviewers can effectively serve as 'gatekeepers'. In turn, inadequate monitoring processes may also facilitate opportunistic behaviours, like the misuse of proceeds for non-sustainability-related expenses (such as general corporate purposes).

4.11 The Challenges of 'Private Hard Law' and Potential Future Avenues

The analysis has highlighted the main challenges of the current market practices, where the parties to securitisation transactions have independently set the rules making the operations 'sustainable'. The reliance on market actors' independent rule-making powers promotes flexibility and stimulates innovation.

However, the analysis of securitisation agreements, additional documentation, external reviews and previous surveys reveals that parties tend to only make limited binding commitments. They also continuously refer to external standards, such as the ICMA GBPs, which serve an 'enabling' function as they prove the reliability of the 'sustainable' profile.

The lack of 'private hard law' is inherently problematic and calls for tailored legal interventions to coordinate 'sustainable' securitisations, following the strategies identified in Chap. 1.

First, further initiatives should promote 'disclosure' to address the general lack of transparency on how sustainability objectives permeate the various aspects of these operations, ranging from the characteristics of the portfolio and the intended use of proceeds, to the governance of SPVs, to monitoring processes. The confidential nature of certain transactional agreements, such as loan purchase agreements, may prevent investors and the broader public from having access to relevant sustainability-related information, like that pertaining to the composition of the portfolio. The formats to disclose the 'sustainable' criteria for the portfolio, the use of proceeds or sustainability-related targets are widely different and inconsistent, complicating investors' due diligence assessments. Originators often do not provide information about their mechanisms to monitor whether the expenses of the operation have been allocated to sustainability objectives, or they only do so to a very limited extent and in fragmented formats.

Ad hoc sustainability-related documentation by originators, such as 'sustainability frameworks', while sometimes vague on the composition of portfolios, tends to provide further details on the use of proceeds or on the sustainability-related targets that the operation pursues, in line with the structure of the ICMA's GBPs. Such frameworks are also sometimes more specific on SPVs' purpose as well as on tracking mechanisms for sustainability-related expenses, when the latter are present. However, there are several constraints to the enforcement of the objectives set out within such documents, unless prospectuses directly refer to ('incorporate') them.

Even the statements issued after external reviews of the issuance can indirectly provide more information around the operation, but they predominantly focus on the alignment of the documentation with external principles, such as the ICMA GBPs. As a result, they may not be specific enough on the characteristics of the asset pool or on the post-issuance progress of the securitised assets against sustainability objectives.

Second, law should also promote the 'shielding' of securitisation funds for the purposes of the operation. It is clear from the analysis of the documentation that the information available to investors about securitisation entities, i.e. SPVs, and about monitoring processes on the use of investors' funds, is very limited. Law could, however, intervene to require SPVs to only make expenses and undertake activities that contribute to sustainability goals, while not undertaking anything additional.

Third, law also has the potential to further regulate the 'gatekeepers' of these operations. The documentation currently fails to provide adequate mechanisms to control and certify 'sustainable' elements.

This is especially true with regard to external service providers. They should review the transactions based on their expertise, built across multiple transactions, and on the application of a consistent methodology. In doing so, they should attest to the reliability and integrity of sustainability commitments. Nevertheless, review statements are generally limited in their scope, focusing on the *planned* use of proceeds *before* the issuance. In doing so, they often overlook or neglect the assessment of the qualities of the portfolio, the progress after the operation and other complementary aspects like the mechanisms for tracking expenses. More importantly, second-party opinions and verifications merely assess the compliance with a series of external principles (such as the ICMA GBPs) or with criteria designed by the providers themselves, rather than the effective sustainability performance. This may result in paradoxical situations, where a transaction receives a positive review statement, but may in practice, not be in line with best practices, thereby potentially deceiving investors. The competition between providers also results in the application of very inconsistent methodologies, which may confuse investors and incentivise originators to engage service providers applying the most convenient (or lenient) criteria.

Fourth, law should also strengthen 'voice' mechanisms for investors to have a say over sustainability matters. It is not clear from the documentation whether it is standard practice for one or more investors to co-design the 'sustainability' characteristics of the documentation together with the originating entity ahead of the issuance. There is no evidence of the involvement of a Lead investor adequately representing the sustainability-related interests of all ABS holders across previous 'sustainable' deals. In particular, it is not common to entrust one investor with the duty to monitor the originator's compliance with the sustainability-related commitments under the documentation, such as internal 'sustainability frameworks'.

Furthermore, there is generally no Lead investor in charge of enforcing the securities on sustainability grounds. While trust deeds generally grant one or a few market actors the discretion to act on behalf of all Noteholders, the same agreements as well as prospectuses often restrict their enforcement avenues through a series of disclaimers and waivers of all liabilities. These clauses, along with the scarcity of provisions effectively binding originators to certain sustainability objectives (in the form of representations, warranties, covenants or events of default), are particularly concerning for the enforcement of 'sustainable' ABS. They may even leave investors with little to no remedies for the circumstances in which issuers make false and misleading claims on their sustainability performance or when they fail to substantiate their commitment (i.e. 'greenwashing' behaviours).

In the absence of clear representation mechanisms, the due diligence costs to assess the 'sustainable' characteristics of these operations are likely to be considerably high and may discourage investments. Excessively high costs may also disincentivise investors from thoroughly monitoring and enforcing the securities after the issuance. However, this may in turn incentivise issuers not to implement sustainability-related projects to the highest standard ('moral hazard'), which may *de facto* translate into 'greenwashing' behaviours. Besides, despite the rise in 'greenwashing' claims globally,[258] there is no record of previous litigation on sustainability grounds across 'sustainable securitisation' deals. This may also confirm the above-mentioned considerable enforcement challenges.

All these shortcomings of 'private hard law' call for a coordination of market practices, which should primarily target the five 'focal points' of sustainable securitisations analysed in this chapter. The following chapters will examine the strengths and weaknesses of the existing coordination initiatives, whilst proposing further regulatory coordination mechanisms resting on private soft law (leading to what will be referred to as 'standardisation') and on EU public regulation (leading to what will be referred to as 'harmonisation').

References

AFME, 'ESG Finance Report – Q2 2021'https://www.afme.eu/Portals/0/DispatchFeaturedImages/AFME%20Sustainable%20Finance%20Report%20-%20Q2%202021-%20v2.pdf

AFME, 'ESG Finance Report – European Sustainable Finance – Q3 2023' (2024) https://www.afme.eu/Portals/0/DispatchFeaturedImages/AFME%20Sustainable%20Finance%20Report%20Q3%202023.pdf

AFME, *European Green Securitisation – Regulatory State of Play* (2022) https://www.afme.eu/Portals/0/DispatchFeaturedImages/AFME_ESGSecuritisation_2022_07_final-2.pdf accessed 9 December 2025

AFME, 'Securitisation Report – European Structured Finance – Q2 2021' (2021) https://www.afme.eu/Portals/0/DispatchFeaturedImages/AFMEQ2th2021SecuritisationReport1.pdf

Agostini F, 'From "Green Bond Principles" to "Green Bond Clauses": Mitigating Greenwashing Through Contract Law' in M Andenas and M Heidemann (eds), *Quo Vadis Commercial Contract?* (Springer2023)

Agostini F, 'One Step Forward, One Step Back: "Sustainable Securitisations" Before and After the EU Green Bond Standard' in M La Torre and S Leo (eds), *Contemporary Issues in Sustainable Finance* (Palgrave Studies in Impact Finance2025)

Alloway T, 'World Bank Backs Crédit Agricole Loans' (FT, 6 June 2014) https://www.ft.com/content/4cbee3c0-ece7-11e3-8963-00144feabdc0

[258] On the rise of this trend and the various types of litigation linked to greenwashing see Joanna Setzer, Catherine Higham, Global trends in climate change litigation: 2023 snapshot (2023) Grantham Research Institute on Climate Change and the Environment and Centre for Climate Change Economics and Policy, 6, 39–41 https://www.lse.ac.uk/granthaminstitute/wp-content/uploads/2023/06/Global_trends_in_climate_change_litigation_2023_snapshot.pdf accessed 30 September 2023.

Alviar-Baquero F, Sweny JP, Nicolaides M, 'The Applications of Structured Finance Techniques to the Cleantech Industry' (2012) Latham & Watkins https://www.americanbar.org/content/dam/aba/publications/nr_newsletters/eemf/201108_eemf.authcheckdam.pdf

Asian Bonds Online, *Green Bond Factsheet 7 – Engaging the External Reviewer* (2022) https://asianbondsonline.adb.org/green-bonds/gbfactsheet7.html

Autorità Garante della Concorrenza e del Mercato, *Delibera AGCM 26173/2016*https://www.agcm.it/dotcmsDOC/allegati-news/PS10211_chiusura.pdf

Awrey D, 'Towards a Supply-Side Theory of Financial Innovation' (2013) 41 *Journal of Comparative Economics*401

Ayres I, Gertner R, 'Filling Gaps in Incomplete Contracts' (1989) 99 *Yale Law Journal*87

Baldwin R, Cave M, Lodge M, 'Self-Regulation, Meta-Regulation, and Regulatory Networks' in id (eds), *Understanding Regulation* (OUP2011)

Banque de France, 'Green Securitisation: A Lever for Financing the Transition' (2025) https://www.banque-france.fr/en/publications-and-statistics/publications/green-securitisation-lever-financing-transition

Bardin Hill Loan Advisors European Funding 2019-1 DAC, 'Listing Particulars' (2019) https://ise-prodnr-eu-west-1-data-integration.s3-eu-west-1.amazonaws.com/legacy/ListingParticulars_b4733386-9ceb-4590-a12e-2b613746ab3c.pdf

Bayfront Infrastructure, 'Bayfront Prices US$402.7 Million Infrastructure Asset-Backed Securities' (Clifford Capital 2022) https://api.cliffordcapital.sg/dev/file/download?id=1862065557811945474

Berg F, Kölbel J, Rigobon R, 'Aggregate Confusion: The Divergence of ESG Ratings' (2019) 26 *Review of Finance*1315

Black J, 'Decentering Regulation' (2001) 54 *Current Legal Problems*103

Black J, Rouch D, 'The Development of the Global Markets as Rule-Makers: Engagement and legitimacy' (2008) 2 *Law and Financial Markets Review*218

Bratton WW, 'Bond and Loan Covenants' (2016) 11 *Capital Markets Law Journal*461

Bureau Veritas Certification, *Assurance Statement* (2020) https://www.climatebonds.net/files/files/Ecoagro-Rizoma-pre-issuance-statement-PUBLIC.pdf https://ecoagro.agr.br/public/storage/Arquivo/1662064095Relat%C3%B3rio%20de%20Verifica%C3%A7%C3%A3o%20P%C3%B3s%20Emiss%C3%A3o%20-%20Bureau%20Veritas%20-%20Ingles%20-%20Ago-2022_compressed.pdf

Cadwalader, Wickersham & Taft LLP, 'ESG Ratings: A Call for Greater Transparency and Precision' (2022) https://www.cadwalader.com/uploads/cfmemos/5d29869eec33512f55a05b5d195e4bea.pdf

CICERO, 'Best Practices' (2022) https://www.spglobal.com/_assets/documents/ratings/green_best_practices_2022.pdf
Climate Bonds Initiative, 'Brazil Sustainable Securitisation – State of the Market Q3 2022' (2022) https://www.climatebonds.net/data-insights/publications/brazil-sustainable-securitisation-state-market-q3-2022
Climate Bonds Initiative, 'China Green Securitisation Report' (2020) https://www.climatebonds.net/data-insights/publications/china-green-Securitisation-state-market-2020-report accessed 9 December 2025
Climate Bonds Initiative, 'Climate Bonds Standard – Version 4.0'https://www.climatebonds.net/files/documents/Climate-Bonds_Climate-Bonds-Standard_V4-0_Apr-2024.pdf
Climate Bonds Initiative, 'Climate Bonds Standard – Version 4.3' (2025) CBI_Standard_V4.3_FINAL_2025-08-20-102147_gbqn.pdf
Climate Bonds Initiative, 'Deep-Dive: Green Synthetic Securitisation Deal' (2017) https://www.climatebonds.net/news-events/blog/deep-dive-green-synthetic-securitisation-deal-frees-usd-2bn-capital-green-investments-cr%C3%A9dit-agricole-mariner-investment-group-take-plunge
Climate Bonds Initiative, 'Green Securitisation: Unlocking Finance for Small-Scale Low Carbon Projects' (2018) https://www.climatebonds.net/data-insights/publications/green-securitisation-unlocking-finance-small-scale-low-carbon-projects
Climate Bonds Initiative, 'Securitisation as an Enabler of Green Asset Finance in India' (2020) https://www.climatebonds.net/data-insights/publications/securitisation-enabler-green-asset-finance-india
Climate Bonds Initiative, 'Sustainable Debt: Global State of the Market 2021' (2022) https://www.climatebonds.net/files/documents/publications/Sustainable-Debt-Global-State-of-the-Market-2021.pdf
Clifford Chance, 'Italy's First Greenwashing Case Between Corporates' (2022) https://www.cliffordchance.com/insights/resources/blogs/business-and-human-rights-insights/2022/01/italys-first-greenwashing-case-between-corporates.html
Coase RH, 'The Problem of Social Cost' (1960) 3 *Journal of Law and Economics* 1
Coffee JC, *Gatekeepers: The Professions and Corporate Governance* (2nd edn, Oxford Univerisity Press 2006)
Coglianese C, 'Environmental Soft Law as a Governance Strategy' (2020) 61 *Jurimetrics* 19
Coglianese C, Mendelson E, 'Meta-Regulation and Self-Regulation' in *The Oxford Handbook of Regulation* (OUP2010)
Cornut St-Pierre P, Gareau-Lindsay J, 'Green Securitisation, a Legal Structure Currently Unfit for Ecological Transition' (2026) 16 *Accounting, Economics, and Law: A Convivium* 2 211.

Credit Suisse, 'Credit Suisse Launches Inaugural Structured Notes with Green Use of Issuance Proceeds' (8 December 2020) https://www.credit-suisse.com/about-us-news/en/articles/media-releases/credit-suisse-launches-inaugural-structured-notes-with-green-use-202012.html

De Gioia Carabellese P, Della Giustina C, *The Law of Securitisations: From Crisis to Techno-Sustainability* (Routledge2023)

Delmas MA, Cuerel Burbano V, 'The Drivers of Greenwashing' (2011) 54 *California Management Review*64

Della Croce R, Kaminker C, Stewart F, 'The Role of Pension Funds in Financing Green Growth Initiatives' (OECD Publishing 2011) https://www.oecd.org/en/publications/the-role-of-pension-funds-in-financing-green-growth-initiatives_5kg58j1lwdjd-en.html

Deutsche Bank, 'ESG and Securitisation: Early Days' (2020) https://structuredfinance.org/wp-content/uploads/2020/02/Deutsche-Bank-Markets-Research-ESG-and-Securitisation-February-2020.pdf

Deutsche Bank, 'European Securitisation Market – Ready for a Comeback?' (2024) EU Monitor – Global Financial Markets https://www.dbresearch.com/PROD/RPS_EN-PROD/European_securitisation_market_%E2%80%93_ready_for_a_comeb/RPS_EN_DOC_VIEW.calias

EBA, 'Developing a Framework for Sustainable Securitisation' (2022) https://www.eba.europa.eu/sites/default/files/document_library/Publications/Reports/2022/1027593/EBA%20report%20on%20sustainable%20securitisation.pdf

EBA, 'Report on the Monitoring of TLAC-/MREL-Eligible Liabilities Instruments of EU Institutions - Update' (2023) https://www.eba.europa.eu/sites/default/files/document_library/Publications/Reports/2023/1061234/Report%20on%20merged%20AT1%20and%20MREL.pdf

Ehlers T, Gao DN, Packer F, 'A Taxonomy of Sustainable Finance Taxonomies' (2021) 118 BIS Papers https://www.bis.org/publ/bppdf/bispap118.pdf

Ehlers T, Packer F, 'Green Bond Finance and Certification' (2017) *BIS Quarterly Review* 89https://www.bis.org/publ/qtrpdf/r_qt1709h.htm

Enel, 'Enel's General Purpose SDG-Linked Bond – Context and Principles' (2019) https://www.enel.com/investors/investing/sustainable-finance/sustainability-linked-finance/sustainability-linked-bonds

Environmental Finance, 'Bond Awards 2022 – BNP Paribas' (2022) https://www.environmental-finance.com/content/awards/environmental-finances-bond-awards-2022/winners/lead-manager-of-the-year-social-bonds-financial-institution;-lead-manager-of-the-year-sustainability-linked-bonds;-and-loan-structurer/arranger/coordinator-of-the-year-bnp-paribas.html

ESMA, 'Sustainable Finance Roadmap 2022–24' (2022) https://www.esma.europa.eu/sites/default/files/library/esma30-379-1051_sustainable_finance_roadmap.pdf

ESMA, 'TRV Risk Report – Trends, Risks and Vulnerabilities' (2023) https://www.esma.europa.eu/sites/default/files/library/ESMA50-165-2438_trv_1-23_risk_monitor.pdf

EU Commission, 'Study on the Potential of Green Bond Finance for Resource-Efficient Investments' (2016) https://op.europa.eu/en/publication-detail/-/publication/5a95e1b0-dc82-11e6-ad7c-01aa75ed71a1/language-en

European Central Bank, 'Eurosystem Reply to the European Commission's Public Consultations' (2020) https://www.ecb.europa.eu/pub/pdf/other/ecb.eurosystemreplyeuropeancommissionpublicconsultations_20200608~cf01a984aa.en.pdf

European Investment Bank, 'CETELEM Enhanced ABS for Climate Action' (2025) https://www.eib.org/en/projects/all/20230971 accessed 9 December 2025

European Investment Bank, 'Environmental and Social Data Sheet' (2025) https://www.eib.org/attachments/registers/241965713.pdf

European Investment Bank, 'DNB Enhanced Support for Climate Action' (2024) https://www.eib.org/en/projects/pipelines/all/20210715

European Investment Bank, 'INBANK Climate Action for Private Individuals' (2024) https://www.eib.org/de/projects/pipelines/all/20240110

Fannie Mae, 'Contractual Representations and Warranties – Overview' (2023) https://selling-guide.fanniemae.com/Selling-Guide/Doing-Business-with-Fannie-Mae/Subpart-A2-Lender-Contract/Chapter-A2-2-Contractual-Representations-and-Warranties/1645977541

Fannie Mae, 'Multifamily Green Bond Framework' (2024) https://capitalmarkets.fanniemae.com/media/20616/display

FCA, 'Enhancing Climate-Related Disclosures' (2021) CP21/18 https://www.fca.org.uk/publication/consultation/cp21-18.pdf

Fitch Ratings, 'Covered Bonds Still a Limited Share of Sustainable Bond Market' (9 December 2021) https://www.fitchratings.com/research/structured-finance/covered-bonds/covered-bonds-still-limited-share-of-sustainable-bond-market-09-12-2021

Fitch Ratings, 'Gemgarto 2021-1 PLC' (2021) https://www.fitchratings.com/research/structured-finance/gemgarto-2021-1-plc-24-02-2021 accessed 9 December 2025

Fitch Ratings, 'More US CLO Managers Considering ESG in investment decisions' (May 2022) https://www.fitchratings.com/research/structured-finance/more-us-clo-managers-considering-esg-in-investment-decisions-27-05-2022

Foley S, 'A Synthetic Path to Green Lending that may just work' (FT, 6 March 2017) https://www.ft.com/content/7042b704-fad1-11e6-bd4e-68d53499ed71

Ford C, *Innovation and the State: Finance, Regulation, and Justice* (CUP2017)

GACI First Investment Company, 'Public Investment Fund Euro Medium Term Note Programme Offering Circular' (2022) https://docs.londonstockexcha

nge.com/sites/default/files/documents/PIF%20Euro%20Medium%20Term%20Note%20Programme%20Offering%20Circular.pdf
Gemgarto 2021-PLC, *Prospectus* (2021) https://ise-prodnr-eu-west-1-data-integration.s3-eu-west-1.amazonaws.com/legacy/Prospectus+-+Standalone_1fe8519a-a55e-4826-8e3d-74f87136927e.PDF
Gibson SD, 'The Case for the Expanded Role of Trustees' (2004) 121 *Banking Law Journal* 387
GlobalCapital, 'Private Securitisation Deal of the Year – Project Boqueron' (2 June 2021) https://www.globalcapital.com/article/28wq98676i4ve6c5xrsxc/polls-and-awards/private-securitization-deal-of-the-year-project-boqueron
Goel R, Gautam D, Natalucci F, 'Sustainable Finance in Emerging Markets: Evolution, Challenges, and Policy Priorities' (2022) IMF Working Papers WP/22/182 https://www.imf.org/-/media/Files/Publications/WP/2022/English/wpiea2022182-print-pdf.ashx
Goetz CJ, Scott RE, 'Principles of Relational Contracts' (1981) 67 *Virginia Law Review* 1089
Gosling T, MacNeil I, 'Can Investors Save the Planet? NZAMI and Fiduciary duty' (2023) 18 *Capital Markets Law Journal* 172
Hannon Armstrong Sustainable Infrastructure Inc, 'Hannon Armstrong Sustainable Infrastructure Inc, 'Hannon Armstrong (HASI) Completes $100,000,000 Asset-Backed 'securitisation of 2.79% Sustainable Yield Bonds' (PR Newswire, 23 December 2013) https://www.prnewswire.com/news-releases/hannon-armstrong-hasi-completes-100000000-asset-backed-'securitisation-of-279-sustainable-yield-bonds-236999811.html
Hart O, Moore J, 'Contracts as Reference Points' (2008) 123 *Quarterly Journal of Economics* 1
Hay J, 'Carbon Permits – Securitisation Finds a Hidden Seam' (Global Capital, 27 September 2021) https://www.globalcapital.com/article/29425pg7njwqa46q9a9kw/securitization/abs/carbon-permits-securitization-finds-a-hidden-seam
Hitachi Capital, 'Green Financing Framework' (2021) https://www.mitsubishihccapital.co.uk/media/pvkl4srm/green-financing-framework-2021.pdf
Hornuf L, Reps M, Schäferling S, 'Covenants in European Investment-Grade Corporate Bonds' (2015) 10 *Capital Markets Law Journal* 345
ICMA, 'Analysis of Amendments to the EuGB Regulation' (2022) https://www.icmagroup.org/assets/ICMA-update-to-its-analysis-of-the-EuGB-Regulation-05012022.pdf
ICMA, 'Green Bond Principles' (2025) https://www.icmagroup.org/sustainable-finance/the-principles-guidelines-and-handbooks/green-bond-principles-gbp/
ICMA, 'Guidelines for External Reviews' (2020) https://www.icmagroup.org/assets/documents/Regulatory/Green-Bonds/June-2020/External-Review-GuidelinesJune-2020-090620.pdf

ICMA, 'Sustainability Bond Guidelines' (2021) https://www.icmagroup.org/assets/documents/Sustainable-finance/2021-updates/Sustainability-Bond-Guidelines-June-2021-140621.pdf
ICMA, 'Sustainability-Linked Bond Principles' (2024) https://www.icmagroup.org/sustainable-finance/the-principles-guidelines-and-handbooks/sustainability-linked-bond-principles-slbp/
ICMA, 'Illustrative Examples for the Selection of Key Performance Indicators (KPIs) for Sustainability-Linked Bond Issuers, Underwriters and Investors' (2023) https://www.icmagroup.org/assets/documents/Sustainable-finance/2022-updates/Registry-SLB-KPIs_Final_2022-06-24-280622.xlsx
Ionova M, 'Obvion Plans Europe's First Green ABS Bond' (Reuters, 24 May 2016) www.reuters.com/article/green-abs/obvion-plans-europes-first-green-abs-bond-idUSL5N18K4NJ
ISS ESG, 'Berlin Hyp Sustainability-Linked Bond Framework' (2021) https://www.icmagroup.org/assets/Uploads/berlin-hyp-sustainability-linked-bond-framework-second-party-opinion-2021.pdf
ISS Oekom, 'Second-Party Opinion' (2019) https://spowidget.iss-corporate.com/icsspoapp/file/documents/spo/spo-berlinhyp-29apr2019.pdf
James K, Parker I, 'New Growth in the Green Securitisation Market' (2019) Simmons & Simmons https://www.simmons-simmons.com/-/media/files/articles/2019/capital%20markets/new%20growth%20in%20the%20green%20securitisation%20market%20january%202019.pdf
Jensen M, Meckling W, 'Theory of the Firm' (1976) 72 *Journal of Financial Economics* 305
KfW, 'Obvion NV Issues EUR 500m Green STORM RMBS' (2016) https://www.kfw.de/nachhaltigkeit/PDF/Nachhaltiges-Investment/Press-Release-Green-Storm.pdf
Latino C, Pelizzon L, Riedel M, 'How to Green the European Auto ABS Market?' (2025) 31 *European Financial Management*627
Legal Community, 'GOP e Nctm nel Finanziamento ESG-Linked' (2022) https://legalcommunity.it/gop-e-nctm-nel-finanziamento-esg-linked-da-1-miliardo-per-marcegaglia/
Lester A, 'Steel Firm Marcegaglia Signs Sustainability-Linked Securitisation' (2022) *Environmental Finance* https://www.environmental-finance.com/content/news/steel-firm-marcegaglia-signs-sustainability-linked-securitisation.html
LMA, 'An Introduction to the Sustainability Coordinator Role' (2022) https://www.lma.eu.com/download?p=5942446308-6796
LMA, LSTA, APLMA, 'Sustainability-Linked Loan Principles' (2025) https://www.lsta.org/content/sustainability-linked-loan-principles-sllp/
LMA, 'Guidance on Sustainability-Linked Loan Principles' (2025) https://www.lma.eu.com/application/files/9416/4873/5604/Sustainability-Linked_Loan_Principles_Guidance_31_March_2022.pdf

MacNeil IR, 'Contracts: Adjustment of Long-Term Economic Relations under Classical, Neoclassical, and Relational Contract Law' (1978) 72 *Northwestern University Law Review* 854

Maddington A, 'Kensington Completes Inaugural £472 Million Social Bond STS Securitisation' (Kensington Mortgages 2021) https://www.kensingtonmortgages.co.uk/blog/article/kensington-completes-inaugural-472-million-social-bond-sts-securitisation

Mariner Investment Group, 'Landmark $3 Billion Socially Responsible Synthetic Securitisation' (PR Newswire, 6 March 2017) https://www.prnewswire.com/news-releases/landmark-3-billion-socially-responsible-synthetic-Securitisation-300418309.html

McBarnet D, Voiculescu A, Campbell T (eds), *The New Corporate Accountability: Corporate Social Responsibility and the Law* (CUP2007)

Miller RT, 'The RMBS Put-Back Litigations And The Efficient Allocation Of Endogenous Risk Over Time' (2014) 34 *Review of Banking and Financial Law* 225

Murphy SD, 'Taking Multinational Corporate Codes of Conduct to the Next Level' (2005) 43 *Columbia Journal of Transnational Law* 389

NatWest, 'GSS Bond Framework' (2020) https://investors.natwestgroup.com/~/media/Files/R/RBS-IR-V2/green-social-and-sustainability-bonds/gss-bond-framework-oct-2020.pdf

NatWest Group, 'Administration Agreement and Cash Management Agreement dated 29 September 2010' (2010) https://investors.natwestgroup.com/~/media/Files/R/RBS-IR-V2/securitisation/reseidential-mortgages/pdf/7-arran-residential-mortgages-2010-1-plc/principal-transaction-documents/2010-1-cash-management-and-administration-agreements.pdf

NatWest Group, 'Arran Residential Mortgages Funding 2010-1 PLC – Account Bank Agreement' (2010) https://investors.natwestgroup.com/~/media/Files/R/RBS-IR-V2/securitisation/reseidential-mortgages/pdf/7-arran-residential-mortgages-2010-1-plc/principal-transaction-documents/014-account-bank-agreement-2010-a1.pdf

NatWest Group, 'Arran Residential Mortgages Funding 2010-1 PLC – Trust Deed' (2010) https://investors.natwestgroup.com/~/media/Files/R/RBS-IR-V2/securitisation/reseidential-mortgages/pdf/7-arran-residential-mortgages-2010-1-plc/principal-transaction-documents/015--trust-deed-2010-a1.pdf

Nedopil C, Dordi T, Weber O, 'The Nature of Global Green Finance Standards – Evolution, Differences, and Three Models' (2021) 13 *Sustainability* 3723 https://www.mdpi.com/2071-1050/13/7/3723

Newmarket Capital, 'Sustainable Investing – Solutions, Impact, Innovation'https://www.newmarketcapital.com/wp-content/uploads/2021/11/ESG-Analysis_v9_2021-11-16_ARM_spreads.pdf

NGFS, 'Enhancing Market Transparency in Green and Transition Finance' (2022) https://www.ngfs.net/en/publications-and-statistics/publications/enhancing-market-transparency-green-and-transition-finance

OECD, *Mobilising Bond Markets for a Low-Carbon Transition* (2017) https://doi.org/10.1787/9789264272323-en

Petit C, Schlosser P, 'Rationale, Potential and Pitfalls of Green Securitisation' (2020) RSCAS Working Paper 35 https://cadmus.eui.eu/bitstream/handle/1814/67018/RSCAS%202020_35.pdf

Pistor K, *The Code of Capital: How the Law Creates Wealth and Inequality* (Princeton University Press, 2019)

Pistor K, 'A Legal Theory of Finance' (2013) 41 *Journal of Comparative Economics* 315

Polinsky AM, Shavell S, 'Economic Analysis of Law' in Vernengo M, Perez Caldentey E, Ghosh J, *The New Palgrave Dictionary of Economics* (Palgrave MacMillan 2008)

Profamily SPV SRL, 'Prospectus: Asset-Backed Fixed Rate Notes due December 2040' (2021) https://www.securitisation-services.com/it/reports/scarica_prospetto.php?f=files/Profamily%20SPV/Profamily%20SPV%20Prospectus.pdf

Ramos Muñoz D, 'In Praise of Small Things: Securitisation and Governance Structure' (2010) 5 *Capital Markets Law Journal*363

Ramos Muñoz D, Ingram S, *The Law of Transnational Securitisation* (OUP2010)

Revak H, 'Corporate Codes of Conduct: Binding Contract or Ideal Publicity?' (2011) 63 *Hastings Law Journal* 1645

S&P Global Ratings, 'North Westerly CLO VI B.V'. (2019) https://www.spglobal.com/_assets/documents/ratings/research/2342405.pdf accessed 30 September 2023

S&P Ratings, 'Second Party Opinion: Sparebank 1 Ringerike Hadeland Green Bond Framework' (2025) https://www.spglobal.com/ratings/en/regulatory/delegate/getPDF?articleId=3496992&type=COMMENTS&defaultFormat=PDF

Scatigna M, Xia Det al, 'Achievements and Challenges in ESG Markets' (2021) *BIS Quarterly Review* 94 https://www.bis.org/publ/qtrpdf/r_qt2112f.pdf

Schwarcz SL, 'The Conundrum of Covered Bonds' (2011) 66 *Business Lawyer*561

Schwartz A, Sepe S, 'Economic Challenges for the Law of Contract' (2020) 38 *Yale Journal of Regulation* 2

Setzer J, Higham C, *Global Trends in Climate Change Litigation: 2023 Snapshot* (2023) https://www.lse.ac.uk/granthaminstitute/wp-content/uploads/2023/06/Global_trends_in_climate_change_litigation_2023_snapshot.pdf

Shirley MM, Ménard C, *Handbook of New Institutional Economics* (Springer2008)

Smoleńska A, 'The Constitutive Role of Law in Sustainable Finance' (2025) *Industrial Law Journal (advance publication)*

Snam SA, 'Euro-Commercial Paper Programme – Information Memorandum Update 2020' (2020) https://www.snam.it/content/dam/snam/pages-attachments/it/investor-relations/documents/euro-commercial-paper/Snam_ECP_-2020_Update-Information_Memorandum.pdf

Société Générale SFH, 'First Supplement to the Base Prospectus Dated 3 June 2020' (2020) https://www.societegenerale.com/sites/default/files/documents/Notations%20Financi%C3%A8res/societe-generale-sfh-emtn-base-prospectus-dated-3-june-2020.pdf

Société Générale, 'Société Générale and Mariner Investment Group Complete a $3.4bn Impact Investment Risk Transfer Transaction' (2019) https://www.societegenerale.com/en/news/newsroom/societe-generale-and-mariner-investment-group-complete-34b-impact-investment-risk

Solana J, 'The Roles of Law in Finance' (2022) Working Paper, on file with the author

Standard Ethics, 'Eco-SISMA Bonus Securitisation Programme' (2022) https://standardethicsrating.eu/component/finances/?view=items&category_id=59

Standard Ethics, 'Security Standard Ethics Rating – Overview' (2020) https://standardethicsrating.eu/SE_Bond_Rating_Overview_2020-1.pdf

Structured Credit Investor, 'Specialist-Squared Solution?' (2022) https://www.structuredcreditinvestor.com/Article.asp?SID=79770

Structured Finance Association, 'ESG Market Sentiment Survey' (2023) https://structuredfinance.org/wp-content/uploads/2023/02/SFA-ESG-2022.pdf

Structured Finance Association, 'RMBS 3.0' (2017) https://structuredfinance.org/wp-content/uploads/2019/05/RMBS-3.0-Sixth-Edition-Final-1109.pdf

Sustainalytics, 'Green Storm 2016 Obvion: Framework Overview and Second Opinion' (2016) https://www.sustainalytics.com/corporate-solutions/sustainable-finance-and-lending/published-projects/project/obvion-n-v/obvion-n.v.-green-storm-2016-cbi-pre-issuance-verification-letter-(2015)/obvion-n.v.-green-storm-2016-cbi-pre-issuance-verification-letter-(2015)

Sustainalytics, 'River Green Finance Green Securitised Bond Framework' (2019) https://mstar-sustops-cdn-mainwebsite-s3.s3.amazonaws.com/docs/default-source/spos/river-green-finance-green-bond-second-party-opinion.pdf

TerraChoice Group Inc, 'The Sins of Greenwashing' (2009) https://www.twosides.info/wp-content/uploads/2018/05/Terrachoice_The_Sins_of_Greenwashing_-_Home_and_Family_Edition_2010.pdf

Tinsley I, Conduit T, 'Implications of Sustainability and ESG for the Securitisation Market' (Allen & Overy 2021) https://www.jdsupra.com/post/fileServer.aspx?fName=698231e7-4302-4500-a87e-f44307e60e3e.pdf

UNPRI, 'ESG Engagement for Fixed Income Investors' (2018) https://public.unpri.org/download?ac=4449

UNPRI, 'ESG Incorporation in Securitised Products: The Challenges Ahead' (2022) https://gsfo.org/resource/esg incorporation securitised products cha llenges-ahead

UNPRI, 'PRI Awards 2019 Case Study: Room2Run' (2019) https://www.unpri.org/showcasing-leadership/pri-awards-2019-case-study-room2run/4848.article

Vigeo Eiris, 'Second-Party Opinion on the Green ABCP Programs Annex to Crédit Agricole's Green Bond Framework' (2020) https://www.ca-cib.com/sites/default/files/2020-10/20200803_VigeoEiris_SPO_Credit-Agricole.PDF

Vigeo Eiris, 'Second-Party Opinion on the Green ABCP Programs – Annex to Crédit Agricole's Green Bond Framework' (2020) https://www.ca-cib.com/sites/default/files/2020-10/20200803_VigeoEiris_SPO_Credit-Agricole.PDF

Wall Chambers, 'Collateralised Loan Obligations (CLOs): Overview' *Practical Law* https://uk.practicallaw.thomsonreuters.com/9-385-8080

Weidemaier MC, Curtis Q, 'Green Bonds, Empty Promises' (2023) *Public Law and Legal Theory Research Paper Series* No 2023-14

Whiley A, 'First Round of UN Green Climate Fund Projects Kick-Starts Green Securitisation Markets in Latin America' (10 November 2015) https://www.climatebonds.net/2015/11/first-round-un-green-climate-fund-projects-kick-starts-green-securitisation-markets-latin

Whiley A, 'Sunny Side Up! FlexiGroup Issues First Australian Green ABS – Proceeds for solar & 5bps pricing benefit for green label! And.... It's Climate Bonds Certified!' (Climate Bonds Initiative, 21 April 2016) https://www.climatebonds.net/news-events/blog/sunny-side-flexigroup-issues-first-australian-green-abs-proceeds-solar-5bps-pricing-benefit-green-label-climate-bonds-certified

Wood P, *International Loans, Bonds, Guarantees, Legal Opinions* vol 5 (3rd edn, Sweet & Maxwell2019)

Yang R, 'What Do We Learn from Ratings about Corporate Social Responsibility?' (2021) Columbia Business School Research Paper No 18–37 https://ssrn.com/abstract=3165783

Zumbansen P, 'The Law of Society: Governance through Contract' (2007) 14 *Indiana Journal of Global Legal Studies* 91

CHAPTER 5

Standardisation of 'Sustainable Securitisations': The Role of 'Private Soft Law'

5.1 Introduction

Chap. 4 has depicted private market actors involved in securitisations as rule-makers, who have begun designing the 'sustainable' component of these transactions on a deal-by-deal basis. Their broad discretion has the potential to promote flexible design and innovation, yet it also entails several shortcomings. A legal response therefore appears imperative, and should arguably consist of initiatives aimed at coordinating 'sustainable' securitisation deals. As previously noted, it follows from the hybridity of financial markets that market actors and state authorities concurrently contribute to streamlining the conduct of market participants and transactions.[1]

In the present chapter, the focus will be on the coordination process and on the related initiatives at market level ('private law'[2]), referred to here as *standardisation*.[3] Traditionally, market actors have tended to adopt

[1] See Chap. 2, Sect. 2.4.

[2] Ibid.; Chap. 2, Sect. 2.2.

[3] In line with academic scholarship, see, among others, Eddy Wymeersch, 'Standardisation by Law and Markets Especially in Financial Services' (2008) 02 Financial Law institute Working Paper 2, 1, 4–5 http://papers.ssrn.com/sol3/papers.cfm?abstract_id=1086987. Kahan and Klausner similarly understand 'standard contract terms' as commonly used contract terms, Marcel Kahan and Michael Klausner, 'Standardisation and Innovation in Corporate Contracting (Or 'The Economics of Boilerplate')' (1997) 83 *Virginia Law Review* 713, 718.

F. Agostini, *The Law of Sustainable Securitisations*, EBI Studies in Banking and Capital Markets Law,
https://doi.org/10.1007/978-3-032-25046-9_5

consistent practices for the design and management of their securitisation operations by replicating non-binding legal tools ('soft law'[4]), such as principles and standard clauses. In line with a general trend across financial markets, the entities responsible for such legal tools have predominantly been financial industry associations,[5] acting as 'intermediaries' between the state and market participants. These associations have also recently taken the lead in a series of initiatives to establish best practices for 'green' or 'sustainable' securities and investments, laying the groundwork for the standardisation of sustainable securitisations.

Against this background, the present chapter will examine how 'private soft law' tools have shaped the 'sustainable' profile of securitisation deals so far, and explore potential additions. It will first sketch out the main dynamics behind standardisation in the financial context, particularly with regard to securitisations, highlighting the responsible actors and their added value. Subsequently, the chapter will dissect the effect of standard legal tools on the most strategic aspects ('focal points') of securitisations identified in Chap. 3. It will also provide normative recommendations about potential adjustments to existing legal tools, or new initiatives that may further standardise these deals. The concluding subsection will highlight the shortcomings that render standard legal tools useful, but insufficient, to regulate these operations. Thus, the complementary role of legislation ('public hard law'), and specifically EU law, will be the subject of Chap. 6.

5.2 Unpacking the Role of 'Standardisation'

According to *Wymeersch*, 'standardisation' is a recurrent trend and a crucial goal of the law supporting financial markets. It is the process (and the outcome of the process) of streamlining the conduct of market participants and the characteristics of transactions around clearly predefined reference points. It can result from a wide array of legal tools.[6] In the following chapter, I will specifically focus on what Wymeersch describes as 'standardisation by the market' as the product of market participants

[4] See Chap. 2, Sect. 2.4.

[5] Julia Black, David Rouch, 'The Development of the Global Markets as Rule-Makers: Engagement and Legitimacy' (2008) 2 *Law and Financial Markets Review* 218, 223.

[6] Wymeersch (n. 3) 1, 4–7. For the standardisation mechanisms in the present work, see below Sect. 5.2.3.

(which I have previously defined as 'private law'[7]) and not of state and international authorities.[8] However, initiatives to *coordinate* the conduct of market actors and transactions can also result in state-designed rules ('public law'), as also discussed in Chap. 6.[9] The following sections will contextualise how standardisation has traditionally affected finance and securitisations. They will also introduce the role and added value of standardisation for transactions that promote environmental, social, and governance goals: notably, they will identify the most promising legal tools as well as providing preliminary remarks about their application and main benefits.

5.2.1 The Main Actors

Academic scholarship has extensively analysed the standardisation trend across financial markets.[10] The main actors behind this process include the associations representing sectors or segments of the market (such as loans, bonds, and derivatives). By way of example, one could mention the Loan Market Association (hereinafterLMA), the International Capital Market Association (hereinafterICMA) and the International Swaps and Derivatives Association (hereinafterISDA)[11] bringing together market

[7] See Chap. 2, Sect. 2.3.

[8] However, it must be noted that *Wymeersch* uses the term 'standardisation' to capture both market-designed and state-designed initiatives. See For example, the reference to national and EU legislation, as well as to International Organisation of Securities Commissions (IOSCO) standards, Wymeersch (n. 3) 2, 5. I will use the term 'harmonisation', instead of 'standardisation', to capture state-designed initiatives, see Chap. 6, s. 6.2.

[9] See Chap. 6, Sect. 5.2.

[10] Bradley, 'Private international law-making for the Financial Markets' (2005) Fordham International law Journal 127, 165–170; Joanne P Braithwaite, 'Standard Form Contracts as Transnational Law: Evidence from the Derivatives Markets' (2012) 75 *Modern Law Review* 5 779; Joanna Benjamin, David Rouch, 'The International Financial Markets as a Source of Global Law: The Privatisation of Rule-Making' (2008) 2 *Law and Financial Markets Review* 78.

[11] LMA represents the loan market in Europe, Middle East and Africa, LMA. 'About us' https://www.lma.eu.com/about-us accessed 9 December 2025; ICMA represents the international debt securities markets (primary, secondary, repo and collateral), see ICMA, About ICMA, https://www.icmagroup.org/About-ICMA/ accessed 9 December 2025; ISDA represents the derivatives market, see ISDA 'About ISDA' https://www.isda.org/about-isda/ accessed 9 December 2025.

participants with the aim to promote constructive discussions among members and to facilitate the circulation of sound market practices.[12]

Their most remarkable contribution has been the draft of model clauses and agreements, which circulated widely and had an almost universal application across less heavily regulated sectors, such as over-the-counter (OTC) derivatives.[13] Even across other segments of the financial markets, the widespread popularity of standard clauses led to a gradual convergence of the contractual terms for financial transactions.[14] In parallel, financial industry associations have designed a number of other tools to standardise transactions, ranging from guidelines[15] to benchmark interest rates.[16]

As a result, several commentators have acknowledged the rule-making function of financial industry associations and the *legal* value of their work ('lex mercatoria').[17] This trend reinforces the previously discussed 'hybridity' of the law governing financial markets, which combines legislation designed at 'public' level with rules designed at 'private' level by market participants.[18]

Abbott et al.'s model can persuasively illustrate the role of industry associations. They identify the existence of several legal 'intermediaries' in various areas of law, such as professional associations, transnational networks or non-governmental organisations (NGOs). Standing between rule-makers at state or international level and the targets of those rules,[19]

[12]For a list of other relevant financial industry associations, see For example, Benjamin, Rouch (n. 10) 79.

[13]For example, ISDA has estimated in 2010 that 90% of credit derivatives transactions across the world were regulated by its Master Agreement, Braithwaite (n. 10) 784.

[14]Dan Awrey, 'The Limits of Private Ordering Within Modern Financial Markets' (2013) *Review of Banking and Financial Law*, Bradley (n. 10) 165–166.

[15]See LMA's Guidelines, LMA, 'Documents & Guidelines' https://www.lma.eu.com/documents-guidelines/documents accessed 9 December 2025.

[16]For a case study of London Interbank Offered Rate (Libor), developed by the British Bankers Association (BBA), bringing together banks, see Awrey, 'Limits of Private Ordering' (n. 114) 31 ff.

[17]Braithwaite (n. 10); Bradley (n. 10) 165–167.

[18]see Chap. 2, Sect. 2.2. Katharina Pistor, 'A Legal Theory of Finance', 41 *Journal of Comparative Economics* 315, 321, 324.

[19]The authors name these actors 'regulators', 'regulatory intermediaries' and 'targets' or 'rule takers'. The terms are adjusted for the purposes of the present analysis to align with the terminology used in the rest of the work. Kenneth W Abbott, David Levi-Faur, Duncan Snidal, 'Theorizing Regulatory Intermediaries: The Rit Model' (2017) 670 *Annals of the American Academy of Political and Social Science* 14.

these actors also contribute to the design, monitoring and enforcement of the law.[20] According to the model, legal 'intermediaries' may serve a law-making function even when their engagement does not derive from an official delegation by the state.[21]

In the financial context, industry associations equally serve an intermediating function between regulators and market participants.[22] On the one hand, the members of professional associations, which draft the various standardisation tools, are themselves active in the respective sector.[23] On the other hand, their work has also had an effect on the law-making process at state level, creating a 'feedback' effect of market practices on legislation. [24] For instance, financial industry associations have engaged in frequent dialogue with regulators, thus influencing several policy developments. As an example, ISDA has produced a Model Netting Act to facilitate the enforcement of close-out netting clauses, which inspired ad hoc legislation in over 40 jurisdictions.[25]

As a result of these mutual feedback effects, the contributions by financial industry associations have gradually shaped market participants' behaviour. Admittedly, the documentation should primarily support the official members of the respective associations, who are also more likely to apply it. However, once standard agreements widely circulate, outsiders are also encouraged to draw from and replicate it.[26]

[20] Ibid., 29; Luc Brès, Sébastien Mena, Marie Laure Salles-Djelic, 'Exploring the Formal and Informal Roles of Regulatory Intermediaries in Transnational Multistakeholder Regulation' (2019) 13 *Regulation and Governance* 127, 128.

[21] Abbott, Levi-Faur, Snidal, ibid., 21–23.

[22] While they do not explicitly mention financial industry associations, Abbott et al recognise that the model is in principle applicable to finance, ibid., 17.

[23] For example, ICMA members include 'firms who actively deal in securities, namely investment banks and issuers, securities dealers and brokers, regional and commercial banks; private banks, and asset managers' along with trading venues and public entities, ICMA, 'Membership' https://www.icmagroup.org/membership/ accessed 9 December 2025; LMA gathers financial institutions and companies active in the loan market, LMA (n. 12); LMA, 'List of Principal Delegates' https://www.icmagroup.org/membership/List-of-principal-delegates-2/ accessed 9 December 2025; ISDA gathers market participants active in the derivatives' market, see ISDA (n. 11).

[24] Abbott, Levi-Faur, Snidal (n. 19) 19, 23, 26–27.

[25] Awrey, 'Limits of Private Ordering' (n. 14) 23–28.

[26] Black, Rouch (n. 5) 218, 223.

5.2.2 *The Standardisation of 'Traditional' Securitisations*

Standardisation dynamics also permeate the securitisation market. However, unlike in other financial segments,[27] there is neither a prevalent industry association nor a unique piece of model documentation tailored specifically to these transactions.

Chap. 3 has already highlighted the multi-layered nature of the documentation for securitisation, reflecting the complexity of these operations.[28] In particular, it is possible to distinguish broadly between 'transactional' agreements, setting out the key legal foundations of the operation (such as the assignment of receivables, the creation of SPVs, and the issuance of ABSs), and other 'intermediation' agreements. The latter set the conditions for the involvement of further intermediaries, such as servicers and underwriters.[29] Depending on the specific characteristics of the transaction, there may also be additional documents, including swap agreements, legal opinions, and documents for the creation of security interests.[30]

It is important to note that, despite the lack of ad hoc documentation, empirical evidence and scholarly analyses point to the convergence of market practices for the design and management of these operations. For instance, Fitch Ratings identified several similarities in some clauses of securitisation documentation.[31] The Structured Finance Industry Group

[27] See above Sect. 5.2.1.

[28] See Chap. 3, Sects. 3.2.–3.3.

[29] In line with the differentiation also outlined in Mohammed S Lahlou, Joseph Tanega, 'Islamic Securitisation: Part 2 – A Proposal for International Standards, Legal Guidelines and.Structures' (2007) 22 *Journal of International Banking Law and Regulation* 7 359, 361, 363.

[30] Wood 'Securitisation Structures', *Project Finance, Securitisations, Subordinated Debt*, Vol. 7, para 7-024, 7-033. For some examples of additional documentation for RMBSs and CMBSs, See LexisNexis PSL, 'Key documents required for a residential mortgage backed securities transaction' https://www.lexisnexis.co.uk/legal/guidance/documents-required-for-a-residential-mortgage-backed-securities-transaction accessed 9 December 2025; see also LexisNexis, 'Precedent signing and closing memorandum for a commercial mortgage – backed securities transaction' https://www.lexisnexis.co.uk/legal/precedents/precedent-signing-closing-memorandum-for-a-commercial-mortgage-backed-securities-transaction, accessed 9 December 2025.

[31] Notably across representation and warranties, see Fitch Ratings, 'Representations, Warranties and Enforcement Mechanisms' in Global Structured Finance Transactions' (2016) https://www.fitchratings.com/research/corporate-finance/representations-warranties-enforcement-mechanisms-in-global-structured-finance-transactions-31-05-2016 accessed 10 December 2025.

For the definition of representations and warranties, see Chap. 3, Sect. 3 and below, Sect. 5.4.1.2.

(hereinafter SFIG), gathering market actors in the sector, has reached a similar conclusion after reviewing the documentation for private transactions backed by residential mortgages (RMBS), whilst also acknowledging persisting differences.[32] Scholars have also noted convergence in the clauses for CLO transactions.[33] These commonalities suggest that, in the absence of ad hoc standard documentation, market actors tend to borrow model agreements from other areas and to adapt them where needed.[34]

The widespread use of standard documentation led Frankel to argue that the rule-making process for securitisation rules is heavily decentralised to market actors.[35]

5.2.3 *'Strategic' Legal Tools for the Standardisation of 'Sustainable' Deals*

In principle, the non-binding ('soft') legal tools driving the standardisation trend could be manifold.[36] For the purposes of the present analysis, I have identified three types of tools that have so far proven relevant, and which could be relevant, to coordinate sustainability-related aspects of securitisations.

First, intermediaries have introduced high-level, non-binding principles,[37] which I will define as 'guidelines' (following the terminology that financial industry associations have at times used[38]). These list general best

[32] i.e. ABS backed by residential mortgages, see Chap. 2. Structured Finance Industry Group (hereinafterSFIG), 'RMBS 3.0' (6th edn 2017) https://structuredfinance.org/wp-content/uploads/2019/05/RMBS-3.0-Sixth-Edition-Final-1109.pdf accessed 10 December 2025.

[33] Zahn Bozanic, Maria Loumioti, Florin P Vasvari, 'Corporate Loan Securitisation and the Standardisation of Financial Covenants' (2018) 56 *Journal of Accounting Research* 45, 65–68. For the definition of CLOs, see Chap. 2, Sect. 3.2.–3.3.1.

[34] Ibid. Tamar Frankel, 'Cross-Border Securitisation: Without Law, But Not Lawless' (1997) *Duke Journal Comp. & Int'l L.*, 8(iv) 255, 275–278.

[35] Ibid.

[36] For some examples, see Wymeersch (n. 3) 4–6.

[37] A contested distinction is the one between 'principles' (ie high-level, vague prescriptions) and 'rules' (ie specific prescriptions), see John Braithwaite, 'Rules and Principles: A Theory of Legal Certainty' (2005) 1–2. http://dx.doi.org/10.2139/ssrn.329400 accessed 10 December 2025.

[38] E.g. LMA categorises its soft law tools as 'documents and guidelines', see LMA, 'Documents and Guidelines' (n. 15); ICMA refers to 'Principles, Guidelines and Handbooks' in the context of sustainable finance, ICMA, 'Principles, Guidelines and Handbooks' https://www.icmagroup.org/sustainable-finance/the-principles-guidelines-and-handbooks/ accessed 10 December 2025.

practices for the design and management of financial transactions and instruments, including securitisations. For instance, the ICMA's Green Bond Principles also provide a definition of 'Secured Green Bonds'.[39]

Second, intermediaries have designed 'metrics' to quantify the performance of companies and assets on various aspects, including sustainability matters.[40] For instance, ICMA has provided a series of 'impact reporting metrics', i.e. indicators to measure the impact of a given project on environmental aspects like water management.[41]

Third, the most remarkable contribution by financial industry associations consists of several examples of 'standard documentation', which range from standard agreements (such as 'master agreements' in the context of derivatives[42]) or standalone standard clauses[43] to other types of templates for the communication of certain information to counterparties or the market. A relevant example is ICMA's Primary Market Handbook,[44] which includes standard clauses as well as templates such as model auditor letters.[45] While the Handbook could be applied to many types of *traditional* securitisations, the chapter will highlight the potential, but also

[39] ICMA, 'Green Bond Principles – –Voluntary Process Guidelines for Issuing Green Bonds' (hereinafterGBPs) (2025), 8 https://www.icmagroup.org/assets/documents/Sustainable-finance/2025-updates/Green-Bond-Principles-GBP-June-2025.pdf accessed 10 December 2025.

[40] Other commentators have referred to the 'norms written in a way to measure performance' as 'standards', Braithwaite (n. 37) 2. I will define them as 'metrics' to avoid confusion with the umbrella category of 'standard legal tools'.

[41] For example, ICMA, 'The GBP Impact reporting working group – –Suggested Impact Reporting Metrics for Sustainable Water and Wastewater Management Projects' (2017) https://www.icmagroup.org/assets/documents/Regulatory/Green-Bonds/Water-Wastewater-Impact-Reporting-Final-8-June-2017-130617.pdf .accessed 10 December 2025. For the other metrics see ICMA, 'Impact reporting for Green Projects' https://www.icmagroup.org/sustainable-finance/impact-reporting/green-projects/accessed 10 December 2025.

[42] See For example, ISDA, 'Master Agreement' (2002) https://www.sec.gov/Archives/edgar/data/1065696/000119312511118050/dex101.htm accessed 10 December 2025.

[43] See For example, LMA's 'bail-in clause', LMA, 'Documents and Guidelines' (n. 15).

[44] Originally conceived for syndicated international bonds and euro-commercial paper programmes in the primary market. ICMA, 'Chap. 1 – Scope' in 'Primary Market Handbook' https://www.icmagroup.org/market-practice-and-regulatory-policy/primary-markets/ipma-handbook-home/ accessed 10 December 2025. It is also possible to apply it to securitisation since ABSs are oftentimes debt securities, see Chap. 3, Sect. 3.3.3.

[45] ICMA, 'Table of contents' in 'Primary Market Handbook' https://www.icmagroup.org/assets/Table-of-contents.pdf accessed 10 December 2025.

the shortcomings of standard documentation as transposed to 'sustainable' transactions.

5.2.4 *The Rule-Making Function of Standard Legal Tools*

These 'soft' legal tools, designed at 'private' level, could serve as a blueprint for market actors to draft their own securitisation documentation. This, in turn, also gives rise to binding legal obligations (what I have previously described as 'hard law'[46]). For instance, parties can incorporate standard clauses within their own securitisation documentation, by which they would need to abide. The binding nature of contractual clauses also translates into a series of self-help remedies when counterparties do not comply, ranging from contractual mechanisms (like 'accelerating', i.e. demanding the immediate repayment of financial obligations) to alternative dispute resolution methods or the power to bring litigation.[47] For instance, market actors may be able to initiate legal proceedings for the incorrectness or inaccuracy of certain reports and statements where market actors have followed standard templates, like the ones to report on green bonds.[48]

It is also worth noting a peculiarity in the application of standard clauses, which *de facto* operate as 'default rules', as intended by contract theorists in economics.[49] In other words, these apply in principle unless market actors explicitly decide to deviate from them, the assumption being that clauses should directly mirror what parties would have wanted anyway.[50] In a similar vein, while model contracts are 'soft law' instruments[51], market actors do not generally make the binary choice between applying them or deviating from them altogether. Drafting documentation independently and anew gives rise to considerable transaction costs.[52] Therefore, market actors are more likely to apply model agreements in

[46] See Chap. 4, Sect. 4.2.

[47] Braithwaite (n. 37) 786–790.

[48] see Sect. 5.5.–5.6.

[49] See the definition by Ayres I, Gertner R, 'Filling Gaps in Incomplete Contracts: An Economic Theory of Default Rules' (1989) 99 The Yale Law Journal 87; Michael Klausner, 'Corporations, Corporate Law, and Networks of Contracts' (1995) 81 *Virginia Law review* 757, 768–769.

[50] Ayres, Gertner, ibid., 97–116. Simon Deakin et al Legal Institutionalism: Capitalism and the Constitutive Role of Law' (2017) 45 *Journal of Comparative Economics* 194–195.

[51] As defined in Chap. 2, Sect. 2.4.

[52] See also Chap. 3, Sect. 3.5.

principle, and to modify or remove certain provisions of them according to their needs. For example, the widely used ISDA 'Master Agreement' for derivatives contains a 'Schedule', which parties use to make amendments to the standard form.[53] Transposing the general 'default rule' logic, these should also in principle reflect best practices, which market actors are likely to want to apply.

Furthermore, standard legal tools, while non-binding, should have a 'normative' effect by facilitating the spontaneous convergence of market practices. There are two main mechanisms which could arguably account for this effect.

On the one hand, they can create 'positive externalities' that can shape the behaviour of other market participants. For instance, contract theorists in economics have already highlighted how the application of standard terms by one company will inspire other companies to do the same ('learning externality'). This is arguably also the case for the above-mentioned 'soft' legal tools due to globalisation and to the inherent interconnectedness of market transactions. The application of uniform standards and clauses by a critical size of market actors also gives rise to a 'network', where all the users reap a certain benefit by adopting it ('network externality'). This will also prompt non-users to do the same. [54]

On the other hand, the circulation of 'soft' legal tools may also foster reluctance among market participants to deviate from the well-known 'path' of existing standards ('path dependence').[55] There has been extensive scholarly analysis around how legal and economic patterns are more likely to be maintained than changed as they become presumptively more efficient in the eyes of market participants. Changes also involve 'switching costs' that market actors are unlikely to bear.[56] This mechanism could also arise across securitisation deals given the popularity of guidelines, metrics, and model documentation.

Thus, standardisation initiatives in the financial sector can act as default rules, while promoting the convergence of market practices through

[53] ISDA, 'Master Agreement' (n. 43) 19–24. Dan Awrey, 'Split Derivatives: Inside the World's Most Misunderstood Contract' (2019) 36 *Yale Journal on Regulation* 495, 518–519.

[54] On 'learning externalities' and 'network externalities', see Klausner (n. 49) 724, 730–733.

[55] Mark J Roe, 'Chaos and Evolution in Law and Economics' (1996) 109 *Harvard Law Review* 641.641, 651, 659; Marcel Kahan and Michael Klausner, 'Standardisation and Innovation' (n. 3) 721.

[56] Kahan, Klausner, ibid., 727–729.

externalities and path dependence mechanisms. These advantages crucially represent the background for the following analysis. Arguably, the design of standard tools specifically leveraging on sustainability-related practices may yield similar benefits for securitisations.

5.2.5 *The Added Value of Standardisation for 'Sustainable' Deals*

In the specific context where securitisations also promote sustainability goals, 'soft' legal tools could serve various functions.

They relieve originators of multiple transaction costs arising in the structuring of these operations. These primarily entail the drafting and negotiation costs for putting together the various agreements of the transactional documentation, including sustainability-related terms.[57] They also consist of outsourcing costs for the involvement of external professionals and legal fees.[58] When securitisations are used to achieve sustainability goals, such costs also arise for the involvement of external reviewers or legal practitioners assessing compliance with principles like ICMA GBPs. Chap. 3 has already highlighted that the magnitude of such costs may currently represent a hurdle, preventing market actors from engaging in sustainable transactions.[59] By contrast, directly applying standard documentation to transactional agreements allows the minimisation of the need for additional negotiation or external support.

Standardisation could also mitigate costs on the investors' side. Chap. 3 has highlighted the high 'information costs' for assessing the characteristics of 'sustainable' ABSs, also potentially requiring outsourcing to analysts (what the Regulation characterises as 'due diligence costs').[60] These may deter investments in sustainable ABSs, stifling the potential growth of the sector as a whole. On the contrary, common guidelines, metrics, and standard clauses should allow market actors to better understand the characteristics of sustainable transactions and make them more comparable. In turn, this should accelerate and facilitate 'price discovery', i.e. the mechanism for market actors to value and price assets, and improve

[57] For a theoretical account of negotiation costs, see Ayres, Gertner (n. 49) 92–93.

[58] See Chap. 3, Sect. 3.5. As also confirmed by SFIG (n. 32) 331–332.

[59] See Chap. 3, Sect. 3.5.

[60] See Chap. 3, Sect. 3.4.4. On the role of analysts for ESG investments in debt securities in general, see UNPRI, 'ESG Engagement for fixed-income investors: managing risks, enhancing returns' 7, 17 https://www.unpri.org/download?ac=4449. accessed 10 December 2025.

liquidity.[61] While the term has different meanings, the reference here is to 'cash-flow liquidity' since standardisation initiatives could ultimately mobilise more investors, thus allowing issuers to raise more capital.[62]

Other costs and challenges arising after the investment relate to the monitoring of originators' performance ('bonding costs') and to the enforcement of their securities ('enforcement costs').[63] In the case of securitisation, the high number and diversity of investors pose the additional challenge of finding a common position to bring forward throughout the engagement activities with originating companies, including on sustainability matters ('coordination costs').[64] On the contrary, standard guidelines and clauses may help delineate more clearly who is responsible for representing investors and the circumstances when these actors are expected to raise their 'voice' with originating companies.

Along with the cost-cutting benefits of standardisation, directly replicating 'default' clauses also allows market actors to avoid errors and gaps inevitably arising if parties negotiate 'sustainable' securitisation agreements on a deal-by-deal basis. As argued by contract theorists in economics, contracts inevitably tend to be incomplete, since parties are unable to foresee all possible contingencies and to contain errors.[65] Further legal intervention would thus be crucial to protect parties effectively.[66] Relatedly, standardisation initiatives could also increase legal certainty in transactions like sustainable securitisation while mitigating legal risks. Chap. 4 has highlighted the ambiguities emerging from the clauses of sustainable securitisation agreements on matters like investors' remedies: For instance, it is unclear whether they could take action if issuers fail to comply with sustainability objectives or fail to report on the final destination of the issuance ('greenwashing' practices).[67] By contrast, standard clauses contribute

[61] Kahan, Klausner, 'Standardisation and Innovation' (n. 3) 724; Awrey, 'Private Ordering' (n. 14) 193–194.

[62] The term also captures the possibility to price and trade securities quickly and with little impact on prices. For the various meanings see Banque de France, 'Financial stability review – special issue on liquidity' (2008) 109-11, 1 https://publications.banque-france.fr/en/february-2008 accessed 10 December 2025.

[63] On due diligence i.e. 'search costs', bonding and enforcement costs, see Michael C Jensen, William H Meckling, 'Theory Of The Firm: Managerial Behaviour, Agency Costs And Ownership Structure' (1976) 3 *Journal of Financial Economics* 305. 338–342.

[64] See also Chap. 3, Sect. 3.5.

[65] Kahan, Klausner, 'Standardisation and Innovation' (n. 3) 720.

[66] Ayres, Gertner (n. 49) 91–93.

[67] For example, see Chap. 4, s. 4.7.

to the creation of legal precedents as well as encouraging legal experts, such as lawyers and consultants, to build expertise in the interpretation and use of the clauses in order to assist market actors in their application.[68] In turn, higher certainty on rights and remedies may encourage investors to raise their 'voice'[69] by bringing legal action vis-à-vis originating entities, thus fostering the creation of a body of case law on sustainability and 'greenwashing' matters.

All in all, standardisation legal tools have the potential to cut costs considerably for issuers and investors whilst addressing the gaps, errors and ambiguities of current market practices.

5.3 Multi-Layered Standardisation Initiatives for 'Sustainable' Securitisations: An Overview

Standardisation initiatives could be a promising solution to mitigate costs and address some of the shortcomings of current market practices, where market actors largely determine sustainability-related clauses on a deal-to-deal basis.

Thus, the following sections will take stock of the existing 'soft' legal tools applicable to the most strategic aspects of securitisation identified in Chap. 3 ('focal points').[70] The purpose will be to examine whether and to what extent they have standardised the 'sustainable' elements of securitisation transactions. Moreover, the following sections will reflect on possible avenues to build on such efforts, namely how intermediaries may fine-tune existing legal tools or design new ones, in order to advance sustainability goals.

While securitisation documentation is complex and multi-layered, the focus will be on the legal tools that show a potential transformative impact.[71] In other words, I will exclusively examine the tools and related adjustments that could arguably provide insight into the sustainability-related goals underpinning securitisations. The following table (Table 5.1) summarises the standard legal tools at the core of the analysis.

[68] Kahan, Klausner, 'Standardisation and Innovation' (n. 3) 726.

[69] On 'voice' as a legal strategy, see Chap. 2, s. 2.6.

[70] See Chap. 3, Sect. 3.4.

[71] On the definition of 'transformative impact' for the purposes of this work, see Chap. 2, Sect. 2.5.–2.6.

Table 5.1 Private Soft Law Initiatives for Sustainable Securitisations

Private 'Soft' Legal Tools for Sustainable Securitisations			
Securitisation Focal Point	Guidelines	Metrics	Standard Documentation
Composition of the asset pool	• LMA, ELFA, Best Practice Guide for Term Sheet Completeness • LMA, APLMA, LSTA: Green Loan Principles, Guidance on Green Loan Principles, Sustainability-linked Loan Principles, Guidance on Sustainability-Linked Loan Principles • LMA, ELFA, Best Practice Guide to Sustainability-Linked Leveraged Loans • Energy-Efficient Mortgage Action Plan (EEMAP) Guidelines	ICMA, Sustainability Indicators for ABSs EEMAP Energy Performance Indicators for Energy-Efficiency Mortgages	• LMA Draft Provisions on Green and Sustainability-Linked Loans • Chancery Lane Project on Green and Sustainability-Linked Loans (discontinued) • RESFARM Project
Terms of issue of ABSs	• ICMA, Green Bond Principles • Social Bond Principles • Sustainability-linked Bond Principles • Guidelines for External Reviews	• ICMA, 'Reporting Metrics' ('Handbook Harmonised Framework for Impact Reporting') • 'Illustrative KPIs Registry'	• LMA – EUROPP, Subscription Agreements for Private Placements • Bond Covenant Group (EUROPP) on covenants • ISDA, Master Agreement • ICMA, 'Templates for Issuers and Reviewers' • Chancery Lane Project, Warranties for Underwriting Agreements
Governance of SPVs	• n/a	• n/a	• n/a
Representation mechanisms for investors	• LMA, on Sustainability Coordinator • UNPRI • ICGN Model Mandate Initiative		• ABA Indenture • IMA Model Discretionary Investment Management Agreement • ICGN Model Mandate Initiative

Table 5.1 (continued)

Private 'Soft' Legal Tools for Sustainable Securitisations	
Monitoring processes	• ICMA, Green and Social Bond Principles • Climate Bonds Standard and Certification Scheme

5.4 Standardisation Initiatives on the Composition of Portfolios

When analysing the transformative potential of extant initiatives, it is important to differentiate between standardisation tools applicable to any type of securitisation and those specifically targeting 'sustainable' financial products or investment practices.

5.4.1 'Traditional' Portfolios

Chap. 3 has identified the clauses of the documentation and rules determining the composition of the portfolio as a strategic step of the securitisation process. They could restrict or disincentivise the inclusion of certain types of credit claims due to their adverse environmental or social impact, as well as require portfolio assets to meet certain sustainability targets or criteria.[72]

It is important to recall here that market actors have previously qualified securitisations as 'sustainable' even when the underlying claims make an arguably neutral or adverse contribution to sustainability goals, such as loans for general corporate purposes.[73] As such, it is essential to investigate what legal tools ensure that, even when portfolios are 'non-sustainable', the underlying assets do not pose significant risks or have a significant adverse impact on environmental and social factors. For instance, while there may be merit in including standard student loans in portfolios, using

[72] See Chap. 3, Sect. 3.4.1.

[73] See Chap. 4, Sect. 4.4.–4.5.

loans for coal or mining projects may run counter to the 'sustainable' label of the transaction.

5.4.1.1 *Origination of Assets*

An interesting example of guidelines that could also apply to the creation of securitisation portfolios is the OECD Guidelines on lending (and underwriting) activities.[74] However, as these represented a trans-governmental effort (i.e. a 'public law' tool[75]), they will be covered in Chap. 6.[76] The Guidelines could apply to any kind of lending activity, regardless of whether lenders extend their finance for sustainability-related projects and targets or not. They list the 'due diligence' measures that lenders should implement to prevent, identify and address sustainability-related risks (such as environmental risks or adverse impacts on human rights across supply chains).[77] In the context of securitisation, market actors who also originate credit claims ahead of their assignment could in principle apply them in order to evidence a minimum level of commitment to sustainability factors in the identification of portfolio assets.

However, alongside these guidelines, limited legal tools support market actors and encourage them to incorporate environmental and social considerations within the terms of their loans when they do not intend to label them as 'sustainable'. For instance, the leading lending association for the EU, Middle East and African markets, the LMA,[78] has provided several standard agreements for loans[79] that can be used for traditional

[74] Specifying how they should implement the OECD Guidelines for Multinational Enterprises. OECD, 'Due Diligence for Responsible Corporate Lending and Securities Underwriting – Key Considerations for Banks Implementing the OECD Guidelines for Multinational Enterprises' (2019) https://mneguidelines.oecd.org/due-diligence-for-responsible-corporate-lending-and-securities-underwriting.pdf accessed 10 December 2025. On their relevance for underwriting activities, see below s. 5.5.

[75] See the distinction in Chap. 2, s. 2.3.

[76] See Sect. 6.4.1.

[77] OECD (n. 74) 16.

[78] On the role of the LMA, See above Sect. 5.2.1.

[79] For example, LMA, 'Standard Terms And Conditions For Par And Distressed Trade Transactions; (Bank Debt/Claims) (2021) in 'Documents and Guidelines' (n. 39) accessed 10 December 2025.

securitisations, with some adjustments.[80] However, these do not contain any explicit reference to the environmental and social risks of loans.

5.4.1.2 *Creation and Assignment of Portfolios*

There is also a lack of standard legal tools guiding market actors on how to consider environmental and social risks when pooling together assets and assigning them to SPVs.

As explained in previous chapters, market actors are likely to embed clauses on the composition of the portfolio within 'loan purchase agreements' or similar contracts for the assignment of the credit claims to SPVs[81], or even within the documentation for the issuance of ABSs, such as prospectuses.[82] Anecdotal evidence and scholarly analyses of traditional securitisation operations suggest that agreements and clauses on the characteristics and financial risks of the securitised assets have largely similar content.[83] However, this is not the case for sustainability-related documents and clauses, as highlighted in Chap. 3. Instead, originators tend to provide insufficient information about the environmental and social impact or risks of the portfolio assets.[84]

Such fragmentation and lack of transparency are likely to depend heavily on the absence of standard legal tools supporting the drafting of these clauses. For instance, standard agreements like LMA's one for 'funded participations'[85], applicable to the contractual assignment of claims across synthetic securitisations[86], make no reference to sustainability-related criteria.

In principle, various types of clauses within loan purchase agreements and prospectuses may provide insight into the composition of the

[80] E.g. market factors need to adjust to the use of portfolio of claims, instead of individual loans. In practice, this may entail additional waivers of liability for originators in case of default over the assets, see LexisNexis PSL, 'Loan portfolio sales – legal documentation' (2023) https://www.lexisnexis.com/uk/lexispsl/bankingandfinance/document/391290/5H0D-CBW1-F185-S1M4-00000-00/Loan_portfolio_sales_legal_documentation accessed 10 December 2025.

[81] On the main content of loan purchase agreements, see Phillip Wood, 'Bond Issue Documentation' in *Project Finance, Securitisations, Subordinated Debt*, Vol. 7 (2019) paras 7–003. More generally on other similar documents underpinning the assignment of proprietary and contractual rights, see Chap. 2, Sect. 3.2.–3.3.

[82] See Chap. 3, Sect. 3.3.3., Chap. 4, 4.6.1.

[83] SFIG (n. 32). Frankel (n. 34) 256.

[84] See Chap. 4, Sect. 4.6.

[85] LMA, Master Funded participation agreement (Par/Distressed), in 'Documents and Guidelines' (n. 79).

[86] On the structure of synthetic securitisations, see Chap. 3, Sect. 3.2.

portfolio. For instance, 'eligibility criteria' clauses sometimes set constraints on portfolio assets;[87] correspondingly, they could also in principle introduce additional sustainability-related criteria or goals. Other clauses amenable to sustainability considerations include 'Representations' and 'warranties', which generally operate in tandem, through which originating entities confirm that the assets meet certain conditions and promise to indemnify SPVs or investors if not.[88] Originators could use them, among other things, to represent that the underlying assets do not pose any material climate-related risks or violate minimum human rights standards.[89] The above-mentioned SFIG review of securitisation clauses[90] has also revealed that many portfolio-related agreements contain a representation that the underlying residential properties or other assets comply with environmental regulation.[91] These clauses could in principle be relevant for 'sustainable' portfolios: For example, originators pooling together 'green' mortgages for residential buildings may represent that they comply with statutory energy efficiency standards. Such clauses could be a particularly promising entry point for investors to hold originators liable due to previous case law on warranties (focusing on share purchase agreements).[92] This suggests that courts are very likely to rule that these clauses have been breached when they do not exhaustively state all the details of the

[87] See Chap. 3, Sect. 3.4.1.1. for instance, Fitch Ratings identifies the following as a possible formulation for transactions under German law: *'Eligibility Criteria: The loan meets the eligibility criteria (...) • The loan is secured by one or more mortgages. • Mortgages are non-accessory mortgages (...) in certified or noncertified form (...). • Mortgages constitute valid mortgage rights and have been entered into the appropriate land register'*, Fitch Ratings, 'Representations, Warranties and Enforcement Mechanisms' (n. 29) 33.

[88] For the definition and an analysis of the most common ones across securitisations, Patricia Ann McCoy and Susan M Wachter, 'Representations and Warranties: Why They Did Not Stop the Crisis' in LA Fennell, BJ Keys (eds.) *Evidence and Innovation in Housing Law and Policy* (Cambridge University Press 2017) 7–11; Wood, 'Bond Issue Documentation' (n. 82) paras 7–003.; Fitch Ratings, ibid.; Chap. 3, Sect. 3.4.1.

[89] On 'do no significant harm criteria' under EU law, see also Chap. 6, Sect. 6.4.1.

[90] See Sect. 5.4.1.

[91] After reviewing several examples of clauses, SFIG recommends the following formulation: 'At the time of origination, each Mortgaged Property was in material compliance with all then-applicable environmental laws pertaining to environmental hazards including, without limitation, asbestos', SFIG (n. 32) 136–137. Fitch ratings also noted its widespread use, see Fitch Ratings (n. 29).

[92] Under English law (often the governing law for most financial transactions), *New Hearts v Cosmopolitan Investments [1997] BCLC 249; Equitix EEEF Biomass 2 Ltd v Fox [2021] EWHC 2531 (TCC).*

circumstances of which counterparties should be aware before the contract. As a result, ABS holders could have the right to indemnification if originators included the above-mentioned representations and the underlying assets turned out to pose climate-related risks.

Alternatively, there may be 'covenants' where originators set out additional commitments.[93] For instance, they may undertake to report on the GHG emissions of the underlying credit claims.

However, while it is common practice to refer to compliance with environmental regulation, there are no standard 'eligibility criteria' clauses, 'representations', 'warranties' and 'covenants' on sustainability-related matters tailored for securitisation-specific agreements. Consequently, market actors need to independently draft such provisions on a deal-by-deal basis or for a series of operations. The ensuing high transaction costs are likely to disincentivise originating entities from being fully transparent about the sustainability risks of their asset pools, as well as from setting restrictive conditions regarding the types of assets they will include.

5.4.2 'Sustainable' Portfolios

Alongside transactions with a 'neutral' portfolio, there have also been cases where originators have exclusively pooled together assets that make a positive contribution to environmental, social, and governance goals ('sustainable portfolios'). Some of the recent 'sustainable finance' initiatives could also be indirectly applicable in these circumstances.[94] They have, however, tended not to be securitisation-specific and their effects are arguably insufficient.

5.4.2.1 Originating 'Sustainable' Assets (I): Guidelines

The LMA[95] has taken the lead in a number of initiatives which could also be useful to market actors originating 'sustainable' assets ahead of securitisation. One of the most popular frameworks to qualify loans as 'green' is the set of Green Loan Principles (GLPs), jointly drafted by the LMA and two other lending industry associations, i.e. the Loan Syndications and Trading Association (LSTA), active in US markets, and the Asia Pacific Loan Market Association (APLMA). The GLPs define 'green loans' as 'any type of loan instruments and/or contingent facilities (such as bonding

[93] Wood, 'Bond Issue Documentation' (n. 82) para 7–003.

[94] For the types of portfolio assets in 'sustainable' transactions, see Chap. 4, 4.4., 4.6.1.

[95] On the role of LMA see Sect. 4.1.2.

lines, guarantee lines or letters of credit) where the proceeds or an equivalent amount shall be exclusively applied to finance, refinance or guarantee, in whole or in part, new and/or existing eligible Green Projects'.[96] The same associations have also drafted a similar set of Principles for Social Loans (i.e. the Social Loan Principles, SLPs), which *'finance, refinance or guarantee, in whole or in part, new and/or existing eligible Social Projects'*.[97]

Following the typology of 'sustainable' securitisations introduced in Chap. 4, the definition of loans under the Principles revolves around the types of *activities* which these loans should finance, i.e. 'Green' and 'Social' Projects. At the same time, both frameworks set out additional best practices that borrowers should comply with before and after extending loans, concerning the identification of the projects, the management of proceeds and external reviews.[98] The associations have also issued more detailed 'guidance', setting out a series of best practices for lending agreements.[99]

However, at least until 2024[100], these did not go as far as to provide a model formulation for provisions that market actors could replicate within their lending agreements. The high-level formulation of the principles, in turn, left room for transaction costs, especially for drafting and negotiating ad hoc contractual clauses which incorporate such principles and guidance.

The same network of associations also issued Sustainability-linked Loan Principles (SLLPs), which characterise such loans by the presence of a *target* (formally denominated sustainability performance target (SPT)). Failing to attain it should affect the financial conditions of loans, resulting in an economic outcome like an adjustment of the interest rate or a reduction of the margin.[101] The Appendix of the SLLPs lists a series of possible targets. Replicating the structure of the GLPs, the SLLPs also provide other high-level recommendations on best practices, such as the

[96] LMA, APLMA, LSTA, 'Green Loan Principles' (2025) 2 https://www.lma.eu.com/application/files/1917/4298/0817/Green_Loan_Principles_-_26_March_2025.pdf accessed 29 March 2025.

[97] LMA, APLMA, LSTA, 'Social Loan Principles' (2025) 2 https://www.lma.eu.com/application/files/1317/4307/3886/Social_Loan_Principles_-_26_March_2025.pdf accessed 29 March 2025.

[98] LMA, APLMA, LSTA, 'Green Loan Principles' (n. 96), 4.

[99] LMA, APLMA, LSTA, 'Guidance on Green Loan Principles', (2025) https://www.lma.eu.com/sustainable-lending/resources accessed 10 December 2025.

[100] On standard documentation, see below, Sect. 5.4.2.2.

[101] Principle 3 LMA, APLMA, LSTA, 'Sustainability-Linked Loan Principles' (2025) 4 https://www.lma.eu.com/application/files/6317/4298/0865/Sustainability-Linked_Loan_Principles_-_26_March_2025.pdf accessed 10 December 2025. See Chap. 3, Sect. 3.3.

identification of the SPTs or the processes for reviewing and reporting on the loans.[102]

The LMA, along with the European Leveraged Finance Association, has also published more detailed guidelines for the issuance of sustainability-linked loans[103] and *leveraged* sustainability-linked loans[104], crystallising the best practices for the issuance of such instruments and for their related documentation.

The LMA and the European Leveraged Finance Association (*hereinafter*ELFA) have also provided preliminary reference to potential sustainability-related clauses in their Guidelines for conventional loan term sheets.[105] For instance, they recommend, among other things, the introduction of ESG 'information reporting requirements', where market actors contractually undertake to regularly report on ESG matters. The Guidelines also recommend incorporating provisions to describe what makes loans 'sustainability-linked' or 'green', such as clauses on KPIs, targets, or the 'use of proceeds'.[106] As discussed later on, these sets of Guidelines are now complemented by standard provisions to assist market actors with the drafting of contractual obligations linked to 'sustainable' lending agreements.

Alongside the LMA initiatives, which should be transversally applicable to 'sustainable' loans, market actors have also introduced guidelines for specific types of mortgages, such as those for energy-efficient upgrades. Under the EU-funded Energy Efficiency Mortgages Action Plan (hereinafterEEMAP), a consortium of mortgage lenders has drafted a definition and compiled a series of best practices for issuing such mortgages.[107]

[102] Principles 1, 2, 4, 5, ibid., p. 3–4.

[103] LMA, APLMA and LSTA, 'Guidance on Sustainability-Linked Loan Principles' (2025) https://www.lsta.org/content/guidance-on-sustainability-linked-loan-principles-sllp/ accessed 10 December 2025.

[104] While the Guide does not define such loans, it specifies that they are not a standalone class, but rather sustainability-linked loans that are also leveraged. The Principles apply on a 'portfolio company level' and not on a sponsor or fund level. LMA, ELFA, 'Best Practice Guide to Sustainability-Linked Leveraged Loans' (2021) 2 https://elfainvestors.com/wp-content/uploads/2023/10/ELFA-Best-Practice-Guide-to-Sustainability-Linked-Leveraged-Loans-1.pdf accessed 3 June 2026.

[105] LMA, ELFA, 'Best Practice Guide for Term Sheet Completeness' (2025) https://www.lma.eu.com/application/files/5816/3843/1629/LMA_Best_Practice_Guide_Term_Sheet_Completeness.pdf accessed 10 December 2025.

[106] Ibid., p. 6. See Sect. 4.4.2.

[107] EEMAP, 'Definition of an Energy Efficient Mortgage' (2018) https://energyefficientmortgages.eu/wp-content/uploads/2021/07/EEMI-Definition-14.11.18.pdf accessed 10 December 2025.

Complying with such (voluntary) Guidelines also allows alignment with a voluntary label which, among others, is meant to encourage market actors to bundle together mortgages and securitise them.[108]

Overall, the guidelines on 'green', 'social' and 'sustainability-linked' loans, as well as on energy-efficient mortgages, represent a welcome first step towards the convergence of market practices. However, their high-level nature leaves room for flexibility and requires effort from market actors to draft their own clauses regarding their sustainability-related commitments and the consequences of non-compliance. In so doing, they fail to address the high transaction costs and may deter market actors from setting ambitious commitments.

5.4.2.2 *Originating 'Sustainable' Assets (II): Standard Documentation*

More recent, preliminary examples of model documentation have sought to fill some of the gaps in the discussed Guidelines.[109] For instance, a network of lawyers, the Chancery Lane Project,[110] designed a wide array of model clauses on sustainability matters. These also include adjustments to LMA's loan agreements to reflect the 'green' element.[111] They entail, among other things, a 'Purpose' clause, through which borrowers identify how they intend to use the proceeds. They also introduce covenants, imposing certain conditions on borrowers,[112] such as the duty to publish a 'Green Loan Framework' to provide further details about the financed projects and to notify lenders if there are cases of non-compliance or ongoing or threatened litigation on environmental matters. There are also representations regarding the truthfulness of sustainability-related information.[113]

[108] EEMAP, 'Energy efficient mortgage label' https://www.energy-efficient-mortgage-label.org/# accessed 10 December 2025.

[109] I would like to kindly acknowledge the valuable research assistance support with regard to standard documentation by Avv. Marco Baio and Dr. Avv. Federico Urbani.

[110] Chancery Lane Project, 'About' https://chancerylaneproject.org/about/ accessed 10 December 2025.

[111] Ibid., Chancery Lane Project, 'Green Loan 'Starter Pack' – –Harrison's clause' (*Chancery Lane Project* 2022) https://chancerylaneproject.org/model-clauses/green-loan-starter-pack/ accessed 10 December 2025. The model provisions have not been maintained, with the last update tracing back to 2021.

[112] Formally denominated 'Undertakings', ibid.

[113] Clauses 25.34 'Green Loan Representations', 26.7–26.8 'Informations: Miscellaneous' and 'Green Loan Reporting', ibid. On the notions of representations and covenants, see also Chap. 3.4.2.1.; Chap. 4.7.1.

Similar clauses are also available for sustainability-linked loans.[114] Importantly, the Chancery Lane Project's model provisions identify non-compliance with SPTs as an 'Event of Default'. In other words, the clauses introduce a regular assessment of the issuer's sustainability performance through the submission of a 'sustainability compliance certificate'. Failing to submit it allows lenders to require immediate repayment of the loan.[115] The model provisions also contemplate alternative stipulations: For instance, parties may agree that failing to achieve sustainability goals triggers a change in the repayment conditions of the loan, such as a margin increase. They may also stipulate a 'Declassification', i.e. a conversion of the instrument into an 'ordinary' loan, which would no longer allow the borrower to portray or advertise it as 'sustainable'.[116]

The novelty and significance of the Chancery Lane Project's model provisions are undeniable. At the time they were introduced, they went significantly beyond the above-mentioned Guidelines by also providing a concrete blueprint for the drafting of these clauses. In so doing, they had the potential to contribute to the standardisation of 'sustainable' lending contracts by also cutting the interrelated transaction costs.

Alongside the Chancery Lane Project's endeavours, the LMA has also started to draft model provisions for 'sustainable' lending agreements since 2023.

For instance, they have created a Model Agreement for sustainability-linked loans, including a series of provisions to articulate sustainability targets. [117] Like the Chancery Lane Project's provisions, this document incorporates covenants where borrowers undertake to regularly disclose how the loans are performing on sustainability matters. They can also undertake to submit reports from external reviewers.[118]

[114] Chancery Lane Project, 'Sustainability-Linked Loans – –Casper's Clause' (*Chancery Lane Project* 2022) https://chancerylaneproject.org/climate-clauses/sustainability-linked-loans/ accessed 10 December 2025. These model provisions have not been maintained either, with the last update tracing back to 2023.

[115] Clause 4.1. [Events of Default]/ [Sustainability-Linked Margin Increase]/ [Declassification], ibid. On the notion of 'Events of default', see Chap. 3, 3.4.2., Chap. 4, Sect. 4.7.1.

[116] Clause 4. ibid.

[117] LMA, 'Draft provisions for sustainability-linked loans' (2023), updated in 2025 https://www.lma.eu.com/documents-guidelines/documents/category/sustainable-finance# accessed 10 December 2025.

[118] In other words, they have to regularly present 'Sustainability compliance certificates', indicating their progress against the targets ('sustainability reports'), and incorporating a third-party review of the truthfulness of such information ('verification report'). see 'External reviewer', clause 26.11, Chancery Lane Project, 'Casper's Clause' (n. 113).

Yet, LMA's model provisions differ from the previously mentioned Chancery Lane Project's provisions by providing less stringent remedies for non-compliance with sustainability-related targets and clauses. Importantly, failing to meet the targets, or making false and misleading statements on sustainability matters ('misrepresentations'), does not constitute an 'Event of Default'. The related clause reads as follows:

> '(i) *Events of Default: (...) [(c)] No Event of Default will occur (...) by reason only of an Obligor's failure to comply with a Sustainability Provision.*
>
> (...) [(b)] *No Event of Default will occur (...) to the extent that the representation or statement concerns, or the document consists of, Sustainability Information'*

In light of the wording of this clause, borrowers essentially cannot demand immediate repayment of the loan for sustainability-related breaches. Instead, such circumstances by default lead to a margin adjustment.[119] Parties also have the alternative option to stipulate that it constitutes a 'declassification event',[120] no longer enabling them to market the loan as 'sustainability-linked'. The related clauses read as follows:

> *[28.41] Declassification Event*
>
> (a) *On and at any time after the occurrence of a Declassification Event the Agent may, and shall if so directed by [the Majority Lenders]/[all the Lenders], by notice to the Parent declassify the Facilities as 'sustainability-linked'.*
>
> (b) *With effect on and from the Declassification Date:*
>
> (i) *Clause [15.3] (Sustainability Margin Adjustment) and each Sustainability Provision shall cease to apply; and*

[119] Clause 15.3 ibid.

[120] This is an option that could be added to the lending agreement, particularly to Clause 2.1, 'Declassification event' (b). However, it is not the default provision, whereby only *ex post* changes to the sustainability targets of the issuance, not agreed by all the parties, can lead to Declassification, ibid., 'Declassification event' (a).

(ii) *no Sustainability Margin Adjustment will apply to any Utilisation.*

(c) *No Facility may be re-classified as 'sustainability-linked' on or after the Declassification Date.*

[28.42] Sustainability publicity

The Parent shall not (and shall ensure that no other member of the Group will) make any disclosure that references any Facility or any Utilisation as 'sustainability-linked' at any time on or after the Declassification Date.

In a similar vein, LMA has created draft documentation for green loans, including a term sheet for use throughout the negotiations and model provisions to add to the main lending agreement ('investment grade document').[121] The latter contains draft clauses concerning the 'Purpose' of the issuance, which, if adopted, create a binding legal commitment to set aside the loan proceeds for the predetermined sustainability-related projects.[122] It introduces a series of 'Representations' concerning the truthfulness of the information concerning sustainability-related elements of the 'green' projects,[123] as well as a series of covenants regarding the monitoring of the development of such projects, along with regular reporting.[124] At the same time, the documentation resonates with that for green loans in designing similar remedies for non-compliance with sustainability-related clauses. This does not lead to a default of the loan, but rather to a mere 'declassification' of the instrument into a regular loan.[125]

All in all, these model agreements for 'green' and 'sustainability-linked' loans could be useful to entities which originate credit claims before securitisation, and less so for those who purchase them from third parties ahead of the assignment to SPVs.[126] The variety in options for the formulation of remedies may also provide welcome flexibility, as I will discuss below.[127] The added value of model clauses, especially compared to the above-mentioned Guidelines, is that market actors could directly replicate them within their agreements without incurring further legal and drafting

[121] LMA, 'Draft provisions for green loans' (2025); LMA, 'Green Loans Term sheet' (2025) https://www.lma.eu.com/documents-guidelines/documents/category/sustainable-finance# accessed 10 December 2025.

[122] Clauses 3.1. 'Purpose', 3.2. 'Management of Proceeds' LMA, 'Green Loans Term sheet'.

[123] Clause 3.2.(3) 'Management of Proceeds'; Clause 21.14 'Green Loans' ibid.

[124] Clause 22.8 'Green Loan reporting'; Clause 22.9 'Green Loan information' ibid.

[125] Clause 24.8. 'Declassification Event' ibid.; Clause 25.3 'Events of Default' (c)-(b).

[126] On this alternative structure for securitisation, see Chap. 3, Sect. 3.3.1.

[127] See Sect. 5.4.3., 5.9.

costs. At the same time, as I will discuss below[128], they are still of limited use for drafting securitisation documentation, especially concerning the assignment of portfolios.

While these initiatives are meant to have a transversal scope of application across 'sustainable' loans, some market-led projects in the EU context have also aimed at standardising specific types of loans, issued in certain sectors or for certain types of assets, with the aim to facilitate the creation of 'sustainable' portfolios. For instance, the RESFARM project, built through the collaboration of unions of farmers and supported by the European Commission[129], aimed to accelerate the renewable energy and energy-efficiency improvements in the agricultural sector through capital market financing. [130] The project culminated, among others, in the design of standard contracts for farmers, like power purchase agreements for the supply of renewable energy. These should foster the practice of pooling together such contracts and their securitisation, which could allow them to have access to capital market funding.[131]

Yet, such projects have also at times highlighted the boundaries of standardisation. For instance, the EU-funded EuroPACE project aimed to provide guidelines to draft contracts for loans towards energy-related and water management projects across commercial and residential properties. According to the initial plans, the standardisation of these contracts should facilitate 'sustainable' securitisations, and specifically the pooling together of the credit claims arising from these agreements for the issuance of RMBS or CMBS.[132] However, the project found standardisation to be unfeasible due to the heterogenous nature of

[128] See Sect. 5.4.3.

[129] Through the Work Programme Horizon 2020, see RESFARM, 'Project' http://resfarmproject.eu/en/project/ accessed 10 December 2025.

[130] RESFARM PROJECT, 'Description of the Investment Instrument – –A Financial Instrument for the Mobilisation of Investments in Sustainable Energy in the Agrarian Sector of Soutern Europe' (2017) http://resfarmproject.eu/en/documentation/ accessed 10 December 2025.

[131] Ibid., specifically 'D.4.2 Standard Contracts' (2017) in ibid.

[132] EuroPACE, 'Refinancing EuroPACE loans in the European fixed income – –Originally: Standardisation protocols for On-Tax Financing' (2020) 7, 22 https://www.eu-peers.eu/knowledge-assets/refinancing-europace-loans-in-the-european-fixed-income-market accessed 10 December 2025.

these lending contracts and of the related enforcement regimes across EU countries.[133]

5.4.2.3 Originating 'Sustainable' Assets (III): Metrics

Alongside guidelines and standard clauses, market actors have also engaged in preliminary work towards the standardisation of metrics to quantify the impact and risks of portfolio assets. For instance, the Asset Management and Investor Council Group within ICMA[134] has provided an illustrative list of key performance indicators (hereinafterKPIs) to measure the performance of loans and leases on ESG matters in the automobile sector.[135] Examples of environmental KPIs include average Co2 emissions, average well-to-wheel Co2 emissions, and the percentage of fuels.[136]

Another example is the previously mentioned EEMAP initiative. The use of the energy-efficient mortgage label requires compliance with a series of market-designed energy performance indicators.[137]

As discussed in Chap. 6, these initiatives, while useful, will also inevitably require coordination with concurrent EU statutory requirements on sustainability indicators.[138]

5.4.2.4 Creation and Assignment of 'Sustainable' Portfolios

The analysed guidelines and standard contracts can support the very preliminary phase of securitisation, when market actors extend 'sustainable' loans. However, no guidelines provide recommendations on how originators could efficiently pool together 'sustainable' assets in order to launch a securitisation deal or programme. There are also no standard documents for the assignment of the portfolio ('loan purchase agreements'[139]), meaning that no model provisions are available for market actors to disclose the sustainability impact of the portfolio ('eligibility criteria' clauses) or to articulate sustainability goals ('representations and warranties' clauses).

[133] Ibid., 24.

[134] On the function of ICMA see above, Sect. 5.2.1.

[135] ICMA, 'Sustainability Indicators for ABS' (2021) https://www.icmagroup.org/assets/documents/Regulatory/AMIC/Sustainability-indicators-for-ABS-181021v2.pdf accessed 10 December 2025.

[136] Ibid., 3. On the EU Taxonomy, see Chap. 6, Sect. 6.4.2.–6.5.2.

[137] EEMAP, 'Master Template – explanatory document' https://energyefficientmortgages.eu/wp-content/uploads/2021/07/EEMI-Master-Template-Explanatory-Document.pdf accessed 10 December 2025.

[138] See Chap. 5, Sect. 5.5.2.–5.5.4.

[139] As discussed in Chap. 3, Sect. 3.3.1.–3.4.1.

Incorporating sustainability-related provisions within these agreements thus gives rise to considerable transaction costs, which may discourage market actors from engaging in 'sustainable' securitisations in the first place.

5.4.3 Normative Recommendations

In light of the thdiscussed shortcomings, there is room for further standardisation initiatives targeting portfolios. When designing further tools, it is essential to ensure that they are securitisation-specific, i.e. that they take the specificities of the creation and assignment of credit claims into account, and transversal, in that they should apply to 'sustainable' and to 'non-sustainable' portfolios.

For instance, there is merit in designing further guidelines to provide a non-exhaustive list of assets eligible for 'sustainable' portfolios, as well as an 'exclusion list' of assets which would not be eligible even for 'non-sustainable' portfolios due to their strongly negative environmental and social impact.[140]

At the same time, going beyond the existing Principles on 'green' and 'sustainability-linked' loans, the Guidelines should also specify best practices for originators after extending the loans. For instance, they should specify how originating entities should monitor the environmental and social impact of the asset pool during the course of the securitisation transaction, and not only the post-lending phase. Building on the OECD Guidelines for lending, they should recommend that originators take specific measures in case of serious environmental or social risks.[141] These may involve following up with clients, or even replacing certain exposures within the portfolio.

One could wonder whether it would be more appropriate to design further legal tools at the state rather than at the market ('private') level, also based on the pre-existing set of OECD guidelines. However, the involvement of market actors like industry associations would arguably ensure that the formulation of ad hoc guidelines reflects existing market practices. It would also build on, and complement, the already existing

[140] As recommended by UNPRI, 'ESG incorporation in securitised products: The Challenges Ahead' (2021) 5, 25 https://gsfo.org/resource/esg-incorporation-securitised-products-challenges-ahead accessed 10 December 2025.

[141] OECD, 'Due Diligence for Responsible Corporate Lending and Securities Underwriting' (n. 74) 51–53.

work towards the design of guidelines for other 'sustainable' assets like bonds and loans.

The guidelines should also provide an illustrative list of metrics for the quantification of the environmental and social risks of the exposures. Market actors should use these metrics in tandem with statutory requirements requiring the measurement of ESG factors on the basis of mandatory technical criteria. Chap. 6 will further elaborate on this point.[142]

Another welcome development would be the design of further standard agreements and clauses. However, it is essential to ensure that model provisions are as accessible as possible to investors. On this note, the SFIG review also noted that publicly available information on the quality and composition of the securitised assets is limited.[143] Prospective investors may not see the loan purchase agreement at all or they may only receive a 'representative sample' of the commonly used clauses in the field, especially when entities assigning the assets to SPVs do not originate them, but rather purchase them from various other market actors.[144] In practice, this means that investors may be unaware of how portfolio 'eligibility criteria' concretely deviate from that list, and of the existence of any sustainability-related constraints.[145] Investors may be able to indirectly obtain such information through rating agencies' reports, which, however, are not always transparent on these matters either.[146]

Due to the limited accessibility of loan purchase agreements,[147] it is essential to create model provisions for the documentation which prospective investors view before the issuance, such as prospectuses.[148] Notably, model provisions should include a standard 'eligibility criteria' clause, allowing originators to disclose about the sustainability-related requirements that credit claims must meet. This should also be applicable to

[142] See Chap 6, Sect. 6.4.1.–6.4.4.

[143] SFIG (n. 32), 331–332, 340.

[144] On the various options for the creation and assignment of securitisation portfolios, see Chap. 3, 3.2., 3.3.1.–3.4.1.

[145] For example, SFIG has tracked the following clause and considered it in line with consolidated market practice. *'The list of Representations and Warranties generally contains the content of certain material representations and warranties which will be provided by the Originators in connection with the [transaction]. The actual list will be Sect.tantially in the same form... but may vary by Originator (...). This list does not purport to constitute a complete list of representations and warranties which are being provided'* SFIG (n. 32), 323–324.

[146] As also discussed in Chap. 3, Sect. 3.4.1.–3.4.4.

[147] See above Sect. 5.5.1.2, and Chap. 4, 4.6.1.

[148] On the role of prospectuses, see Chap. 3, Sect. 3.3.3.

transactions regardless of whether the underlying portfolio is 'sustainable' or not. To this end, an option may be to provide two alternative formulations: First, there should be a provision allowing *all* originators to disclose about how they prevent the inclusion of assets posing significant environmental or social *risks.* [149] Second, an additional clause could allow originators to disclose about the *impact*, i.e. the positive contribution, of the assets on environmental, social, and governance matters when the portfolio is fully 'sustainable'.

Model provisions should also include 'representations', where originating entities confirm that they have assessed that the portfolio does not pose material sustainability risks. Incorporating such clauses in loan purchase agreements or in the documentation for the offer of ABSs would also have the crucial implication of expanding the set of remedies especially in case of 'greenwashing' practices. The breach of representations ('misrepresentation') would allow investors to take action against originating entities and to receive monetary compensation. This right would also arise in case of false or misleading statements on the sustainability qualities of the issuance in the prospectus or other documentation. Moreover, if market actors included such a term in the prospectus, there would also be additional remedies under securities legislation, discussed in the following chapter.[150]

It would also be desirable to design standard 'covenants', through which originating entities make commitments vis-à-vis investors regarding the composition and qualities of the portfolio. For instance, certain multilateral development banks (MDBs) have started making their finance conditional to 'climate covenants', i.e. contractual clauses requiring prospective borrowers to make various commitments to align with the Paris agreement, like regularly disclosing to the bank about the impacts of climate change on their activities.[151] Similarly, there is room for standard clauses for prospectuses, requiring originating entities to report on the

[149] For examples of sustainability risks, see above 4.3.1.

[150] See Chap. 6, Sect. 6.5.1.

[151] Among others, see the European Bank for Reconstruction and Development (EBRD)'s Paris agreement methodology, whose application requires financed entities to comply with some covenants, among others. EBRD, 'Methodology to determine the Paris Agreement alignment of EBRD investments' (2024) Para 3.11, 30 https://www.ebrd.com/content/dam/ebrd_dxp/assets/pdfs/green/Methodology%20to%20determne%20the%20Paris%20Agreement%20alignment%20of%20EBRD%20investments.pdf accessed 10 December 2025.

sustainability-related risks and impact of the portfolio on a regular basis. It would be preferable for such clauses to also entail a quantitative element, where originators specifically undertake to measure the environmental and social impact of their 'portfolios' according to given metrics, or the compliance with specific targets.

Lastly, model provisions should delineate the types of remedies available to investors if originating entities fail to comply with 'eligibility criteria' clauses or with reporting covenants. As explained above, LMA model provisions and other initiatives such as the Chancery Lane Project clauses have provided various options, such as Events of Default, margin adjustments or the 'declassification' of loans, where market actors lose their ability to qualify loans as 'green' or 'sustainability-linked'.[152] It would be particularly desirable for the parties negotiating the documentation (such as the originating entities and underwriters[153]) to choose from various options depending on the characteristics of the transaction and on the breaches. For instance, if the credentials on the use of proceeds are particularly strong, it may be sufficient to tie breaches to margin adjustments, rather than to Events of Default, which would entail the termination of the relationship with investors.[154] The clauses clarifying investors' remedies would play a crucial role, increasing legal certainty, and thereby encouraging ABS holders to raise their 'voice' on sustainability matters, whilst ensuring some level of flexibility to the parties drafting the documentation.[155]

Along with the proposed transversal standardisation tools, there may also be merit in the design of further sectoral initiatives exclusively targeting specific types of assets. Industry associations are arguably the best placed to evaluate whether some claims, like commercial or consumer mortgages, require additional ad hoc guidelines or model documentation with a view to incentivising securitisation. The above-mentioned EU projects for the design of ad hoc contracts further evidence this point, since there have been cases where they have culminated in useful standardisation tools like contracts, and instances where they have concluded that such effort would be impractical.[156]

Overall, the proposed guidelines, metrics, and model provisions are likely to yield a twofold effect. First, creating standard legal tools could

[152] See above, Sect. 5.4.2.3.

[153] On underwriters, see Chap. 3, Sect. 3.3.3.; see below Sect. 4.4.1.

[154] See above Sect. 4.4.1.

[155] On investors' 'voice', see Chap. 2, Sect. 2.6.

[156] See Sect. 5.4.2.3.

facilitate the creation of securitisation portfolios. They may be especially beneficial for smaller-sized market actors such as SMEs, for which transactional costs are disproportionately higher than for larger and frequent market players in the securitisation sector.[157]

Second, standardisation tools accessible to investors and to the market more broadly, such as standard clauses for prospectuses, could make sustainability-related qualities of portfolios more transparent and comparable. Disclosure may promote investors' confidence in the reliability of ABSs. In the long run, it may mobilise more investments in sustainable securitisations.

5.5 The Terms of Issue of ABSs

5.5.1 Traditional Documentation

Chap. 3 has illustrated how the documentation ahead of and contextual to the issuance of ABSs represents the primary reference point for originating companies to articulate sustainability-related targets and projects connected with the issuance. However, existing standardisation initiatives offer very limited insight into how originators could incorporate such goals within their documentation, thus making them binding.

5.5.1.1 Underwriting Agreements

'Underwriting' or 'subscription' agreements, which 'underwriters' and SPVs or originators of the credit claims conclude, represent the first relevant pre-issuance document. They delineate underwriters' obligation to market ABSs to investors and to retain the unsold ones.[158]

The above-mentioned OECD Guidelines, covering lending and underwriting activities, outline best practices for underwriters to assess and mitigate the environmental and social risks of the portfolios and of the issuance. They recommend that prospective investors to evaluate how their counterparties (i.e. the issuers of securities) identify and manage sustainability-related risks related to their operations in general (such as their overall climate risks or the impact of their supply chain).[159] While not

[157] On costs and size as obstacles to 'green' securitisation, see Petit, Schlosser (n. 34) 12.

[158] LexisNexis, 'residential mortgage-backed securities' (n. 30).

[159] The Guidelines clarify that the assessment relates to issuers' activities in general and not to specific projects or assets. OECD, 'Due Diligence for Responsible Corporate Lending and Securities Underwriting' (n. 74) 38.

a 'private law' initiative[160], these Guidelines provide a starting point for understanding the role that underwriters could play in 'sustainable' securitisations. In particular, they demonstrate the great 'leverage' that these actors could have on sustainability matters, especially ahead of the issuance. This power considerably decreases after the sale of securities, as they cease to owe any contractual duties to issuers or noteholders.[161] The Guidelines also encourage a proactive approach whereby underwriters react to threatened or current adverse impacts on sustainability matters by demanding further information or engaging with their clients.[162] Therefore, applying the Guidelines to 'sustainable' securitisations would elevate underwriters to 'gatekeepers', who provide an external check on sustainability qualities before the issuance.[163] While they encourage financial institutions to incorporate their policies and expectations from counterparties in their contracts[164], their main shortcoming is that they fail to specify how issuers and underwriters should articulate sustainability-related commitments across the clauses of their 'underwriting agreements'.

Standard documents which may assist parties in such endeavour include, among others, the model 'subscription agreements' designed by the LMA[165] and by the European Private Placement Group (EuroPP), bringing together several industry associations, including ICMA.[166] Both sets of

[160]On the role of OECD as a trans-governmental network and as a rule-maker, see Jacint Jordana, 'Transgovernmental Networks as Regulatory Intermediaries: Horizontal Collaboration and the Realities of Soft Power' in Regulatory Intermediaries in the Age of Governance (2017) 670 *The Annals of the American Academy of Political and Social Science* 245, 247–248.

[161]Except for the circumstances where underwriters are interested in maintaining the relationship with issuers for the purposes of future transactions. OECD, 'Due Diligence for Responsible Corporate Lending and Securities Underwriting' (n. 74) 36–37.

[162]Ibid., 36–37.

[163]On the notion of 'gatekeepers', see Chap. 2, Sect. 2.6.

[164]Ibid., 49.

[165]LMA, 'Pan-European Private Placement Sect.cription Agreement' (2020) https://www.lma.eu.com/documents-guidelines/documents; as summarised by LexisNexis, 'Key_features_in_LMA_PEPP_documentation' (2023) https://www.lexisnexis.co.uk/legal/guidance/key-features-in-lma-pepp-documentation, accessed 10 December 2025.

[166]Particularly the 'Sample Sect.cription Agreement and Bond Terms and Conditions' within EuroPP, 'EuroPP Charter' (2019) 1, 31 ff. https://eact.eu/Core/Images/Documents/Euro%20PP%20Charter%20(October%202019).pdf; https://www.euro-privateplacement.com/wp-content/uploads/2024/01/Euro_PP-Note-EN-2019.pdf accessed 10 December 2025.

model provisions were originally designed for private placements of debt securities, but there is no reason to rule out their applicability to ABSs[167]

While such model agreements contain no reference to sustainability goals, they may lend themselves to strategic adjustments.

For instance, some model clauses grant underwriters some control over issuers' activities. One example is 'conditions precedent' which originating entities have to meet, such as the submission of specific reports or compliance with certain legal requirements.[168] They are generally a prerequisite for the settlement of the operation: In other words, no duty to market the securities arises if issuers do not comply. [169] There are also model 'representations',[170] where issuers confirm facts, such as their unawareness of any event making the information in the offering material untrue or incorrect. As previously discussed, the incorrectness of these statements could expose issuers to liability vis-à-vis investors for misstatements. [171]

Parties willing to incorporate sustainability considerations may, among others, adjust model conditions precedent to compel issuers to report on sustainability matters. They may draw on model representations to require issuers to confirm that they comply with sustainability standards such as the Green Bond Principles.[172] In so doing, the clauses of underwriting agreements elevate underwriters to 'gatekeepers' of issuers' sustainability practices. The fact that they subscribe the issuance would attest to the reliability and integrity of the 'sustainable' label of the issuance.

[167]On the differences between ABSs traded on capital markets and private placements, and on ABSs as debt securities see Chap. 3, Sect. 3.3.3.

[168]Wood (2019) 5-011.

[169]See LexisNexis, '*Residential mortgage backed securities*'; Wood (2019) *25-040, For example, 'The undertaking by the Sect.criber[s] (...) to Sect.cribe and pay for the Notes on the Settlement Date is subject to satisfaction of the following conditions precedent: (a) (i) the absence, on the Settlement Date, of any event or change rendering any one of the undertakings made or any one of the representations made or warranties given, under the terms of Articles 3 and 5(A) of this Agreement untrue or incorrect, in the same terms as if they had been formulated, given or made on such Settlement Date [...] (b) the submission to the Sect.criber[s], at the latest by the Settlement Date, of the following documents: (i) [a statement of the debts of the Issuer [and those of its Sect.idiaries/Material Sect.idiaries];] [...]'* Clause 6 'Conditions Precedent', EuroPP (n. 166).

[170]For examples of representations in underwriting agreements, see also Geoffrey Fuller, *The law and practice of international capital markets* (3rd edn, Lexisnexis 2012) paras 7.220–7.224.

[171]LexisNexis, 'Key features in LMA PEPP documentation' (n. 165).

[172]On the Green Bond Principles, see Sect. 5.4.2.1. and 5.5.2.

Another possibility for market actors may be 'warranties' for underwriting agreements, which the Chancery Lane Project has designed. [173] They allow issuers to reassure underwriters that they have, among other things, implemented the TCFD, complied with minimum human rights and labour standards, and that there are no pending claims or litigation concerning social and labour abuses. For instance, one warranty reads as follows: *'Additional Warranties. (...) The Company has implemented in full the recommendations of the Task Force on Climate-related Financial Disclosures in relation to the governance, strategy, risk management and disclosure of climate risks and opportunities'.* [174]

As previously mentioned, the breach of a warranty would entitle underwriters to compensation.[175]

5.5.1.2 Documentation for the Offer of ABSs

The documentation for the offer of securities to investors also serves a strategic function. As previously discussed, this may include prospectuses or more general 'terms and conditions' and 'offering circulars'. It is not only relevant to disclose key financial characteristics of the securities, but also to define the perimeter of issuers' liability.[176] Misstatements in prospectuses can give rise to two types of remedies: Civil liability vis-à-vis investors, which is strictly dependent on the terms of the documentation, and regulatory sanctions, which will also be discussed in the following chapter.[177]

The content of prospectuses tends to be uniform across EU Member States as a result of maximum harmonisation efforts.[178] The 2017 Prospectus Regulation, applicable to most securitisation transactions with some exceptions, prescribes the items that market actors must include to trade ABSs within the EU market.[179] Some of the mandatory elements may provide information on sustainability matters: For instance, prospectuses should include a clause on the 'use of proceeds'.[180] As explained in

[173] See 1.5, 1.17, 1.18 Chancery Lane Project, 'Dotties' Clause – –Climate Underwriting Sponsor Warranties' (*Chancery Lane Project,* 2021) https://chancerylaneproject.org/climate-clauses/climate-underwriting-sponsor-warranties/ accessed 10 December 2025. The clause was however discontinued.

[174] clause 1.5 ibid.

[175] See above Sect. 5.4.2.3.

[176] See Chap. 3, Sect. 3.3.3.

[177] See Sect. 6.5.1.

[178] i.e. on the notion of maximum harmonisation, see Chap. 6, Sect. 5.2.1.

[179] Regulation (EU) 2017/1129. See Chap. 6, Sect. 6.5.1.

[180] Annex I, IV, C. Regulation (EU) 2017/1129.

Chap. 3, 'sustainable' ABSs (especially 'green' and 'social') allow issuers to finance specific environmentally friendly activities. Thus, the 'use of proceeds' clause is a valid reference point to list these objectives, as also evidenced by previous 'sustainable securitisation' transactions. [181]

In addition, the requirement to disclose the 'risk factors'[182] of the issuance may leave room for issuers to provide information about the sustainability-related risks related to the portfolio or about the planned projects.

Beyond the mandatory requirements, there are other contractual conditions for the offer of ABSs (in prospectuses or similar documents) that may serve a strategic function.

For instance, parties may add additional conditions that have to be met after the issuance of ABSs ('covenants').[183] To this end, they may draw from model provisions such as those drafted by the Bond Covenant Group[184] for issuances in euro-sterling.[185] For instance, they may adjust model covenants on reporting[186] to require issuers to regularly inform investors about sustainability matters.

They may also adjust the above-mentioned LMA model provisions for private placements[187] or ISDA documentation for derivatives[188] to introduce new remedies, allowing investors to take action when issuers do not comply with sustainability-related targets or do not successfully implement

[181] See Chap. 4, Sect. 4.7.1.

[182] Annex I, IV, D Regulation (EU) 2017/1129. On the relevance of risk factors clauses, see also Joseph Tanega, *Securitisation law* (n. 38) 2.69.

[183] The LMA – EuroPP group documentation does not specifically include these clauses, but it recommends their use as a best practice in the documentation for the issuance of bonds, LexisNexis, 'Key features in LMA PEPP documentation' (n. 165). Clause 2.2. The Investment Management Association and the Bundesverband Investment und Asset Management, The Bond Covenant Group, 'Model Covenants in Sterling and Euro Bond Issues' (2010) 2 https://www.theia.org/sites/default/files/press-releases/document/20100621-01.pdf accessed 3 June 2026.. UNPRI has also noted their strategic role for ESG bond issuances, see UNPRI, 'ESG Engagement For Fixed Income Investors' (n. 62) 34.

[184] Bringing 4 investment associations together, i.e. the Association of British Insurers, The National Association of Pension Funds, The Investment Management Association and the Bundesverband Investment und Asset Management, The Bond Covenant Group, ibid., 2.

[185] Ibid.

[186] For example, *'So long as any of the Bonds are outstanding, the Issuer shall [...] post on the Website: (i) as soon as they are available, but in any event within 120 days after the end of each of its financial years, its audited consolidated financial statements for that financial year'* Clause 2.5.2. a) i) ibid.;

[187] LexisNexis, 'Key features in LMA PEPP documentation' (n. 158).

[188] This is especially common across synthetic transactions, see David Ramos Muñoz, Steven Ingram, *The law of transnational Securitisations* (Oxford University Press 2010) 104

'sustainable' projects. These events may, among others, represent an 'Event of Default'[189], allowing ABS holders to accelerate the repayment of their securities. Alternatively, they may stipulate a 'Declassification Event', such as the one proposed by LMA for sustainability-linked loans[190], where issuers would need to cease qualifying ABSs as 'sustainable' should they fail to meet the related targets.

However, all these possible adjustments, while promising, largely represent theoretical options which do not find a widespread application in market practice. Chap. 4 has already compared various examples of documents for the offer of 'sustainable' ABSs, where clauses like conditions precedent, covenants and Events of Default did not make any reference to sustainability objectives.[191] Further, there is empirical evidence that market documentation for sustainability-linked bonds (and, in turn, for ABSs) is very inconsistent in relation to sustainability-related targets, KPIs and penalties, with frequently inadequate penalty mechanisms.[192] This further illustrates that the lack of sustainability-related standard clauses and the need for market actors to adjust other legal tools for this purpose result in high transaction costs which they are unlikely to incur.

5.5.2 Sustainability-Related Documentation

Sustainable finance initiatives have recently gone further in the standardisation of the documents that specifically characterise 'sustainable' ABSs.

In line with LMA's work on loans,[193] the leading financial association operating in capital markets, ICMA,[194] has designed various guidelines on what makes debt securities 'sustainable',[195] which consequently

[189] In securitisation market practice, they can also have different denominations like 'credit events' (especially across synthetic securitisations) or 'Early Amortisation Events', see Ramos Muñoz, ibid.

[190] See above Sect. 5.4.2.

[191] See Chap. 4, 4.7.1.–4.7.2.

[192] ESMA, 'TRV Risk Monitor – –ESMA Report on Trends, Risks and Vulnerabilities' (2023) 36 https://www.esma.europa.eu/sites/default/files/library/ESMA50-165-2438_trv_1-23_risk_monitor.pdf. Accessed 10 December 2025.

[193] See above Sect. 5.4.2.1.–5.4.2.2.

[194] See Sect. 5.2.2.

[195] Other guidelines outside the scope of the present analysis are, among others, those on Climate Transition Bonds. ICMA, 'Climate Transition bond guidelines' (2025) https://www.icmagroup.org/assets/documents/Sustainable-finance/2025-updates/Climate-Transition-Bond-Guidelines-CTBG-November-2025.pdf accessed 10 December 2025.

apply to securitisation.[196] The most popular ones include the Green Bond Principles, Social Bond Principles and Sustainability-Linked Bond Principles (the GBPs, the SBPs and the SLBPs, respectively).[197] Chap. 4 has highlighted the wide reference to these Principles across 'sustainable' securitisation deals.[198] This evidences their fundamental contribution to the standardisation of market practices. Chap. 4 has also illustrated the different focus of the sets of Principles, which replicates the parallel distinction between 'green' or 'social' and 'sustainability-linked' loans.[199] While the 'green' and 'social' labels tie to the types of activities that issuers implement through the finance raised from investors ('use of proceeds'), the 'sustainability-linked' qualification follows from the connection between specific sustainability-related targets (SPTs) and the financial conditions of ABSs (such as the interest rate). All sets of Principles also recommend additional best practices for issuers on matters like the steps for the identification of projects, reporting and external reviews of 'sustainable' qualities.[200] As a complement, ICMA has also designed more detailed guidelines on some of these aspects, including external reviews and reporting.[201]

Moreover, ICMA has also issued more succinct Sustainability Bond Guidelines, which more generally define 'sustainability bonds' as those that contribute *both* to environmental and social goals, while also cross-referring to the other above-mentioned sets of principles.[202]

[196] On ABSs as debt securities, see Chap. 3, Sect. 3.3.3.
[197] ICMA, 'GBPs' (n. 39); ICMA, 'SLBPs' (n. 39). The last versions of GBPs and SBPs explicitly clarify that they apply to securitisations ('Secured Green/social collateral Bonds' and 'Secured Green/social standard bonds') ICMA, GBPs, 8. ICMA, SLBPs, 7.
[198] See Chap. 4, 4.3.
[199] See Chap. 4, Sect. 4.
[200] The SLBPs name the latter process 'verification'. For clarity, I will use the term 'review' as an umbrella term, encompassing second party opinions, verifications, ratings and certifications. ICMA has also used the term 'review' in the same vein, see ICMA, 'Guidelines for Green, Social and Sustainability Bonds External Reviews' https://www.icmagroup.org/assets/documents/Regulatory/Green-Bonds/June-2020/External-Review-GuidelinesJune-2020-090620.pdf accessed 10 December 2025; see also Chap. 4, Sect. 4.6.4., 4.7.3.
[201] Ibid. See also ICMA, 'Handbook Harmonised Framework for Impact Reporting' 1 (2022) 22–30 https://www.icmagroup.org/assets/documents/Sustainable-finance/2022-updates/Harmonised-Framework-for-Impact-Reporting-Green-Bonds_June-2022-280622.pdf accessed 10 December 2025.
[202] IMCA, 'Sustainability Bond Guidelines' (2021) https://www.icmagroup.org/assets/documents/Sustainable-finance/2021-updates/Sustainability-Bond-Guidelines-June-2021-140621.pdf accessed 10 December 2025.

ICMA's work also went beyond the design of high-level guidelines. Remarkable contributions have been various templates for the various phases of the issuance of these securities. For instance, a model information template should facilitate the *ex ante* disclosure of information about the identified projects, and an illustrative summary template should facilitate regular *ex post* reporting. There is also a review form that should point out the aspects that all service providers should evaluate when reviewing bonds.[203] Information templates certainly represent a step ahead in facilitating the 'disclosure' of what makes ABSs 'sustainable', which may improve the current lack of transparency surrounding the market observed in Chap. 3. It may also strengthen the role of external reviewers as effective 'gatekeepers' of these transactions. In particular, Chap. 4 highlighted how the current divergences in methodologies across various reviews or reviewers may not ensure in-depth control over the 'sustainable' credentials of securitisation programmes.[204] By contrast, model review templates may provide insight into the steps that they should follow and the elements that they should assess. As I will further discuss in Chap. 6, complementary statutory measures are, however, necessary to ensure the consistency of reviews.[205]

Moreover, ICMA has engaged in preliminary work towards the identification of promising metrics for green and sustainability-linked securities. These consist of indicators within the previously mentioned guidance on reporting and within the SLBPs, which should allow issuers to quantify the contribution of their green, social and sustainability-linked bonds to sustainability objectives.[206] They should facilitate the drafting of reports on green bonds' impact. Besides, ICMA has developed a registry of KPIs on environmental and social matters for the structuring of sustainability-linked bonds, also illustrating their application so far. [207] These KPIs could

[203] ICMA, 'Templates for issuers and reviewers' https://www.icmagroup.org/sustainable-finance/sustainable-bonds-database/#Templatesforissuers accessed 10 December 2025.

[204] See Chap. 4, Sect. 4.6.4., 4.7.3.

[205] See Chap. 6, Sect. 6.5.4.

[206] ICMA, 'Illustrative Examples for the Selection of Key Performance Indicators (KPIs) for Sustainability-Linked Bond Issuers, Underwriters and Investors' https://www.icmagroup.org/assets/documents/Sustainable-finance/2022-updates/Registry-SLB-KPIs_Final_2022-06-24-280622.xlsx accessed 10 December 2025; ICMA, 'Core indicators' in 'Impact reporting' (n. 201) 10–43.

[207] ICMA, 'Illustrative KPIs registry' (2022) https://www.icmagroup.org/assets/documents/Sustainable-finance/2022-updates/Registry-SLB-KPIs_Final_2022-06-24-280622.xlsx accessed 10 December 2025.

also act as standard metrics from which originators may draw when they set the targets of the issuance. It is, however, important to note that legislation plays a complementary role to these standardisation tools by providing a yardstick to quantify the impact of ABSs. For instance, market actors can label their securities as being in line with the EU Green Bond Standard as long as they allocate the proceeds to activities complying with the technical criteria of the Taxonomy Regulation.[208] A centralised process for external reviews will ensure that all issuers align with the Taxonomy criteria.[209] This example illustrates how 'hard law' requirements can more effectively foster consistent market practices than 'soft' legal tools when they are linked to a 'public' system of external oversight.[210]

Despite the abundance of standardisation tools, which also go as far as providing templates that market actors can directly apply, model clauses for the underwriting and offer of 'sustainable' ABSs are still missing. I have argued elsewhere that the lack of standard clauses represents a persistent barrier to the growth of 'sustainable' instruments.[211] Similarly, the lack of model provisions to articulate the sustainability-related conditions for the offer of ABSs (such as representations, covenants, Events of Default) leaves room for high transaction costs and inconsistent practices.

5.5.3 Normative Recommendations

The gaps in previous initiatives highlight the large scope for further standardisation tools tailored to the issuance of 'sustainable' ABSs.

First, further guidelines, building on the already existing ones by the OECD, should clarify the role of underwriters prior to the offering of 'sustainable' securities. It would be particularly advisable for private actors, such as industry associations, to engage in the design of these guidelines due to their previous work on 'sustainable' ABSs and their general ability to bring together market best practices. ICMA and LMA's previous involvement in similar initiatives also suggest that these associations would

[208] See Chap. 6, 6.4.4, 6.5.2.

[209] On enforcement powers as a corollary of EU harmonisation, See Chap. 6, Sect. 6.2.

[210] Yet, on the flaws of the EUGBS, See Chap. 6, 6.4.4, 6.5.2. On the boundaries of standardisation, see Sect. 5.9.

[211] Federica Agostini, 'From 'Green Bond Principles' to 'Green Bond Clauses': Mitigating Greenwashing Through Contract Law' in Andenas, M., Heidemann, M. (eds) *Quo vadis Commercial Contract?* (Springer 2023) LCF Studies in Commercial and Financial Law, Vol. 1 https://doi.org/10.1007/978-3-031-14105-8_6

have the interest and expertise to engage. These guidelines should elevate underwriters to 'gatekeepers' of the 'sustainable' qualities of the issuance. For instance, they should recommend that these actors have a say over the documentation, negotiating the types of projects that the proceeds will finance (in the case of 'green' or 'social' ABSs) or the sustainability targets (in the case of 'sustainability-linked' ABSs[212]) together with issuers. A valid reference point could be a separate set of LMA guidelines on 'sustainability coordinators'. While these cater to loans, they also recommend the direct involvement of a market actor in arranging the sustainability-related elements of the documentation with the issuer.[213] Similarly, the proposed guidelines should clarify how underwriters should intervene ahead of the issuance of ABSs.

Another aspect to take into consideration is the benefits for underwriters to act as 'gatekeepers' of 'sustainable' issuances. There is ample anecdotal and empirical evidence of the involvement of banks and large investment funds in structuring previous green bond and green securitisation issuances.[214] This demonstrates market actors' wide interest in acting in this capacity, also due to reputational advantages and profits.

Additionally, a standard underwriting agreement could assist in negotiating sustainability-related elements. It should explicitly introduce sustainability-related conditions, requiring underwriters to perform robust due diligence assessments of issuers' environmental and social practices, in line with OECD guidelines. At the same time, it would be unreasonable to demand that underwriters evaluate the sustainability profile of the portfolio assets in the absence of exhaustive information about them. As a result, standard underwriting agreements should include 'conditions precedent', compelling originating entities to provide reports on the environmental and social risks and impact of the portfolio assets or of the projects to which the proceeds of 'sustainable' ABSs will be allocated. These clauses would allow underwriters to 'screen' the existence of severe risks and also to intervene by performing additional assessments at asset level or project

[212] On the distinction, see above, Sect. 4.5.1., and Chap. 3, Sect. 3.2.

[213] LMA, 'Introduction to the sustainability coordinator role', 3 https://www.lma.eu.com/sustainable-lending/documents#; .Also cited by LMA, Loan Market Association, 'Sustainability-Linked Loan Principles' (n. 101) 3.

[214] See, inter alia, the top 15 lead managers in green and sustainability-linked bond issuances in 2022, mainly including large financial institutions. Environmental Finance, 'Sustainable Bonds Insight 2023' (2023) 27. https://www.environmental-finance.com/assets/files/research/sustainable-bonds-insight-2023.pdf accessed 10 December 2025.

level, or by directly engaging with issuers. The agreement should also link these conditions to remedies for underwriters. For instance, underwriters should have the right to refrain from subscribing to the securities if originating entities fail to comply with the conditions precedent.

Furthermore, there is room for additional standard clauses targeting the documentation linked to the offer of ABSs. It is, however, essential to ensure their compatibility with the other statutory requirements for certain 'sustainable' ABSs under the EUGBS, which I will discuss more extensively in Chap. 6.[215]

For instance, there is merit in designing standard 'Use of Proceeds' and 'Risk Factors' clauses, which market actors could replicate in order to establish commitments about the allocation of investors' funds and to disclose the adverse environmental and social risks of the portfolio. In the case of 'sustainability-linked' ABSs, standard clauses could notably provide standard formulations for sustainability targets, KPIs to measure them, and the changes to financial conditions in case of non-compliance.

Building on Chancery Lane's model 'warranties' on TCFD, there is also room for the design of standard 'representations and warranties', where originating entities confirm that they have assessed all climate-related and other sustainability risks linked to the portfolio and to the issuance. [216] As previously discussed, the inclusion of these clauses would expose issuers to liability vis-à-vis investors for misstatements. Therefore, including them in the prospectus would give rise to additional statutory remedies.[217]

Standard clauses should also address the ambiguity currently arising from the documents for the offer of ABSs regarding investors' remedies when originators do not meet sustainability-related goals. Building on ICMA's and Chancery Lane's model provisions, a promising option may be to provide alternative clauses from which market actors may choose. For instance, there may be a model 'Event of Default' clause, allowing investors to demand early repayment of their ABSs if originators do not comply with 'use of proceeds' or 'representation' clauses, or if they fail sustainability-related targets. [218] Another option is a standard 'Declassification Event' clause, similar to the one proposed by the

[215] See Sect. 6.4.4, 6.5.2.

[216] The Chancery Lane's Dotties' clause (n. 174) is more focussed on the implementation of TCFD at corporate level. The proposal in the present chapter would focus on the risk assessment at portfolio and/or asset level. See above Sect. 4.5.2.

[217] See Chap. 6, Sect. 6.5.1.

[218] See above Sect. 4.6.1.

Chancery Lane Project for 'sustainable' loans, which would require issuers to stop categorizing their ABSs as 'green' or 'sustainability-linked'.

Offering various alternative clauses for parties to choose from may also address some of the objections that the expansion of Events of Default may raise.

For instance, ICMA itself and another industry association commented on one of the versions of the proposal for an EU Green Bond Standard, raising concerns that broadening the circumstances triggering default may deter market actors from issuing 'green' securities.[219] EBA also warned against the systemic threats that may derive from linking ESG objectives to 'Events of Default', especially when bonds are issued by banks and they also serve to absorb other losses on the balance sheets.[220] It is important to note that the need to advance sustainability objectives through ABSs does not justify disproportionate legal tools. As a result, it is arguably not advisable to design clauses qualifying the non-compliance with sustainability targets as an event triggering 'Cross-default'. Such standard provisions, common in bond documentation, would provide for the default in all other obligations towards other creditors,[221] potentially opening the door to insolvency and broader systemic risks. Nevertheless, 'Events of Default', not affecting other obligations in banks' balance sheets, could introduce a good deterrent mechanism to encourage the compliance with sustainability-related clauses. Therefore, allowing market actors to choose between those and 'Declassification Events' may be an appropriate compromise, allowing market actors to shape remedies based on their preferences and the characteristics of the transaction.

[219] ICMA, 'Updated analysis of the proposals for the EuGB Regulation' (22 June 2022) 2–3. https://www.icmagroup.org/assets/EU-GB-Updated-ICMA-commentary_220622.pdf; accessed 10 December 2025; Federation for European Securities Exchanges, 'Position Paper on Green Bond Prospectus'(2019) 1–2. https://www.fese.eu/app/uploads/2019/04/190429-fese-position-paper-on-green-bond-prospectus.pdf accessed 3 June 2026.

[220] Specifically in relation to Additional Tier 1 instruments (AT1), even though the argument could also be transposed to ABSs more generally. EBA, 'Report on the Monitoring of Additional Tier 1 (AT1) European Union Institutions – Update', Vol. EBA/REP/20 (2021) 35-37 https://www.eba.europa.eu/publications-and-media/press-releases/eba-updates-monitoring-additional-tier-1-tier-2-and-tlacmrel-eligible-liabilities-instruments accessed 3 June 2026.

[221] See Chap. 3, Sect. 3.4.2.1. Lars Hornuf, Markus Reps, Stefan Schäferling, 'Covenants in European Investment-Grade Corporate Bonds' (2015) 10 *Capital Markets Law Journal* 345, 356; William W Bratton, 'Bond Covenants and Creditor Protection: Economics and Law, Theory and Practice, Sect.tance and Process' (2006) 7 *European Business Organisation Law Review* 39, 18.

As I will also discuss in Chap. 5, it is essential to coordinate such standard clauses with the requirements to publish standalone documentation before and after issuing 'sustainable' ABSs under the upcoming EU Green Bond Standard. This comprises, among others, a 'factsheet', describing the planned sustainability-related projects, as well as regular 'allocation' and 'impact' reports. When drafting such documentation, market actors and reviewers have to follow statutory templates.[222] As a result, there will be very limited margin for further standardisation initiatives on the terms of issue for 'sustainable' ABSs across EU countries.

Overall, the proposed standard clauses would provide significant support to market actors willing to issue 'sustainable' ABSs. One may argue that they may be reluctant to apply them due to their perceived higher litigation risk if they do not comply with the provisions. In fact, there is anecdotal evidence of US companies shying away from issuing green bonds under the assumption that these securities would expose them to investors' claims.[223] However, issuers of 'green' ABSs will already be subject to a series of disclosure requirements under EU law, already exposing them to liability risks, as I will discuss in Chap. 5. For this reason, the marginal costs of introducing additional constraints, like warranties or events of default, drawing from model documentation, would be marginal. Conversely, these additional commitments may lead to potential reputational advantages for issuers since they would testify to their commitment to sustainability objectives, attracting prospective underwriters and investors. [224]

The other advantage of incorporating 'sustainability-related' clauses within prospectuses is to expand the remedies available to investors, encouraging them to raise their 'voice'[225] and to hold issuers accountable for their sustainability goals. The following chapter will also discuss additional legal avenues to do so, based on the documentation for the offer of ABSs. 925

[222] Annex I – II – III EUGBS.

[223] Kate Allen, 'Strict US Market Rules Limit Corporate Sellers of Green Bonds' (20 February 2018 *Financial Times*) https://www.ft.com/content/baa217c4-157c-11e8-9376-4a6390addb44 accessed 20 July 2025.

[224] Indeed, UNPRI noted that an increasing number of investors is pressing for a wider incorporation of information on ESG factors within the documentation of securitised products, UNPRI, 'ESG Incorporation in Securitised Products' (n. 140) 5.

[225] See Chap. 2, Sect. 2.6.

5.6 (The Case for) Standardisation Initiatives Concerning the Governance of SPVs

Chap. 3 has described the role of special purpose vehicles (SPVs) as intermediating entities, especially created for the purposes of issuing ABSs and managing these operations. One of their distinctive features is their specialisation in these transactions, which also results in limited business activities on any matter other than securitisations. There is anecdotal evidence that the documentation for the constitution of these entities generally reflects such purpose of SPVs, while introducing constraints on their sphere of action.[226] As previously mentioned, this documentation entails articles of association when SPVs are companies, trust deeds when they hold securitisation assets as trustees, or management regulations when they are investment funds.[227] However, there have not been any remarkable standardisation initiatives that have created uniform guidelines or standard provisions for the constitution and governance of SPVs. A 2009 study by the Bank for International Settlements (BIS) also confirmed the divergences across market practices: For instance, it highlighted only limited similarities between the constitutional provisions of SPVs across different transactions.[228]

Self-evidently, the lack of standard legal tools for SPVs in general also results in a lack of guidance on setting up these entities for 'sustainable' transactions and on their governance.

Admittedly, the lack of standard legal tools reflects the discretion traditionally granted to boards to manage companies. However, the case of securitisations promoting a 'sustainable' purpose is inherently different. This also translates into a need for additional legal tools to ensure that such entities focus their activities on the allocation of investors' funds towards sustainability-related projects and targets.

As a result, there is arguably room to develop standard 'constitutional' documentation, such as model Articles or management regulations, for SPVs within 'sustainable' transactions. A point of reference for the design of standard provisions could be, among others, the 'Example Objects' clauses, available for UK charity companies. These provide various

[226] See Chap. 3, Sect. 3.3.2. see also Pierre De Gioia Carabellese, Camilla Della Giustina, *The law of securitisations: From crisis to techno-sustainability* (Routledge 2023) 63.

[227] See Chap. 3, Sect. 3.2.–3.3.2.

[228] BIS, 'Report on special purpose entities' (2009) 44 https://www.bis.org/publ/joint23.pdf accessed 20 December 2025.

examples of how market actors could draft constitutional documents articulating what charitable purpose these companies pursue.[229] While these model clauses represent the result of guidance by a public body (i.e. the UK Charity Commission), there is no reason to rule out the possibility of similar initiatives at market level (through 'private law'). Therefore, financial industry associations or other professional bodies could design standard 'Object' clauses to incorporate within SPVs' constitutional documents. These could clarify that the implementation of sustainable projects or expenditure on sustainability-related targets is the sole purpose of their constitution and existence. A model covenant should also introduce a restriction on SPVs' activities, so that they do not engage in parallel transactions or incur any further expenditure other than that for 'sustainable' securitisation operations.

It is important to note that such clauses would not prevent SPVs from acting outside their scope or render any other acts invalid.[230] Instead, they would lay out the parameters to assess the scope of directors' duties and their liability standard. Chap. 2 has highlighted how courts have hardly ever established that SPVs' directors were liable vis-à-vis investors for breaches of their duties.[231] However, standard documentation arguably has the power to indirectly shape such duties, since acts and decisions deviating from the main 'object' may signal a breach.[232] 'Greenwashing' practices, such as allocating investors' proceeds to different goals from the ones identified before the issuance of ABSs, might signal that directors are not acting in line with SPVs' purpose and thereby breaching their duties.

The advantage of such standardisation tools would be their easy replicability across any type of constitutional document for SPVs. They would

[229] See, among others, Charity commission, 'Example objects – sustainable development' https://www.gov.uk/government/publications/example-charitable-objects/example-objects-sustainable-development (2013) accessed 10 December 2025.

[230] As the history of object clauses in UK suggests. They are no longer a requirement for the incorporation of companies under the Companies Act 2006, but still virtually possible and common. They also do not constrain directors' decisions and transactions, which still bind companies even when they go beyond the perimeter of object clauses. See Art 39 Companies Act 2006 (UK). Cristopher Hare, Sarah Worthington, PL Davies, *Gower: Principles of Modern Company Law* (11th edn, Sweet & Maxwell 2021) para 8-029. See also Chap. 3, Sect. 3.4.3.

[231] Ibid. See Chap. 3, Sect. 3.4.3.

[232] For example, under UK law, directors must act within the purposes of their powers. See SS 171, 172 Companies Act 2006 (UK). For a similar provision from an EU jurisdiction, see Art 2380-bis (1) cc (Italy).

considerably cut transaction costs, thus incentivising the creation of these entities and, in turn, 'sustainable securitisations' in general.

Despite the potential of the proposed model clauses, they would still leave room for market actors not to adopt them. Therefore, the complementary legal tools that could ensure a uniform application, holding directors accountable for their decisions on sustainability matters, are arguably statutory requirements. Chap. 6 will further illustrate how EU law can complement the proposed standardisation tools.

5.7 Investors' Representation Mechanisms

5.7.1 For 'Traditional' Securitisations

Other elements of securitisations showing a strong transformative impact are a series of mechanisms for the representation of investors in ABSs. As explained in Chap. 3, a lead investor is generally in charge of protecting their interests (raising their 'voice'). Legally speaking, they generally act as investment managers or trustees, depending on the surrounding legal framework.[233]

Various parts of the documentation shape the duties of investors' representatives as well as their powers vis-à-vis originating entities. For instance, ABS holders may use the clauses of the contract for their appointment (an 'Intermediation'[234] or 'Investment agreement'[235]). If representatives act as trustees, a trust deed (also defined as 'Indenture', especially in US) is likely to delineate their function[236]: this gives rise to a duty to act in strict accordance with its terms.[237]

Alternatively, securitisation documentation also sometimes includes an additional inter-creditor agreement, which differentiates the rights of various tranches and categories of investors.[238] It is also sometimes common

[233]See Chap. 2, Sect. 3.4.4.

[234]Lahlou, Tanega (n. 29) 361, 363.

[235]Ramos Muñoz (n. 188).

[236]For an example by Natwest, see Arran Residential Mortgages Funding 2010-1 Plc As Issuer-Citicorp Trustee Company Limited As Trustee, Trust Deed, https://investors.natwestgroup.com/~/media/Files/R/RBS-IR-V2/securitisation/reseidential-mortgages/pdf/7-arran-residential-mortgages-2010-1-plc/principal-transaction-documents/015--trust-deed-2010-a1.pdf accessed 10 September 2025.

[237]See For example, under UK law *Clough v Bond* (1838) 3 My & Cr 490, 496-7

[238]As anecdotally reported by David Ramos Muñoz, 'Bankruptcy-Remote Transactions and Bankruptcy Law-a Comparative Approach (Part 1): Changing the Focus on Vehicle

across issuers of debt instruments (e.g. across master trust structures) to set out investors' rights in advance through the documentation for the offer of ABSs: this means that prospectuses or 'terms and conditions' may also designate representatives and articulate their rights.[239]

Various standardisation tools, albeit not securitisation-specific, may support market actors in the design of the clauses delineating investors' representation mechanisms. For instance, LMA has developed various types of standard inter-creditor agreements.[240] A popular standard 'indenture' for parties to appoint trustees is the one by the American Bar Association (hereinafterABA).[241] Even the Investment Association (hereinafterIMA), gathering different investment firms across the UK and Europe,[242] has drafted a model discretionary investment agreement. [243] There is, however, no evidence of a wide preference for any of these agreements among securitisation market actors. On the contrary, a closer analysis of market practices reveals a wide discrepancy. For instance, the clauses of securitisation agreements are not consistent in the extent to which representatives must monitor ABSs in the course of the transaction and identify whether Events of Default have materialised.[244] They also differ in the level of discretion granted to representatives to enforce ABSs. Upon a review of RMBS documentation, the above-mentioned SFIG network has confirmed that trustees can sometimes independently decide to accelerate ABSs, or they sometimes need to receive instructions from a third-party

Shielding' (2015) 10 *Capital Markets Law Journal* 239, 20. For an analysis of the function of intercreditor agreements when securities are tranched and secured by a pool of assets, while not securitisation-specific, see KE Belln, PS Burke, Michael Girondo et al, 'Comparing Intercreditor Arrangements' (2015) LSTA Loan Market Chronicle 2015, 139 https://webstorage.paulhastings.com/Documents/PDFs/comparing_intercreditor_agreements.pdf accessed 10 December 2025.

[239] Steven L Schwarcz, 'Commercial Trusts as Business Organisations: Unraveling the Mystery' (2003) 58 Business Lawyer 559, 9–10. For the various types of documents for the offer of ABSs, See Chap. 3, Sect. 3.3.3.

[240] All available to LMA Sect.cribers at: https://www.lma.eu.com/documents-guidelines/documents accessed 10 December 2025

[241] ABA, 'Revised Model Simplified Indenture' (2000) 55 *the Business Lawyer* 3 1115, particularly 1139–1144 https://www.jstor.org/stable/40657063?seq=1 accessed 10 December 2025.

[242] IMA, 'About us' https://www.theia.org/about-us (2021) accessed 10 December 2025

[243] IMA, 'Model Discretionary Investment Management agreement' . https://www.theia.org/sites/default/files/2019-08/Model_IMA_website_version.pdf accessed 31st March 2021.

[244] SFIG (n. 32) 257, 258, 263. Ramos Muñoz (n. 188) 384–385. Chap. 3, Sect. 3.4.4.

reviewer or from a majority of investors.[245] There are also divergences in the rights of other noteholders: In some transactions, they may retain their concurrent right to take action on their ABSs. In others, a 'no-action' clause prevents noteholders other than the representative from bringing parallel lawsuits.[246]

As discussed in Chap. 4, the frequent confidentiality of agreements such as investment agreements also prevents an accurate assessment of the extent to which market actors have incorporated sustainability-related clauses.[247] It could be reasonable to assume that one promising entry point could be the clause within investment agreements where 'investment objectives' are set out.[248] Correspondingly, an analysis by the United Principles for Responsible Investment (UNPRI) network has highlighted that investment management agreements between asset managers and asset owners are in most cases the key source for articulating sustainability-related objectives.[249] Therefore, the report has recommended amending the terms of trust deeds and other intermediation agreements in order to require them to pursue a sustainability impact.[250] These documents may also introduce decision-making fora for investors or set out the circumstances in which representatives may act on investors' behalf, including for the enforcement of ABSs.[251]

The inconsistencies in the formats for the appointment of these representatives are particularly problematic for 'sustainable' transactions. Chaps. 2 and 3 have already highlighted the coordination costs and challenges among investors, which restrict their ability to raise their 'voice'.[252]

[245] SFIG, RMBS 3.0 (n. 77) 200, 201, 205, 346.

[246] For various examples of clauses on their rights, see ibid., 252–258. On the 'No-action' clause, see Chap. 3, Sect. 3.4.4.

[247] See Chap. 4, Sect. 4.6.

[248] See, among others, clause 6., 'investment objectives and guidelines', Investment Association, 'Model Discretionary Investment Management Agreement' 6126. (2015) https://www.theia.org/sites/default/files/2021-11/Model%20IMA%202021.pdf accessed 20 December 2025; for a concrete example, see clause 6, Barclays, 'Barclays Discretionary Investment Agreement' (2015) https://privatebank.barclays.com/content/dam/privatebank-barclays-com/en-gb/private-bank/documents/support-and-information/india-important-information/discretionary-pms-agreement.pdf

[249] UNPRI, 'A legal framework for impact: Sustainability impact in investor decision-making' (2021) 435–437, 487 https://public.unpri.org/policy/a-legal-framework-for-impact accessed 20 December 2025.

[250] Ibid., 487–492.

[251] For some examples, see Chap. 3, Sect. 3.4.4.

[252] See Chap. 2, Sect. 2.3.4.; Chap. 3, Sect. 3.4.5.

Yet, the 'sustainable' purpose behind the transaction should justify closer scrutiny of how companies pursue these goals and allocate investors' funds. The fact that investors have intentionally chosen ABSs with a 'sustainable' profile should also highlight their broader interest in originators' behaviours and in holding them accountable.

5.7.2 *For Representation on Sustainability Matters*

The momentum for 'sustainable finance' has culminated in several guidelines to advance sustainability goals within investment practices, which may also indirectly affect sustainable securitisations. For instance, the UNPRI has formulated high-level guidelines on strategies to integrate ESG considerations and to deliver an environmental or social impact. These include, among others, embedding such factors in decision-making processes, investment analyses, and demanding disclosure from investee companies on these matters.[253] Investors applying these guidelines already signal their interest in sustainability objectives: Therefore, they could be more likely to act as representatives. However, the scope is self-evidently too broad to capture how certain investors should protect the interests of ABS holders.

A welcome development, albeit not securitisation-specific, is the Model Mandate Initiative by the investor-led networks International Corporate Governance Network (ICGN) and Global Investors for Sustainable Development Alliance (GISD).[254] It has more specifically targeted the relationship between asset owners and asset managers, with the aim of enabling the former to better clarify their expectations across investment agreements. Investment approaches on environmental, social, and governance matters are one of the core themes of the initiative.[255] Importantly, unlike other high-level initiatives, it combines detailed guidelines and draft clauses to put them into action. [256] Using the recommendations to adjust standard terms for the appointment of asset managers is encouraged[257]: as

[253] UNPRI, 'What are the principles for responsible investment?' (2023) https://www.unpri.org/about-us/what-are-the-principles-for-responsible-investment accessed 20 December 2025.

[254] About the networks, see ICGN, 'About us' https://www.icgn.org/about accessed 20 December 2025. GISD, 'The Alliance' https://www.gisdalliance.org/about accessed 20 December 2025.

[255] ICGN, GISD, 'Model Mandate' (2022) 4 https://www.icgn.org/model-mandate accessed 20 December 2025.

[256] Ibid., 15–19, 21–30.

[257] Ibid., 6.

a result, it seems promising to use ICGN draft clauses throughout trust deeds and investment agreements for noteholders' representatives.

For instance, a model provision requires investors to regularly monitor investee companies, and also ESG factors more specifically.[258] The clauses also require investors to regularly assess sustainability-related risks[259] and to inform clients about possible remedies in case of significant concerns related to performance or to investee companies. Possible ways to take action for ABS holders, considering the legal restrictions on their powers,[260] include engagement, collaborative engagement, divestment or litigation.[261] The clauses also require investors to undertake all activities 'in the best interests of the client' in order to 'influence' the management of their counterparties on sustainability matters.[262] Lastly, investors should provide regular reports to noteholders on their sustainability impact and about their engagement activities.[263]

The application of UNPRI Principles and ICGN model clauses would considerably strengthen the role of investors' representatives by empowering them to raise investors' 'voice' on sustainability matters through engagement activities and litigation vis-à-vis originating companies. At the same time, as these guidelines and standard clauses focus on 'sustainable' investment practices *in general*, their application to securitisation will inevitably vary across transactions.

5.7.3 Normative Recommendations

Facing the general scope of existing standardisation tools, additional securitisation-specific initiatives are necessary to ensure the adequate representation of ABS holders on sustainability matters.

The starting point could be guidelines, providing recommendations on the role of investors' representatives, their duties as well as the skills and professional standards that these entities should meet. As recommended

[258] 'The Manager shall monitor its counterparties on an ongoing basis and manage the Portfolio's overall exposure to any one counterparty such that risks of default or failure are appropriately controlled' ICGN, Model Mandate Initiative (n. 256) 29, more generally see ibid., Clause 13, 20.
[259] Clause 16. P. 22.
[260] See Chap. 2, Sect. 2.3.4.
[261] Clause 16 ibid.
[262] Clause 13 ibid. A model clause also specifically invites to engage with counterparties on their net-zero commitments, ibid., p. 22.
[263] Clauses 37–39, p. 30.

by the Model Mandate Initiative, it may be preferable to appoint entities with expertise in financial and sustainability matters, which should also demonstrate their compliance with best practices, such as UNPRI.[264]

There is also room for a standard 'investment agreement' for these representatives. This should comprise an 'investment objective' clause explicitly referring to environmental, social, and governance goals.[265] Therefore, market actors who are more likely to perform this function should be investment funds that place sustainability issues at the forefront of their mandates, such as SRI, ethical and impact funds.[266] These are more likely to identify and mention sustainability objectives within fund agreements.[267]

Model provisions should also regulate how these representatives should promote the 'voice rights' of ABS holders on sustainability matters after the issuance. To this end, they should, among other things, require representatives to monitor how originators progress towards sustainability-related goals and targets, as well as whether significant environmental and social risks arise. There should also be model 'covenants' through which investors' representatives would undertake to engage with originating companies on sustainability matters through bilateral meetings and to report to end investors.[268] Given the high coordination costs, especially ahead of litigation, standard clauses should also empower such representatives to bring legal proceedings against originating companies in sustainability-related circumstances, such as false statements regarding sustainability goals ('greenwashing') or the failure to achieve the targets.

The advantage of these model provisions would be their easy replication as well as their flexibility: for instance, investors and their representative may decide that only some classes of ABS holders should receive regular notice about sustainability matters or participate and decide during investors' meetings, as already occurs in some EU RMBS issuances.[269] It may also be possible to provide different standard formulations for the clauses concerning enforcement rights: These may give representatives full

[264] Global Investors for Sustainable Development Alliance (GISD) International Corporate Governance Network (ICGN), 'Model Mandate' (n. 256) 20 https://www.icgn.org/model-mandate

[265] See above Sect. 4.7.1.

[266] See Chap. 1.

[267] Irene-marié Esser, Iain MacNeil, 'From a Financial to an Entity Model of ESG' (2021) *European Business Organisations Law Review* 1, 17–18.

[268] Clauses 37–39 International Corporate Governance Network (ICGN) (n. 256).

[269] SFIG, RMBS 3.0, 205.

discretion to act or require them to give investors notice. There may also be the option to require a decision within dedicated 'general meetings' or by way of written resolutions.[270] This may come together with a clause that restricts the enforcement rights of all investors other than the representative ('no action clause').[271] The ability to choose from various model clauses would allow the tailoring of the powers of representatives to the investor base: For instance, having a say over enforcement may be attractive to qualified institutional investors. By contrast, noteholders with a short-term horizon and little interest in sustainability goals may prefer to delegate such powers to their representative.

Nevertheless, the possibility to adjust and to deviate from standard clauses (as 'soft law' tools[272]) also represents their inherent limitation. While it may be desirable to adjust the breadth of enforcement powers to the characteristics of the transaction, it is also imperative to ensure that investors can oversee how originating companies use their funds and hold them to account if they deviate from their sustainability-related commitments. Therefore, legislation should also intervene in tandem with the proposed standardisation initiatives to grant all investors in 'sustainable' securitisation transactions a minimum level of protection and representation, as I shall discuss in Chap. 6.[273]

All in all, standard investment agreements could be a significant step forward to strengthen investors' 'voice' rights on sustainability matters and to address the costs and challenges of coordinating their interests. In turn, it would also make originating companies more accountable for sustainability matters, while leaving market actors with a margin of flexibility.

[270] See For example, 'Collective decisions are adopted at a general meeting (the 'General Assembly') or by unanimous decision following a written consultation (the 'Unanimous Written Decision'). [...] The General Meeting can only validly deliberate based on the first call if the Bondholders present or represented hold at least one-fifth (1/5) of the principal amount of the Bonds in circulation. [...] At the initiative of the Issuer or the Representative of the Mass, Collective Decisions may also be taken by Unanimous Written Decision. Such a Unanimous Written Decision shall be signed by or on behalf of all Bondholders without having to comply with the formalities [...]' Clause 11, EuroPP Group, 'Sample Sect.cription Agreement and Bond Terms and Conditions' within the EuroPP charter (n. 169).

[271] see also Chap. 3, Sect. 3.4.4.

[272] See above Sect. 5.2.4.

[273] See Chap. 6, Sect. 6.8.3.

5.8 Monitoring Processes

5.8.1 For 'Traditional' Securitisations

As mentioned in Chap. 2, there are various options for originators and investors to monitor securitisation expenses, ensuring the proper segregation of funds towards sustainability goals ('shielding'). Legislators and market actors can introduce statutory or contractual provisions providing for the appointment of external entities to oversee the circulation of cash flows, or introducing mechanisms for the segregation of funds. [274] However, previous standardisation efforts by financial industry associations have failed to cover these aspects.

Therefore, originators or SPVs can appoint third parties, like servicers or management companies[275] through 'servicing' or 'administration' agreements. While some market actors have made publicly available a sample that they use across all their operations,[276] these agreements tend to be highly specific to each transaction, depending on the conditions, size of the operation and parties involved.

Alternatively, or in addition to those, parties may agree on additional mechanisms towards the 'ring-fencing' of investors' proceeds, such as segregated bank accounts or trusts. The main purpose is to prevent enforcement actions from other creditors, which may dilute the funds. In order to do so, parties need to sign contractual provisions such as 'bank account agreements', which again tend to be highly individual and less amenable to standardisation efforts by associations.[277] Only large issuers are likely to design their own sample agreements.[278]

[274]See Chap. 3, Sect. 3.4.5.

[275]The latter, especially when the transaction is 'revolving'. See Chap. 3, Sect. 3.2., 3.4.5.

[276]See For example, Mayer Brown, 'Servicing agreement' https://www.mayerbrown.com/-/media/files/perspectives-events/publications/2012/12/german-standards-for-loan-transactions/files/agreement-english/fileattachment/mayerbrownservicingagreementform-englishl.pdf accessed 10 December 2025.

[277]Chap. 3, Sect. 3.4.5.

[278]see, among others, Natwest, which has published one of their bank account agreements as an example of its own documentation for regular issuances. see Natwest Group, 'Arran Residential Mortgages Funding 2010-1 PLC – –Account Bank Agreement' (2010) https://investors.natwestgroup.com/~/media/Files/R/RBS-IR-V2/securitisation/reseidential-mortgages/pdf/7-arran-residential-mortgages-2010-1-plc/principal-transaction-documents/014-account-bank-agreement-2010-a1.pdf accessed 10 December 2025.

Thus, even parties willing to appoint external entities or to implement additional mechanisms for the proper allocation of funds towards sustainability-related goals will not be able to rely on standard provisions. This results in high transaction costs, which may represent a considerable deterrent.

5.8.2 *On Sustainability-Related Costs*

As part of the recent initiatives on 'green' financial products, a noteworthy development is a series of preliminary guidelines around the 'tracking' of expenses, which should ensure that investors' funds are channelled towards sustainability goals.

One of the core pillars of ICMA's GBPs and SBPs is the 'management of proceeds'. Both frameworks recommend issuers to develop a 'formal internal process' to evidence their mechanisms to track expenses[279] and communicate them to investors. They also recommend the implementation of 'earmarking' mechanisms to set aside investors' funds, such as 'a sub-account, a sub-portfolio', while also allowing issuers to design other instruments, as long as proceeds are 'tracked in an appropriate manner'. Issuers should also rely on 'an external auditor, or other third party, to verify the internal tracking method and the allocation of funds from the Green Bond proceeds'.[280] Therefore, the two sets of Principles are grounded on two legal strategies discussed throughout this work, i.e. the 'disclosure' of tracking mechanisms to investors and the external verification by a third party ('gatekeeping').

Another interesting aspect which market actors may consider when drafting servicing and 'bank account' agreements for sustainability goals is the use of unallocated proceeds. ICMA's GBP allow temporary alternative uses as long as issuers predetermine the criteria and conditions for all non-sustainability-related expenses and as long as they inform investors.[281]

By contrast, the Climate Bonds Standard replicates the minimum tracking requirements under ICMA's GBP,[282] while also going beyond. It

[279] ICMA, 'GBPs' (n. 39) 8. ICMA, 'SLBPs' (n. 39) 4.
[280] ICMA, 'GBPs' (n. 39), 6.
[281] Ibid., 8; ICMA, 'SLBPs' (n. 39) 7.
[282] It also refers to a 'sub-account', a 'sub-portfolio' or to other adequate tracking mechanisms. See clause 3.1.1., 7.1. Climate Bonds Initiative, 'Climate Bonds Standard – Version 4.3' (2025) A.2.3. 'Climate Bonds Standard – version 4.3' (2025) CBI_Standard_V4.3_FINAL_2025-08-20-102147_gbqn.pdf accessed 20 December 2025.

recommends spending the unallocated balance only towards obligations of a revolving nature so that funds could be easily withdrawn in the future to meet sustainability-related expenses. Alternatively, issuers may invest it in money market instruments (like treasury bonds) as long as these do not promote carbon and gas intensive projects,[283] which would be at odds with the purposes of 'sustainable' ABSs.

5.8.3 Normative Recommendations

Market actors using securitisation instrumentally to achieve sustainability goals should put in place additional safeguards to ensure that securitisation funds are used for that purpose, which would not be as necessary in traditional securitisations. Therefore, there is room for the design of standard clauses for 'servicing' or 'administration' agreements, delineating the day-to-day duties of external entities. This should include conditions such as covenants where they contractually undertake to oversee the management of funds. They should also keep track of the payments on the underlying loans and of the distributions to investors.

In order to ensure the effective 'shielding' of the funds, it would be useful to draft a model 'bank account agreement' designating a bank account for the expenses of the operation. The peculiarity for 'sustainable' transactions could also be the segregation of the account, so that it may not be subject to any other claims from other creditors.[284] Specific covenants could then require the exclusive allocation of the funds for sustainability-related purposes.

At the same time, prospective originators may also perceive a segregated bank account as very restrictive to their business activities. These limitations may discourage the widespread adoption of such clauses in securitisation documentation. Therefore, as previous sections have already recommended in relation to other proposed standard clauses,[285] it may be advisable to include alternative formulations. Following ICMA's principles on tracking and 'earmarking' of investors' funds, the standard bank account agreement could also provide for alternative standard covenants, requiring issuers to put in place tracking mechanisms for securitisation expenses. These could, among others, require regular reporting on all expenses of the proceeds. An additional covenant, based on the Climate

[283] Clauses 3.1.1.–3.1.3.; 7.3.1.–7.3.3 ibid.

[284] See Chap. 3, Sect. 3.4.5.

[285] See above Sect. 5.4.2.3.

Bonds Standard, could also set out the criteria to spend unallocated proceeds. For instance, this may allow the covering of urgent expenses and exclude activities with an adverse environmental impact. There should also be a time limit for issuers to reallocate such funds to the outstanding balance. If issuers do not meet this deadline, they should be liable vis-à-vis investors. Such standard clauses would encourage a well-pondered use of the expenses as well as strengthen transparency and issuers' accountability.

Relatedly, another way to strengthen the 'shielding' of the funds for the purposes of the operation could be to ensure the appointment of a third party overseeing the circulation of cash flows. Therefore, financial industry associations may also design a model 'servicing agreement' introducing covenants where servicers undertake to keep track of the expenses.

5.9 The Boundaries of Standardisation Mechanisms

The analysis of the shortcomings of previous initiatives has shown the potential for market actors, such as representatives of the financial industry, to continue taking a leading role in the design of guidelines, metrics, and model clauses, to standardise sustainable securitisations. Private 'soft' legal tools could play a crucial role in reducing costs and promoting liquidity. They could mitigate legal risks and incentivise securitisation parties to adopt consistent behaviours. Certain market actors, acting as 'intermediaries' between parties and regulators, could create a 'feedback effect', where they gather best practices through standardisation initiatives, and may also inspire further regulatory measures on sustainable securitisations. [286]

However, the standardisation tools analysed in this chapter bring a series of limitations inherent in their *soft law* and *private* nature.

First, 'soft' tools like guidelines, metrics, and standard clauses are likely to widely circulate among market participants, but their non-prescriptiveness inevitably leaves room for divergent behaviours. As explained in Chap. 1, the distinctive feature of 'soft law' is the lack of formal sanctions, actioned by state, or other public authorities, when entities and individuals do not comply with rules. This is also the case for the members of industry associations when not applying model clauses, metrics, or guidelines for securitisations. Overall, the decision to apply 'soft' legal tools is likely to depend on cost and reputational factors. The latter could include peer pressure or even the threat of internal sanctions linked to the membership within

[286] See Sect. 4.1.2., 4.1.4.

certain associations. [287] For instance, the LMA may suspend or withdraw membership under exceptional circumstances like conduct which the Board considers to be 'commercially, reputationally or otherwise prejudicial to the interests of the Association'.[288] However, some have even questioned the effectiveness of these powers in practice since the associations have a direct interest in a broad membership base, especially when involving influential market actors.[289]

As previously mentioned, other dynamics which may promote the circulation of 'soft' legal tools involve market actors inspiring and attracting their peers ('learning' and 'network externalities') as well as replicating their past behaviours ('path dependence' mechanisms).[290]

Despite these potential effects, the lack of formal sanctioning procedures inherently allows market actors to deviate from these legal tools altogether or to 'cherry-pick' certain guidelines or model provisions. These practices are particularly problematic for transactions with a 'sustainable' profile since market actors may have leeway to only apply the least restrictive clauses. For instance, they may decide not to introduce any remedies for breaches of sustainability-related provisions. In turn, this may result in 'greenwashing' practices.

Besides, the possibility for market actors to independently determine the 'sustainable' profile of their transactions inherently gives rise to 'market failures'. In neoclassical economic terms, these arise in circumstances where individuals in private transactions fail to produce, allocate, and consume goods efficiently[291], like asymmetric information, negative externalities, the presence of public goods, imperfect competition and biases in

[287] On a systematic analysis of soft law disciplining and enforcement mechanisms within the financial regulation arena, see Christopher Brummer, 'A Compliance-Based Theory of International Financial Law' in *Soft Law and the Global Financial System: Rule-Making in the twenty-first century* (2nd ed., Cambridge University Press 2020) 140, 154–155. On the enforcement of market-based rules like private soft law clauses see also Joanna Benjamin, Paul Bowden, David Rouch, 'Law and Regulation for Global Financial Markets: Markets as Rule-Makers – Enforcement, Dispute Resolution and Risk' (2008) 2 *Law and Financial Markets Review* 321.

[288] Art 7.3.2. LMA, 'Byelaws relating to membership of the association' (2022) https://www.lma.eu.com/application/files/5316/6964/5469/LMA_Byelaws_-_November_2022_changes.pdf accessed 10 December 2025.

[289] Brummer (n. 88) 154–156. Awrey, 'The Limits of Private Ordering' 29–30.

[290] See above Sect. 4.1.4.

[291] As defined, among others, by Ménard, Shirley (n. 39) 838. Armour, Awrey et al (n. 15) 52–61.

decision-making processes. [292] According to the law-and-economics theorists of the Chicago and neo-institutional school, market failures justify legal intervention.[293]

For the purposes of the present analysis, sustainable securitisations inherently give rise to an asymmetry between issuers of financial instruments and investors, where the former have more information on certain conditions and quality of the transaction[294] (like the environmental and social risks of the portfolio assets or the projects financed through the issuance). Therefore, as the present chapter highlighted, originators could be incentivised to strategically withhold information around how their portfolio contributes to, or negatively affects, sustainability goals and targets. They also lack the incentives to introduce covenants or additional remedies for investors if originators miss sustainability-related targets.

Furthermore, Chap. 2 has highlighted how information on sustainability matters acts as a 'public good' since market actors not fully disclosing it to the public can 'free-ride', obtaining additional benefits. Conversely, the disclosure of such information yields 'non-excludable' benefits beyond each transaction for *all* market participants. These include, among others, encouraging the allocation of funds towards sustainability objectives and strengthening originators' accountability. [295]

Facing these market failures, the proposed non-binding legal tools may provide inadequate incentives for issuers to provide all information on sustainability-related matters. As a result, statutory measures with binding effect ('hard law') are equally crucial to require the dissemination of such information.

SECOND, OTHER challenges arise from the design of legal tools at market level ('private law'). Despite their intermediating function between regulators and market participants, financial industry associations represent a specific segment of the market. For instance, the LMA identifies itself as the 'authoritative voice' of the syndicated loan market.[296] This inevitably gives rise to an accountability concern: in whose interest are

[292] Ibid.

[293] Ménard, Shirley (n. 39) 43, 44, 209, 838.

[294] Hill. 'Securitisation: A low-cost sweetener' 1086.

[295] On the 'Positive externalities' effects of disclosure obligations, Enriques L, Gilotta S, 'Disclosure and Financial Market Regulation' in Niamh Moloney, Eilís Ferran, and Jennifer Payne, *The Oxford Handbook of Financial Regulation* (Oxford University Press 2015), 7; see also Chap. 2, Sect. 2.6.

[296] LMA, 'About us' (n. 11).

they acting and in what ways do they respond for their conduct?[297], Thus, there is the risk that the design of standard legal tools reinforces their self-interests[298], rather than promoting the general interests and the global sustainability agenda.

Overall, the soft law and private law nature of standardisation tools may not suffice to foster the convergence of market practices, while giving rise to market failures and accountability challenges. These shortcomings call for complementary measures with binding effect ('hard law') and at state level (which I will define 'public law').

5.10 Conclusions

The present chapter has illustrated the role that certain market actors, especially financial industry associations, have played towards the 'standardisation' of securitisation market practices through the design of non-binding guidelines, metrics, and model clauses. These tools yield several benefits: They cut transaction costs and attract liquidity, enhance legal certainty while mitigating the risks of errors in the documentation and they have the potential to inspire consistent behaviours among market actors as a result of 'network externalities' and 'path dependence' mechanisms. All these advantages could partially address some of the shortcomings of current market practices, where actors negotiate sustainability-related goals on a deal-by-deal basis, as discussed in Chap. 3.

As a result of the momentum for 'sustainable' financial products and investment practices, some industry associations, including LMA and ICMA, have increasingly started to design guidelines and preliminary examples of metrics and contract clauses. These initiatives could also lay the first step towards the standardisation of 'sustainable' securitisations. However, as their scope is generally broader, they fail to cover some strategic aspects of these operations, such as the types of assets which can be part of portfolios, or the optimal arrangements to set up and govern SPVs. Only recently have industry associations have started to move away from

[297]On accountability as the responsibility to justify a certain conduct ex ante and as the possibility to hold a certain actor into account ex post, see Thomas Schillemans and Mark Bovens, *The Challenge of Multiple Accountability: Does Redundancy Lead to Overload?* (1st edn, Routledge 2011) 4–5. On the accountability challenges of the initiatives by financial industry associations, see Black, Rouch (n. 5) 2255–227.

[298]As already argued in relation to standard clauses in general, see Black and Rouch (n. 4) 223, 225; Benjamin, Bowden, Rouch (n. 287)., 83.

high-level guidelines, by also providing model provisions for integrating sustainability-related elements into the clauses of transactional documentation, thus making them binding.

Against this background, I have proposed the introduction of further standardisation tools, revolving around the sustainability-oriented legal strategies explored in Chap. 1.

First, guidelines, metrics, and model clauses could promote the 'disclosure' of strategic aspects of the transaction, which currently remain ambiguous or under-disclosed. For instance, they could contain a provision requiring issuers to publish information about the qualities of the portfolio assets ('eligibility criteria') or about the mechanisms to monitor sustainability-related expenses.

Second, standardisation tools could designate 'gatekeepers' to oversee the 'sustainable' profile of these operations. Actors who generally negotiate the clauses of the documentation with originators, such as underwriters, are especially well placed to act in this role. Therefore, securitisation-specific guidelines could shape their role, referring to their need to negotiate the terms of the issuance together with SPVs and/or originators as well as make due diligence assessments. Standard clauses could clarify the minimum sustainability-related conditions that the operation would need to meet for underwriters to subscribe to it.

Third, it is desirable to strengthen the rights of ABS holders to raise their 'voice' on sustainability matters. The first step could be model articles for the constitution of SPVs, incorporating sustainability matters within the purpose of these entities. These would represent the point of reference, guiding the respective directors to focus decisions and business activities on the implementation of sustainability-related projects and targets. Such provisions may in turn allow investors to hold directors accountable if the directors fail to act in line with such sustainability purposes. There is also room for further model clauses, requiring a representative of all investors to monitor how originators are performing on their sustainability goals and empowering ABS holders to bring legal proceedings if originators or SPVs do not comply with their commitments.

Fourth, standardisation tools can ensure the 'shielding' of the funds for the purposes of the operation. For instance, a model 'bank account agreement' can provide standard terms for parties to segregate the funds of the operation. By putting the funds outside the reach of creditors, a segregated bank account can ensure that all cash flows and expenses contribute to sustainability-related purposes.

Despite the merits of standardisation tools, their non-binding effects (as 'soft law' instruments) and their design at market level (as 'private law' instruments) undermine the effectiveness of these measures to address 'market failures'. These features also pose accountability concerns. The following chapter will discuss how complementary initiatives to coordinate sustainable securitisations at EU level may address these shortcomings.

References

ABA, 'Revised Model Simplified Indenture' (2000) 55 *The Business Lawyer* 3, 1115https://www.jstor.org/stable/40657063?seq=1

Abbott KW, Levi-Faur D, Snidal D, 'Theorizing Regulatory Intermediaries: The RIT Model' (2017) 670 *Annals of the American Academy of Political and Social Science* 14

Agostini F, 'From "Green Bond Principles" to "Green Bond Clauses": Mitigating Greenwashing Through Contract Law' in Andenas M and Heidemann M (eds), *Quo Vadis Commercial Contract?* (Springer2023) LCF Studies in Commercial and Financial Law, vol 1 https://doi.org/10.1007/978-3-031-14105-8_6

Allen K, 'Strict US Market Rules Limit Corporate Sellers of Green Bonds' (*FT*, 20 February 2018) https://www.ft.com/content/baa217c4-157c-11e8-9376-4a6390addb44

Awrey D, 'The Limits of Private Ordering Within Modern Financial Markets' (2013) 34 *Review of Banking and Financial Law* 83

Awrey D, 'Split Derivatives: Inside the World's Most Misunderstood Contract' (2019) 36 *Yale Journal on Regulation* 495

Ayres I, Gertner R, 'Filling Gaps in Incomplete Contracts: An Economic Theory of Default Rules' (1989) 99 *Yale Law Journal* 87

Banque de France, 'Financial Stability Review – Special Issue on Liquidity' (2008) https://publications.banque-france.fr/en/february-2008

Barclays, 'Barclays Discretionary Investment Agreement' (2015) https://privatebank.barclays.com/content/dam/privatebank-barclays-com/en-gb/private-bank/documents/support-and-information/india-important-information/discretionary-pms-agreement.pdf

Belln KE, Burke PS, Girondo Met al, 'Comparing Intercreditor Arrangements' (2015) *LSTA Loan Market Chronicle*https://webstorage.paulhastings.com/Documents/PDFs/comparing_intercreditor_agreements.pdf

Benjamin J, Bowden P, Rouch D, 'Law and Regulation for Global Financial Markets: Markets as Rule-Makers – Enforcement, Dispute Resolution and Risk' (2008) 2 *Law and Financial Markets Review* 321

Benjamin J, Rouch D, 'The International Financial Markets as a Source of Global Law: The Privatisation of Rule-Making' (2008) 2 *Law & Financial Markets Review* 78

Black J, Rouch D, 'The Development of the Global Markets as Rule-Makers: Engagement and Legitimacy' (2008) 2 *Law and Financial Markets Review* 218
Bond Covenant Group, 'Model Covenants in Sterling and Euro Bond Issues' (2010) https://www.theia.org/sites/default/files/press-releases/document/20100621-01.pdf
Bozanic Z, Loumioti M, Vasvari FP, 'Corporate Loan Securitisation and the Standardisation of Financial Covenants' (2018) 56 *Journal of Accounting Research* 45
Bradley C, 'Private International Law-Making for the Financial Markets' (2005) 29 *Fordham International Law Journal* 127
Braithwaite J, 'Rules and Principles: A Theory of Legal Certainty' (2005) http://dx.doi.org/10.2139/ssrn.329400
Braithwaite JP, 'Standard Form Contracts as Transnational Law: Evidence from the Derivatives Markets' (2012) 75 *Modern Law Review* 779
Bratton WW, 'Bond Covenants and Creditor Protection' (2006) 7 *European Business Organisation Law Review* 39
Brès L, Mena S, Salles-Djelic ML, 'Exploring the Formal and Informal Roles of Regulatory Intermediaries' (2019) 13 *Regulation and Governance* 127
Brummer C, *Soft Law and the Global Financial System: Rule-Making in the twenty-first century* (2nd ed., Cambridge University Press 2020)
BIS, 'Report on Special Purpose Entities' (2009) https://www.bis.org/publ/joint23.pdf
Della Gioia Carabellese P, Della Giustina C, *The Law of Securitisations: From Crisis to Techno-Sustainability* (Routledge 2023)
Chancery Lane Project, 'About'https://chancerylaneproject.org/about/
Chancery Lane Project, 'Climate Underwriting Sponsor Warranties – Dotties' Clause' (2021) https://chancerylaneproject.org/climate-clauses/climate-underwriting-sponsor-warranties/
Chancery Lane Project, 'Green Loan 'Starter Pack' – Harrison's Clause' (2022) https://chancerylaneproject.org/model-clauses/green-loan-starter-pack/
Chancery Lane Project, 'Sustainability-Linked Loans – Casper's Clause' (2022) https://chancerylaneproject.org/climate-clauses/sustainability-linked-loans/
Charity Commission, 'Example Objects – Sustainable Development' (2013) https://www.gov.uk/government/publications/example-charitable-objects/example-objects-sustainable-development
Climate Bonds Initiative, 'Climate Bonds Standard – Version 4.3' (2025) CBI_Standard_V4.3_FINAL_2025-08-20-102147_gbqn.pdf
Deakin Set al, 'Legal Institutionalism: Capitalism and the Constitutive Role of Law' (2017) 45 *Journal of Comparative Economics* 194
EBA, 'Report on the Monitoring of Additional Tier 1 (AT1) EU Institutions' (2021) EBA/REP/20 https://www.eba.europa.eu/publications-and-media/press-releases/eba-updates-monitoring-additional-tier-1-tier-2-and-tlacmrel-eligible-liabilities-instruments

EBRD, 'Methodology to Determine the Paris Agreement Alignment of EBRD Investments' (2024) https://www.ebrd.com/content/dam/ebrd_dxp/assets/pdfs/green/Methodology%20to%20determne%20the%20Paris%20Agreement%20alignment%20of%20EBRD%20investments.pdf

EEMAP, 'Definition of an Energy Efficient Mortgage' (2018) https://energyefficientmortgages.eu/wp-content/uploads/2021/07/EEMI-Definition-14.11.18.pdf

EEMAP, 'Energy Efficient Mortgage Label'https://www.energy-efficient-mortgage-label.org/#

EEMAP, 'Master Template – Explanatory Document'https://energyefficientmortgages.eu/wp-content/uploads/2021/07/EEMI-Master-Template-Explanatory-Document.pdf

Enriques L, Gilotta S, 'Disclosure and Financial Market Regulation' in Moloney N, Ferran E, Payne J (eds), *Oxford Handbook of Financial Regulation* (OUP2015)

Environmental Finance, 'Sustainable Bonds Insight 2023' (2023) https://www.environmental-finance.com/assets/files/research/sustainable-bonds-insight-2023.pdf

ESMA, 'TRV Risk Monitor' (2023) https://www.esma.europa.eu/sites/default/files/library/ESMA50-165-2438_trv_1-23_risk_monitor.pdf

Esser IM, MacNeil I, 'From a Financial to an Entity Model of ESG' (2021) 23 *European Business Organisations Law Review* 9

EuroPACE, 'Refinancing EuroPACE Loans in the European Fixed Income Market' (2020) https://www.eu-peers.eu/knowledge-assets/refinancing-europace-loans-in-the-european-fixed-income-market

EuroPP, 'EuroPP Charter' (2019) https://eact.eu/Core/Images/Documents/Euro%20PP%20Charter%20(October%202019).pdf; https://www.euro-privateplacement.com/wp-content/uploads/2024/01/Euro_PP-Note-EN-2019.pdf

Federation for European Securities Exchanges, 'Position Paper on Green Bond Prospectus' (2019) https://www.fese.eu/app/uploads/2019/04/190429-fese-position-paper-on-green-bond-prospectus.pdf

Fitch Ratings, 'Representations, Warranties and Enforcement Mechanisms' (2016) https://www.fitchratings.com/research/corporate-finance/representations-warranties-enforcement-mechanisms-in-global-structured-finance-transactions-31-05-2016

Frankel T, 'Cross-Border Securitization: Without Law, but not lawless' (1997) 8 *Duke Journal of Comparative & International Law* 255

Fuller G, *The Law and Practice of International Capital Markets* (3rd edn, LexisNexis 2012)

GISD, 'The Alliance'https://www.gisdalliance.org/about

Hare C, Worthington S, Davies PL, *Gower: Principles of Modern Company Law* (11th edn, Sweet & Maxwell 2021)

Hill CA, 'Securitisation: A Low-Cost Sweetener for Lemons' (1997) 10 *Journal of Applied Corporate Finance* 64

Hornuf L, Reps M, Schäferling S, 'Covenants in European Investment-Grade Corporate Bonds' (2015) 10 *Capital Markets Law Journal* 345
ICGN, 'About Us'https://www.icgn.org/about
ICGN, GISD, 'Model Mandate' (2022) https://www.icgn.org/model- mandate accessed 3 June 2026.
ICMA, 'About ICMA', https://www.icmagroup.org/About-ICMA/
ICMA, 'Climate Transition bond guidelines' (2025) https://www.icmagroup.org/assets/documents/Sustainable-finance/2025-updates/Climate-Transition-Bond-Guidelines-CTBG-November-2025.pdf
ICMA, 'Green Bond Principles – Voluntary Process Guidelines for Issuing Green Bonds' (2025), https://www.icmagroup.org/assets/documents/Sustainable-finance/2025-updates/Green-Bond-Principles-GBP-June-2025.pdf
ICMA, 'GBP Impact reporting working group — Suggested Impact Reporting Metrics for Sustainable Water and Wastewater Management Projects' (2017) https://www.icmagroup.org/assets/documents/Regulatory/Green-Bonds/Water-Wastewater-Impact-Reporting-Final-8-June-2017-130617.pdf .
ICMA, 'Handbook Harmonised Framework for Impact Reporting' (2022) https://www.icmagroup.org/assets/documents/Sustainable-finance/2022-updates/Harmonised-Framework-for-Impact-Reporting-Green-Bonds_June-2022-280622.pdf
ICMA, 'Illustrative Examples for the Selection of Key Performance Indicators (KPIs) for Sustainability-Linked Bond Issuers, Underwriters and Investors' (2023) https://www.icmagroup.org/assets/documents/Sustainable-finance/2022-updates/Registry-SLB-KPIs_Final_2022-06-24-280622.xlsx
ICMA, 'Illustrative KPIs registry' (2022) https://www.icmagroup.org/assets/documents/Sustainable-finance/2022-updates/Registry-SLB-KPIs_Final_2022-06-24-280622.xlsx
ICMA, 'Impact reporting for Green Projects'https://www.icmagroup.org/sustainable-finance/impact-reporting/green-projects/
ICMA, 'Membership'https://www.icmagroup.org/membership/
ICMA, 'Primary Market Handbook'https://www.icmagroup.org/market-practice-and-regulatory-policy/primary-markets/ipma-handbook-home/ and https://www.icmagroup.org/assets/Table-of-contents.pdf
ICMA, 'Principles, Guidelines and Handbooks'https://www.icmagroup.org/sustainable-finance/the-principles-guidelines-and-handbooks/
IMCA, 'Sustainability Bond Guidelines' (2021) https://www.icmagroup.org/assets/documents/Sustainable-finance/2021-updates/Sustainability-Bond-Guidelines-June-2021-140621.pdf
ICMA, 'Sustainability Indicators for ABS' (2021) https://www.icmagroup.org/assets/documents/Regulatory/AMIC/Sustainability-indicators-for-ABS-181021v2.pdf
ICMA, 'Templates for issuers and reviewers'https://www.icmagroup.org/sustainable-finance/sustainable-bonds-database/#Templatesforissuers

ICMA, 'Updated analysis of the proposals for the EuGB Regulation' (22 June 2022) https://www.icmagroup.org/assets/EU-GB-Updated-ICMA-commentary_220622.pdf;

IMA, 'About Us' (2021) https://www.theia.org/about-us

IMA, 'Model Discretionary Investment Management Agreement' (2019) https://www.theia.org/sites/default/files/2019-08/Model_IMA_website_version.pdf

Investment Association, 'Model Discretionary Investment Management Agreement' (2015) https://www.theia.org/sites/default/files/2021-11/Model%20IMA%202021.pdf

ISDA, 'About ISDA'https://www.isda.org/about-isda/

ISDA, 'Master Agreement' (2002) https://www.sec.gov/Archives/edgar/data/1065696/000119312511118050/dex101.htm

Jensen MC, Meckling WH, 'Theory of the Firm: Managerial Behaviour, Agency Costs And Ownership Structure' (1976) 3 *Journal of Financial Economics* 305

Jordana J, 'Transgovernmental Networks as Regulatory Intermediaries' (2017) 670 *Annals of the American Academy of Political and Social Science* 245

Kahan M, Klausner M, 'Standardisation and Innovation in Corporate Contracting (Or 'The Economics of Boilerplate')' (1997) 83 *Virginia Law Review* 713

Klausner M, 'Corporations, Corporate Law, and Networks of Contracts' (1995) 81 *Virginia Law Review* 757

Lahlou MS, Tanega J, 'Islamic Securitisation: Part II – A Proposal for International Standards, Legal Guidelines and.Structures' (2007) 22 *Journal of International Banking Law and Regulation* 359

LexisNexis PSL, 'Key documents required for a residential mortgage backed securities transaction'https://www.lexisnexis.co.uk/legal/guidance/documents-required-for-a-residential-mortgage-backed-securities-transaction

LexisNexis, 'Key_features_in_LMA_PEPP_documentation' (2023) https://www.lexisnexis.co.uk/legal/guidance/key-features-in-lma-pepp-documentation,

LexisNexis PSL, 'Loan portfolio sales – legal documentation' (2023) https://www.lexisnexis.com/uk/lexispsl/bankingandfinance/document/391290/5H0D-CBW1-F185-S1M4-00000-00/Loan_portfolio_sales_legal_documentation

LexisNexis, 'Precedent signing and closing memorandum for a commercial mortgage – backed securities transaction'https://www.lexisnexis.co.uk/legal/precedents/precedent-signing-closing-memorandum-for-a-commercial-mortgage-backed-securities-transaction,

LMA, 'About us'https://www.lma.eu.com/about-us

LMA, 'Byelaws relating to membership of the association' (2022) https://www.lma.eu.com/application/files/5316/6964/5469/LMA_Byelaws_-_November_2022_changes.pdf

LMA, 'Documents & Guidelines' https://www.lma.eu.com/documents-guidelines/documents

LMA, 'Draft Provisions for Green Loans' (2025) https://www.lma.eu.com/documents-guidelines/documents

LMA, 'Draft Provisions for Sustainability-Linked Loans' (2023) https://www.lma.eu.com/documents guidelines/documents/category/sustainable-finance#
LMA, ELFA, 'Best Practice Guide to Sustainability-Linked Leveraged Loans' (2021) https://elfainvestors.com/wp-content/uploads/2021/07/ELFA-Best-Practice-Guide-to-Sustainability-Linked-Leveraged-Loans.pdf accessed 3 June 2026.
LMA, ELFA, 'Best Practice Guide for Term Sheet Completeness' (2025) https://www.lma.eu.com/application/files/5816/3843/1629/LMA_Best_Practice_Guide_Term_Sheet_Completeness.pdf
LMA, 'Green Loans Term Sheet' (2025) https://www.lma.eu.com/documents-guidelines/documents/category/sustainable-finance#
LMA, APLMA, LSTA, 'Green Loan Principles' (2025) https://www.lma.eu.com/application/files/1917/4298/0817/Green_Loan_Principles_-_26_March_2025.pdf
LMA, APLMA, LSTA, 'Guidance on Green Loan Principles' (2025) https://www.lma.eu.com/sustainable-lending/resources
LMA, 'Introduction to the Sustainability Coordinator Role'https://www.lma.eu.com/sustainable-lending/documents#
LMA, 'List of Principal Delegates' https://www.icmagroup.org/membership/List-of-principal-delegates-2/
LMA, 'Pan-European Private Placement Subscription Agreement' (2020) https://www.lma.eu.com/documents-guidelines/documents
LMA, APLMA and LSTA, 'Guidance on Sustainability-Linked Loan Principles' (2025) https://www.lsta.org/content/guidance-on-sustainability-linked-loan-principles-sllp/
LMA, APLMA, LSTA, 'Social Loan Principles' (2025) https://www.lma.eu.com/application/files/1317/4307/3886/Social_Loan_Principles_-_26_March_2025.pdf
LMA, 'Standard Intercreditor Agreements'https://www.lma.eu.com/documents-guidelines/documents
LMA, APLMA, LSTA, 'Sustainability-Linked Loan Principles' (2025) https://www.lma.eu.com/application/files/6317/4298/0865/Sustainability-Linked_Loan_Principles_-_26_March_2025.pdf
Mayer Brown, 'Servicing agreement'https://www.mayerbrown.com/-/media/files/perspectives-events/publications/2012/12/german-standards-for-loan-transactions/files/agreement-english/fileattachment/mayerbrownservicingagreementform-englishl.pdf
McCoy PA, Wachter SM, 'Representations and Warranties: Why They Did Not Stop the Crisis' in Fennell LA, Keys BJ, *Evidence and Innovation in Housing Law and Policy* (CUP 2017)
Natwest Group, 'Arran Residential Mortgages Funding 2010-1 PLC – Account Bank Agreement' (2010) https://investors.natwestgroup.com/~/media/Files/R/RBS-IR-V2/securitisation/reseidential-mortgages/pdf/7-arran-resi

dential-mortgages-2010-1-plc/principal-transaction-documents/014-account-bank-agreement-2010-a1.pdf

Natwest, 'Arran Residential Mortgages Funding 2010-1 Plc As Issuer-Citicorp Trustee Company Limited As Trustee – Trust Deed' (2010) https://investors.natwestgroup.com/~/media/Files/R/RBS-IR-V2/securitisation/reseidential-mortgages/pdf/7-arran-residential-mortgages-2010-1-plc/principal-transaction-documents/015--trust-deed-2010-a1.pdf

OECD, 'Due Diligence for Responsible Corporate Lending and Securities Underwriting' (2019) https://mneguidelines.oecd.org/due-diligence-for-responsible-corporate-lending-and-securities-underwriting.pdf

Pistor K, 'A Legal Theory of Finance' (2013) 41 *Journal of Comparative Economics*315

Ramos Muñoz D, 'Bankruptcy-Remote Transactions and Bankruptcy Law-a Comparative Approach (Part 1): Changing the Focus on Vehicle Shielding' (2015) 10 *Capital Markets Law Journal*239

Ramos Muñoz D, Ingram S, *The law of transnational Securitisations* (Oxford University Press2010)

RESFARM Project, 'Description of the Investment Instrument – A Financial Instrument for the Mobilisation of Investments in Sustainable Energy in the Agrarian Sector of Soutern Europe' (2017) http://resfarmproject.eu/en/documentation/

RESFARM Project, 'Project'http://resfarmproject.eu/en/project/

Roe MJ, 'Chaos and Evolution in Law and Economics' (1996) 109 *Harvard Law Review* 641

Schillemans S, Bovens M, *The Challenge of Multiple Accountability: Does Redundancy Lead to Overload?* (1st edn, Routledge 2011)

Schwarcz SL, 'Commercial Trusts as Business Organisations: Unravelling the mistery' (2003) 58 *Business Lawyer* 559

Structured Finance Industry Group (SFIG), 'RMBS 3.0' (6th edn, 2017) https://structuredfinance.org/wp-content/uploads/2019/05/RMBS-3.0-Sixth-Edition-Final-1109.pdf

UNPRI, 'A Legal Framework for Impact' (2021) https://public.unpri.org/policy/a-legal-framework-for-impact

UNPRI, 'ESG Engagement for Fixed-Income Investors: managing risks, enhancing returns'https://www.unpri.org/download?ac=4449

UNPRI, 'ESG Incorporation in Securitised Products: The challenges ahead' (2021) https://gsfo.org/resource/esg-incorporation-securitised-products-challenges-ahead

UNPRI, 'What Are the Principles for Responsible Investment?' (2023) https://www.unpri.org/about-us/what-are-the-principles-for-responsible-investment

Wall, Chamberset al, *Collateralised Loan Obligations (CLOs): Overview* (Practical Law) https://uk.practicallaw.thomsonreuters.com/9-385-8080

Wood P, *Project finance, securitisations and subordinated debt* vol 7 (3rd edn, Sweet & Maxwell2019)

Wymeersch E, 'Standardisation by Law and Markets Especially in Financial Services' (2008) 02 Financial Law institute Working Paper 2, http://papers.ssrn.com/sol3/papers.cfm?abstract_id=1086987

CHAPTER 6

EU Harmonisation of 'Sustainable Securitisations': The Role of 'Public Hard Law'

6.1 Introduction

Chap. 5 has shed light on the potential of market-designed non-binding legal tools ('private soft law' instruments) to coordinate 'sustainable' securitisations. At the same time, it revealed their shortcomings, calling for complementary coordination mechanisms at state level.

The theoretical framework presented in Chap. 2[1] already highlighted the 'hybrid' nature of the law governing financial markets. Market participants as well as state and international authorities jointly contribute to the design of the law applicable to financial transactions and services ('private law' and 'public law', respectively[2]). Public law operates on a spectrum: it may occupy a broader or narrower regulatory space depending on the subject matter, but it never goes as far as to rule private law out entirely.

Within the international regulatory landscape, several legislators and policy-making authorities have increasingly been involved in the design of rules to promote sustainability goals in financial markets (what has been referred to as 'sustainable finance'). In the following chapter, I will

[1] See Chap. 2, Sect. 2.2.; also consistently with Katharina Pistor, 'A Legal Theory of Finance' (2013) 41 Journal of Comparative Economics 315, 318.

[2] See Chap. 2, Sect. 2.3., also in line with the denomination used by Gortsos, see Christos Gortsos, *The Evolution of European (EU) Banking Law under the Influence of (Public) International Banking Law: A Comprehensive Overview* (4th edn) (January 31, 2022) https://ssrn.com/abstract=3334493.

F. Agostini, *The Law of Sustainable Securitisations*, EBI Studies in Banking and Capital Markets Law,
https://doi.org/10.1007/978-3-032-25046-9_6

specifically focus on the EU-level initiatives, contributing to a phenomenon which I will call 'harmonisation'.

Traditionally, harmonisation has encompassed both the process leading to and the final outcome of bringing different national rules and standards into accordance with one another, while preserving unsurmountable country-specific differences.[3] While this has represented a common regulatory approach across several international organisations, it plays a broader and peculiar role in the EU context. EU authorities, unlike other transnational bodies, have the power to introduce legally binding harmonised rules. The treaties also provide a concrete legal basis for EU institutions to enact ad hoc rules on sustainability matters, placing EU law in a unique position to coordinate sustainable securitisations.

Despite the proliferation of EU law-making initiatives targeting securitisation, especially in the aftermath of the 2007 Global Financial Crisis, they have largely failed to regulate sustainability aspects. Even the EU sustainable finance agenda, internationally well-known for its wide-ranging scope, has only culminated in initiatives which have an indirect and limited application on securitisations.

Against this background, Sect. 2 will conceptualise the distinctive meaning of harmonisation within the EU legal framework and discuss its added value.

The following subsections will examine the EU framework for 'sustainable' securitisations. Sect. 3 will provide an overview of the most salient initiatives. Subsections 4–8 will critically assess the extent to which such initiatives enable market actors to promote sustainability goals through securitisations. Following the structure of Chaps. 3 and 4, I will focus on the effects of the existing legal framework and forthcoming initiatives on the most strategic aspects of securitisation law for sustainability purposes ('focal points'[4]).

Along with the analysis of the existing legal framework, the chapter will also provide normative considerations about potential targeted reforms to prompt sustainability-related practices in securitisations. These could improve the design and management of 'sustainable' securitisation deals as well as encouraging more market actors to originate or invest in these transactions.

[3] Michael Dougan, 'Minimum Harmonisation and the Internal Market' (2009) 46 *Common Market Law Review* 3, 854.
[4] See Chap. 3, Sect. 3.4.

6.2 The Role of EU Harmonisation for Sustainable Securitisation

Throughout the present chapter, I will refer to the term 'harmonisation' as a legal technique to coordinate the laws for sustainable securitisations. It has the potential to address the shortcomings of the current market practice discussed in Chap. 3, where market actors largely articulate sustainability-related elements of these transactions on a deal-by-deal basis. However, unlike 'standardisation', harmonisation does not arise from the efforts of specific market participants, like financial industry associations, but rather from the initiatives of state and supranational authorities. The following sections will explore the peculiar connotation of the concept under EU law and the available legal tools to achieve it. They will describe the added value of harmonisation compared to standardisation initiatives. They will also justify the choice of EU harmonisation as the focus of the analysis, by examining its legal bases and contextualising it within the broader spectrum of coordination initiatives at the transnational level.

6.2.1 *The Distinctive Meaning of 'Harmonisation' Under EU Law*

The concept of 'harmonisation' has a recurrent use in various legal areas, like international trade law and international commercial arbitration.[5] It broadly captures both the *process* and the final *outcome*[6] of bringing rules, standards and policies of different jurisdictions and legal systems into accordance, which fundamentally leads to the creation of new law.[7] The treaties and some examples of EU legislation have inconsistently used similar terms, like 'approximation' and 'coordination', to refer to the same concept.[8]

[5] As discussed by Eva J Lohse, 'The meaning of harmonisation in the context of European union law – a process in need of definition' in Mads Andenas and Camilla Baasch Andersen (eds), *Theory and Practice of Harmonisation* (Edward Elgar 2012) 309.

[6] Academic scholarship does not converge on whether 'harmonisation' should only capture the process or the outcome, ibid. The term will be used with reference to both in the present work.

[7] Marcus Klamert, 'What We Talk about When We Talk about Harmonisation' (2015) 17 *Cambridge Yearbook of European Legal Studies* 360, 360–361. Lohse in Andenas and Baasch Andersen (n. 5) 573.

[8] For example, art 113, 191 (2) Consolidated version of the Treaty on the Functioning of the European Union (2016) OJ C 202/47. As noted by Klamert, ibid., 361. Inconsistent translations, like 'assimilation', have exacerbated this ambiguity, see Lohse in Andenas, Baasch Andersen (n. 5) 284.

The harmonisation process rests upon a fundamentally hierarchical relationship between supranational and national rule-making authorities, yet the domestic competence is still crucial for the incorporation of harmonised rules into national legal orders.[9]

Harmonisation has acquired a distinct meaning under EU law, reflecting the *sui generis* features of the EU in the context of international organisations. This is evident at two levels. First, EU legal tools, approved following the relevant legal procedure[10], have the power to harmonise domestic legislation with *binding* effect, without the need for further consent by Member States.[11] Second, EU harmonisation does not always derive from the alignment of existing domestic statutory provisions; rather, it may also lead to the creation of *new* rules, originating at EU level.[12]

Regarding the first aspect, EU's competence to harmonise national laws emanates from the powers that Member States have conferred to EU institutions through the Treaties. Therefore, it is generally possible on all subject matters where the EU shares law-making powers with Member States.[13] There are also provisions in the Treaties providing specific powers to introduce harmonisation measures functional to the internal market.[14] This has led commentators to argue that there would be a more general 'harmonisation' power for all measures that further EU objectives.[15]

As the following paragraph will illustrate, various legal instruments can introduce binding rules, which are either directly applicable or require transposition in all EU Member States.[16] The harmonisation process does not lead to a complete unification of national laws,[17] but rather leaves room

[9] Ibid., 309, 310.

[10] See below Sect. 6.2.2.

[11] Lohse in Andenas and Baasch Andersen (n. 5) 286.

[12] Ibid., 291.

[13] See Art 2(2) TFEU; 4(2) TFEU lists the subject matters where the competence is shared; see also Paul Craig, Gráinne De Búrca, *EU Law: Text, Cases and Materials* (7th edn, Oxford University Press) 118–119.

[14] See For example, art 114, art 50 TFEU, see below Sect. 5.2.3. As pointed out by Lohse in Andenas, Baasch Andersen (n. 5) 284.

[15] As argued by ibid., 301, 310, in relation to art 114 TFEU.

[16] See below Sect. 6.2.2.

[17] Piet Jan Slot, 'Harmonisation' (1996) 21 European Law Review 378, 379. As also argued by Baasch Andersen, 'Applied Uniformity of a Uniform Commercial Law: Ensuring Functional Harmonisation of Uniform Texts through a Global Jurisconsultorium of the CISG' in Andenas, Baasch Andersen (n. 5), 30–31. The author refers to the term 'uniformity'. For clarity, I will exclusively refer to 'harmonisation' for the purposes of the present chapter.

for concurrent national legislation and some freedom to market actors.[18] EU efforts can also have different levels of intensity. Commentators have compellingly distinguished between 'maximum harmonisation' initiatives, leaving no room for Member States to derogate from the EU standard, and 'minimum harmonisation', leading to the creation of an EU minimum benchmark, but also allowing Member States to introduce more stringent measures.[19]

Member States significantly contribute to the harmonisation process too when implementing EU rules. Their legislative, executive and judiciary authorities, transposing and interpreting EU rules, allow the shaping of uniform, EU-specific (i.e. harmonised) legal concepts.[20]

The peculiarities of national legal systems make an exhaustive codification of a certain subject area at EU level practically infeasible. For instance, an exclusively EU law-centred regime for securitisation would be unconceivable due to the interlinkages to national provisions on property and contract.[21] Therefore, EU harmonisation will be intended, for the purposes of the present analysis, as an inherently 'partial' phenomenon.[22]

The second distinctive feature of EU harmonisation is that, unlike other examples of legal harmonisation, it has not only resulted in EU institutions bringing national laws together. Throughout the history of EU legislation, this rule-making exercise has often been 'preventive', culminating in rules which were not formerly part of national legal frameworks. [23] The European Court of Justice (*hereinafter* CJEU) has held this practice to be compatible with EU law when in line with the general objectives of the EU and when retaining national rules may create potential obstacles to

[18] As also discussed in Chap. 3.

[19] For a review of these concepts see Klamert (n. 7) 362–363–364; Slot (n. 16) 382–389.

[20] Lohse in Andersen, Baasch Andersen (n. 5) 295, 301, 313.

[21] See For example, Chap. 3 Sect. 3.4.1., providing different examples of national provisions concerning the assignment of securitisation portfolios.

[22] Some commentators have differentiated between 'total' and 'partial' harmonisation, see For example, Klamert (n. 7) 362–363. For critiques to the distinction, see C Tietje in E Grabitz et al (eds), *Das Recht der Europäischen Union: EUV/AEUV*, 56th ed (Beck 2015), art 114, para 36

[23] Andenas, Baasch Andersen (n. 5) 291. Moloney also defined EU's law-making exercise as one of 'regulatory harmonisation' which has stretched the mandates of EU institutions, Niamh Moloney, 'Capital Markets Union: 'Ever closer union' for the EU financial system' (2016) 41 European Law Review 3 307, 321.

trade.[24] It would only exceed EU competence if EU legislation leads to new standards that co-exist with previous national ones.[25]

6.2.2 *Harmonisation Legal Tools*

The present work will examine different instruments which, regardless of their legal form, can practically yield the harmonising effects across EU jurisdictions described above.[26]

Throughout the history of EU integration, the traditional instrument for harmonisation has been directives, which only bind Member States to a certain result, while leaving some freedom as to the means.[27] This narrow interpretation has stemmed from the wording of the Treaties. For example, Art. 115 Treaty on the Functioning of the European Union (hereinafter TFEU), providing a legal basis for the harmonisation of laws on the internal market, specifically empowers EU institutions to issue directives.[28]

Conversely, regulations are directly applicable across EU Member States.[29] This has led commentators and the CJEU to describe their effect as one of 'unification' of national laws. Considering regulations as harmonisation tools is not incompatible with the wording of the Treaties: for instance, Art. 115, while making express reference to directives, does not prevent the use of regulations altogether.[30] The practice of the CJEU and EU legislation has further blurred the distinction between legislative instruments: It was held that directives can have direct effects[31] and there have been examples of regulations whose titles referred to 'approximation'

[24] See For example, C-377/98, where the applicants claimed that a new directive, introduced to address ambiguities in the interpretation of national provisions, did not justify harmonisation. The CJEU held that harmonisation is possible 'if the aim is to prevent the emergence of future obstacles to trade resulting from multifarious development of national laws provided that the emergence of such obstacles is likely and the measure in question is designed to prevent them' Case C-377/98 *Netherlands v Parliament and Council* [2001] ECR I-7079, para 15; similarly, Case C-376/98 *Germany v Parliament and Council* [2000] ECR I-7079, para 6.

[25] Case 436\03 *European Parliament v Council* [2006] I-03733, para 41–44.

[26] In line with ibid., 311; Slot (n. 28) 378.

[27] Art 288 (3) TFEU; Lohse in Andenas, Baasch Andersen (n. 5) 291–293; Klamert (n. 7) 365.

[28] Consolidated Version of the Treaty on the Functioning of the European Union (2012) C 326/49.

[29] Art 288 (2) TFEU.

[30] Ibid. 293.

[31] Case C-91/92 Faccini Dori (1994) ECR I-03325.

or 'harmonisation'.[32] For instance, the most comprehensive regulatory effort on securitisation is a regulation,[33] but it also leaves a margin of freedom to Member States regarding property, company, and contractual law aspects.[34] Regulations and directives can take a legislative or non-legislative form depending on the procedure for their approval.[35]

The present work will also consider several non-legislative instruments[36], i.e. delegated and implementing acts.[37] These are also directly applicable across all Member States. Delegated acts supplement or amend specific and non-essential elements of an existing legislative act.[38] Implementing acts lay down the general rules for the implementation of EU law across Member States.[39]

Lastly, even decisions by EU institutions like the CJEU or the ECB can indirectly lead to the harmonisation of national laws due to their binding nature.[40] Court decisions are a notable example since they require Member States to modify their existing laws or to interpret EU provisions in accordance with decisions.[41]

Some have contended that soft law instruments like opinions, recommendations and non-binding guidelines by EU institutions and authorities may also contribute to harmonisation.[42] While the present chapter also

[32] Lohse (n. 5) 310; see For example, the title of Regulation (EU) No 305/2011 of the European Parliament and of the Council of 9 March 2011, laying down *harmonised* conditions for the marketing of construction products and repealing Council Directive 89/106/EEC (2011) OJ L 88.

[33] Regulation 2017/2402 of the European Parliament and of the Council of 12 December 2017 laying down a general framework for securitisation and creating a specific framework for simple, transparent and standardised securitisation, and amending Directives 2009/65/EC, 2009/138/EC and 2011/61/EU and Regulations (EC) No 1060/2009 and (EU) No 648/2012 (*hereinafter*Securitisation Regulation).

[34] See For example, the very flexible wording on SPVs, Sect. 5.6.1.

[35] Craig, De Búrca (n. 13) 150–158.

[36] 'Non-legislative' acts follow a different procedure than 'legislative' acts, which require the approval by the Parliament and the Council generally following a proposal by the EU Commission, see Artt 289-290-291 TFEU; for an analysis of the different procedures see Craig, De Búrca (n. 13).

[37] See art 290–291 TFEU. Ibid., 152–154; some will be listed within Sect. 5.3.

[38] Art 290 TFEU.

[39] Art 291 TFEU.

[40] Art 298 TFEU.

[41] Principle of 'harmonious interpretation', Case 14/83 *Von Colson and Kamann v Land Nordrhein-Westfalen* [1984] ECR 1891, para 26–28; Lohse in Andenas, Baasch Andersen (n. 5) 298.

[42] For example, Lohse in Andenas, Baasch Andersen (n. 5) 293.

mentions guidelines by EBA[43], it will focus on the other above-mentioned legal tools due to their binding effects. Their 'hard law' [44] nature is arguably crucial for the regulation of sustainability-related elements of securitisation deals.

6.2.3 The Rule-Making Function of EU Harmonisation for Sustainable Securitisation

EU harmonisation, like market-driven 'standardisation' analysed in Chap. 4, can also facilitate the coordination of sustainability-related aspects of securitisation. Yet, the prescriptiveness of EU rules represents an added value, since they can more effectively ensure a uniform and consistent application among market actors.

Notably, Chap. 4 has discussed the inherent dynamics which enable the circulation of non-binding legal tools across market actors, who tend to replicate the behaviour of their peers ('learning' and 'network' externalities) or past behaviours ('path dependence' mechanisms). Other mechanisms indirectly encourage individuals, companies and investors to comply with 'soft law', including reputational threats and internal sanctioning mechanisms within industry associations. However, their effectiveness is inherently limited, since divergent behaviours are ultimately always possible. These could occur when market actors interpret and apply guidelines inconsistently, or if they fail to adopt standard clauses.[45]

By contrast, EU rules are accompanied by stronger enforcement mechanisms. EU Treaties and legislation grant EU and national authorities general or ad hoc enforcement powers: In the case of securitisation, these include, among other powers, the power to impose administrative sanctions if issuers provide misleading information, or fail to provide mandatory information, in their prospectuses.[46] Besides, while soft law mechanisms like Guidelines[47] may also leave room for inconsistent interpretations, the involvement of regulatory authorities and courts in overseeing and adjudicating securitisation law should ensure a consistent interpretation and application of the rules.

[43] Preparatory document for one of the key legal initiatives on sustainable securitisations, see Sects. 6.3.–6.4.

[44] For the distinction between 'hard law' and 'soft law', see Chap. 2, s. 2.4.

[45] As explained in Chap. 5, Sect. 5.9.

[46] See below Sect. 6.5.1.

[47] For the typology of 'soft' legal tools see Chap. 5, Sect. 5.2.3.

Harmonisation initiatives should also address some of the accountability concerns that standardisation at 'private' level may pose. EU rules derive from the competence that Member States have conferred under the Treaties for the fulfilment of the objectives of the Union.[48] In exchange for such rule-making powers, 'public' rules must follow extensive procedures, also including consultations with Member Countries and the civil society.[49] These external constraints, which mark the greatest distinction from 'privately' designed legal tools, should put EU institutions in the best position to protect the public interest, and not merely the interest of specific groups. As discussed in Chap. 5, this risk can materialise when financial industry associations design guidelines, metrics, or standard clauses that would apply for their own sector.[50]

These characteristics of EU harmonisation initiatives also cast light on the role that they could play for the law of 'sustainable' securitisations.

As discussed in Chap. 3, the starting point for securitisation law *in general* is the set of rules defining the contractual and proprietary rights over the portfolio assets and over the securities (serving a 'constitutive' function).[51] On the contrary, the legal framework which governs and which should govern the sustainability-related elements of these deals arguably serves three different functions.

First, law enables economic activities, like securitisation operations, by conferring *positive rights* to the parties involved ('enabling' function).[52] In other words, it could set the prerequisites which *allow* companies and financial institutions to name their securitisation operations as 'sustainable'. For instance, the EUGBR requires those willing to use of the 'EU Green Bond' (EUGB) label to abide by a series of rules (on matters like reporting). It also empowers public authorities to oversee the compliance with the Regulation and to enforce it.[53]

Second, EU legislation can serve a 'framing' function by drawing boundaries to the behaviours of the parties involved, establishing 'the rules of the game' [54] Chap. 5 discussed how standardisation efforts can

[48] See above, Sect. 6.2.1.

[49] On the role of consultations and the processes for the introduction of harmonisation tools (For example, under art 352 TFEU) see below Sect. 6.2.4.

[50] See Chap. 5, Sect. 5.9.

[51] See Chap. 3, Sect. 3.3.

[52] As defined in Chap. 2, Sect. 2.5; see also Pistor, 'Legal Theory of Finance' (n. 1) 318, 325; Javier Solana, 'The Roles of Law in Finance' (2022) on file with author, 3.

[53] See below Sect. 6.4.2.–6.5.2.

[54] Chap. 2, Sect. 2.5.

arguably also result in recommendations 'framing' parties' behaviour. For instance, guidelines on 'green' securities may, among others, recommend issuers to only allocate the proceeds to specific types of projects. However, their power to prohibit certain behaviours is inherently limited as it depends on parties' willingness to replicate such 'soft' legal tools within their documentation, thus making them binding. On the contrary, the prescriptive nature of EU legislation means that the applicability of its rules is mandatory and overseen by public authorities. For the purposes of the present analysis, they may, among others, prevent originators from adding certain types of assets to portfolios[55] or SPVs from concluding certain types of transactions.[56]

Third, 'public' laws can also consist of high-level provisions, crystallising the objectives and values which should drive specific legislation ('programmatic' function[57]). In the case of 'sustainable' securitisations, such rules are especially enshrined in the EU Treaties. The following section will delve into these provisions.

6.2.4 EU Treaties as Legal Bases for Harmonisation

As a prominent part of the 'primary law' of the EU, several provisions of the Treaty on European Union (hereinafter TEU) and of the TFEU represent the starting point for the EU's rule-making power for sustainable securitisation. They have a twofold nature.

Treaty provisions have a 'programmatic' value by articulating the goals and ideals which should inspire the above-mentioned harmonisation legal tools. For instance, key legal bases include Art. 114(1) and Art. 115 TFEU, which empower EU authorities to introduce legislative and non-legislative measures instrumental to the EU internal market. 'sustainable securitisations' could be a crucial example thereof since they increase market actors' liquidity through the issuance of ABSs and also allow banks to release part of their capital in order to comply with capital requirements.[58]

[55] See s. 6.4.

[56] See s. 6.5.

[57] Javier Solana, 'Market Efficiency as a Directive Principle of EU Monetary Policy' in Anna Chadwick, Eleonora Lozano-Rodríguez, Andrés Palacios-Lleras, Javier Solana (eds.), *Markets, Constitutions, and Inequality* (2022) Routledge, 7–8.

[58] See e.g. Sect. 5.4.3.; on the functions of securitisation more generally see Claire Hill, 'Securitisation: A low-cost sweetener' for Lemons' (1997) 10 Journal of Applied Corporate Finance 64.

Therefore, they are crucial to the functioning of the EU internal market and instrumental to its integration, as the recitals to the main piece of EU legislation on securitisation confirm.[59]

Additionally, Art. 352 TFEU, i.e. the general 'flexibility' clause for all the measures necessary for the objectives of the EU,[60] is another legal basis for the power to harmonise national laws on securitisation. Measures that promote sustainability-related goals notably result from the combination of Art. 352 with the Treaty provisions that acknowledge certain environmental and sustainability-related principles as general principles of the EU.

For example, Art. 3 of the TEU mentions 'sustainable development' as well as 'high level of protection and improvement of the quality of the environment' as factors towards the creation of the internal market.[61] Even Art. 11 TFEU refers to 'environmental protection' as a key principle underpinning EU policies. Thus, the combination of Art. 352 TFEU and these provisions have an aspirational value, encouraging EU institutions to reconcile the goal of supporting the internal market with environmental and social considerations when designing new policies on securitisation. It should encourage the CJEU and national courts to interpret and apply the securitisation regime with a view to promoting sustainability goals.

Next to the 'programmatic' value, these provisions also serve a stronger, 'framing' function by integrating sustainability-related principles within the mandate of EU institutions. According to Art. 11 TFEU, environmental principles 'must be integrated into the definition and implementation of the Union's policies and activities, in particular with a view to promoting sustainable development'. Arguably, the wording of the provision elevates 'environmental protection' to a principle of 'integration',[62] giving rise to a procedural and substantive obligation. In other words, it not only inspires, but also *requires* EU institutions to take environmental protection into account at the decision-making level when designing laws.

[59] See Regulation 2017/2402/EU(Securitisation Regulation), which mentions art 114 TFEU as the main legal basis. on the Regulation, see below s. 6.3.

[60] It involves a stricter legislative procedure, see Art 352 (1)-(2) TFEU; Craig, De Búrca (n. 12) 125–126.

[61] Consolidated version of the Treaty on European Union (2016) OJ C 202/13.

[62] Originally formulated as '*Environmental protection requirements shall be a component of the Community's other policies*', article 130r(2) Single European Act [1987] OJ L169/1 74; Andrea Lenschow, *Environmental Policy Integration: Greening Sectoral Policies In Europe* (Earthscan 2002), 7 ff.; Julian Nowag, 'The sky is the limit' in Sjåfjell B, Wiesbrock A (eds) *The Greening Of European Business Under EU Law: Taking Art 11 TFEU Seriously* (1st ed, Routledge 2015) 19 ff.

It also empowers EU institutions to introduce specific rules that promote environmental goals in non-environmental areas.[63]

Such general principles should bind all EU institutions, including the CJEU and the European Central Bank (ECB).[64] As I shall explain in Sect. 6.4.3., securitisation law is also inextricably linked to the rules governing market actors' operations with central banks. Further evidence of the relevance of these general principles for the ECB is art 127 TFEU, which identifies price stability as the primary objective of the European System of Central Banks, whilst making reference to Art. 3 TEU. Therefore, the legal framework governing ECB's operations as well as its decisions,1,[65] also indirectly affecting sustainable securitisations in turn[66], would need to be in line with Art. 3, and thereby with the sustainable development principle.

All in all, Art. 11 TFEU and Art. 3 TEU have substantive policy implications since they encourage but also require EU institutions, including law-making authorities, the CJEU and the ECB, to act on environmental and sustainability-related measures even in the securitisation remit.

6.2.5 EU Harmonisation as (the Preferable) International Regulatory Coordination Mechanism

The present analysis focuses on EU harmonisation, which only constitutes one example of a broader range of coordination mechanisms in the 'public law' sphere. A large body of scholarship has extensively analysed the function of international financial fora, such as the International Organisation of Securities Commissions (IOSCO), the Basel Committee on Banking Supervision (BCBS) and the International Accounting Standards Board (IASB).[67] Scholarly accounts have recurrently shed light

[63] For an analysis of the two dimensions and their interpretation by the CJEU, see Javier Solana, 'The Power of the Eurosystem to Promote Environmental Protection' (2019) 30 *European Business Law Review* 4 547, 557–560. On a more nuanced view of ECB's mandate on climate change and its limitations, see Chiara Zilioli, Michael Ioannidis, 'Climate change and the mandate of the ECB: Potential and limits of monetary contribution to European green policies' (2022) 59 *Common Market Law Review* 2 363, 371–383.

[64] On the horizontal dimension of Art. 11 under EU law, see Nowag (n. 62) 19; Lenschow (n. 62) 7 ff; on its value for ECB, see Solana, ibid. 10–14, 23; Zilioli, Ioannidis, ibid.

[65] Binding as a result of 132(1) TFEU.

[66] See below Sect. 6.4.3.

[67] Within the umbrella of the International Financial Reporting Standards (IFRS) Foundation, See Christopher Brummer, *Soft Law and the Global Financial System: Rule Making in the twenty-first century* (2nd ed., Cambridge University Press, 2020) 82.

on their rule-making function. In other words, these bodies have introduced a wide range of non-binding principles, practices and guidelines[68] in the financial and corporate law remit that have a global scope of application.[69] Securitisation has also fallen under the radar of such standard-setters after the 2007 global financial crisis, becoming the subject of IOSCO Principles to identify 'simple, transparent and standardised' (STS) transactions and of BCBS principles, providing for a dedicated prudential regime for these deals.[70]

Some of the above-mentioned transnational bodies have also been active within the sustainable finance debate, contributing to the coordination of global market practices. For instance, the International Sustainability Standards Board (ISSB) arose as a 'sister board' of the IASB during the 2021 United Nations Climate Change Conference (COP26) in order to introduce global sustainability-related reporting standards.[71]

Despite the merits of these and of other similar initiatives, the present work identifies EU harmonisation as the preferable regulatory coordination mechanism for sustainable securitisation for various reasons.

[68] The FSB has proposed this categorisation, which differentiates financial standards according to their specificity, see FSB, 'Key Standards for Sound Financial Systems' (2001) https://www.fsb.org/work-of-the-fsb/about-the-compendium-of-standards/key_standards/ accessed 10 December 2025; for an overview of these standards, see Gortsos (n. 1) 67–75.

[69] For example, Brummer, defining them as 'international financial standard setters' to differentiate them from 'international agenda setters' like G20 and the FSB, who only set high-level policy objectives. Brummer, C. (2015). The Architecture of International Financial Law. In *Soft Law and the Global Financial System* (n. 67) 68. https://doi.org/10.1017/cbo9781316423875.004 Armour, Awrey, et al, 'International Regulatory Coordination' in *Principles of Financial Regulation* (2016) 619–620; Gortsos (n. 1) 89–93; Mariana Pargendler, 'The Rise of International Corporate Law' (2021) 98 *Washington University Law Review*. 32–35; Rosa M Lastra, Marco Bodellini, 'Background paper on soft law and sovereign debt' (2018) 3–4 https://unctad.org/system/files/non-official-document/SOFT%20LAW%20AND%20SOVEREIGN%20DEBT.pdf accessed 10 December 2025.

[70] BCBS, Board of IOSCO, 'Criteria for Identifying Simple, Transparent and Comparable Securitisations' (2015) https://www.iosco.org/library/pubdocs/pdf/IOSCOPD494.pdf accessed 10 December 2025; under the current Basel III framework, see CRE 40.66, 40.73, Basel III framework (2023) https://www.bis.org/basel_framework/chapter/CRE/40.htm?inforce=20230101&published=20201126 accessed 10 December 2025.

[71] See IFRS, 'Who we are' (2023) https://www.ifrs.org/about-us/who-we-are/ accessed 20 December 2025; the main standards include the IFRS S1, supporting corporate sustainability-related disclosures, and the IFRS S2, supporting climate-related disclosures. ISSB, 'Frequently Asked questions' (2025) https://www.ifrs.org/groups/international-sustainability-standards-board/issb-frequently-asked-questions/ accessed 20 December 2025.

As previously discussed,[72] the 'programmatic' provisions of the Treaties provide for a legal basis for supporting and regulating sustainability-related elements of securitisations, which would be in line with the general objectives of the EU.

Furthermore, the insights from financial regulation and corporate governance theorists also arguably confirm the preferability of coordination at the EU level. A large body of scholarship has extensively appraised the merits of designing uniform laws at the supranational level (what could be referred to as 'uniformisation') or compare it with the opposite 'regulatory competition' trend, where regulators unilaterally choose their rules within each legal system.[73] Many have highlighted how globalisation directly affects the laws of the respective states, leading either towards the 'persistence' of national standards[74] or towards their spontaneous 'convergence'.[75]

It is important to note that retaining 'competition' between the laws of EU countries on securitisation would not be realistic or desirable. As previously discussed, sustainable securitisations contribute to the general objectives of the EU, EU institutions have a prevailing competence over Member States for their regulation in the light of EU Treaties.[76]

Rather, the most compelling question that may arise pertains to whether it would be preferable to introduce a *globally* uniform regime on sustainable securitisation or to retain competition between EU law and the legal frameworks of non-EU countries. While it is not possible for space constraints to consider all the angles of the 'competition' v. 'uniformisation'

[72] See above Sect. 5.2.4.

[73] Pargendler (n. 69) 2–3; Armour et al, 'International Regulatory Coordination 28.1', 616–628. On 'centralisation' see also Howell E Jackson, 'Centralisation, Competition, and Privatisation in Financial Regulation' (2001) 2 *Theoretical Inquiries in Law* 4.

[74] Parglender, ibid., 2–3; John C Coffee, 'The Future as History: The Prospects for Global Convergence in Corporate Governance and Its Implications' (1999) 93 Northwestern University Law Review 641, 679–681.; contrary to this view, see Marcel Kahan, Ehud Kamar, 'The Myth of State Competition in Corporate Law' (2002) 55 *Stanford Law Review* 3, 747.

[75] Jackson noted the existence of parallels between the two areas, Jackson (n. 73) 4,; in favour of competition, see, inter alia, Roberta Romano, 'The Need for Competition in International Securities Regulation' (2009) 2 Theoretical Inquiries in Law 0. 389; Wolf-Georg Ringe, 'Regulatory Competition in Global Financial Markets – The Case for a Special Resolution Regime' (2016) 1 *Annals of Corporate Governance* 175. Conversely, for a critique against competition, see Kamar, Kahan (n. 74) 727.

[76] See above Sect. 6.2.4.

debate[77], advocates for competition have rightly emphasised its power to foster innovation and yield cross-border effects. Market participants, choosing one jurisdiction or another based on their preference for certain rules, give a signal to regulators. A wide migration from a certain jurisdiction could naturally spur a 'feedback mechanism', compelling the respective regulators to adjust and innovate their regimes.[78] As a result, competition favours the spontaneous convergence of legal regimes towards an efficient outcome (race to the top).[79]

These arguments highlight the merits of an EU framework on sustainable securitisation, which would bring the EU legislation in accordance ('harmonisation'), whilst retaining competition with non-EU countries. This solution would also have virtuous 'race to the top' effects, since dedicated rules could attract foreign investment, arguably making EU law the 'global gold standard' for sustainable securitisations. This also seems to be the goal behind EU sustainable finance initiatives like the EU Green Bond Standard, also (partially) regulating securitisations.[80] The attractiveness of the EU framework would also trigger a virtuous circle among non-EU countries, since competition dynamics may foster 'feedback mechanisms', inspiring similar regulatory initiatives across other jurisdictions.

The main objection that commentators have raised against competition is the potential risk of a 'race to the bottom' effect, materialising when retaining diverse laws leads to suboptimal, rather than efficient, rules.[81] Relatedly, there may also be a risk of 'forum shopping', where market actors choose the legal framework that provides the least stringent enforcement regime.

[77] For a literature review on merits and demerits of regulatory competition, see Kahan, Kamar (n. 74), notes 4–16, 30; see also Armour et al (2016), 635–639, summarising the main arguments.

[78] Romano (n. 75) 6–9.

[79] Coffee (n. 74) 679.

[80] Alexander Lehmann, 'The EU green bond standard: sensible implementation could define a new asset class' (13 July 2021 Brugel) https://www.bruegel.org/blog-post/eu-green-bond-standard-sensible-implementation-could-define-new-asset-class . accessed 28 March 2025; see also Federica Agostini, 'Gold' standards for 'Greening' Financial Flows? The 'Brussels Effect' Of The Eu Sustainable Finance Agenda' (2025) In E. Baroncini, P. Manzini, & G. Tellarini (Eds.), *Sustainable Development as Fundamental Pillar of Economic Governance and Public Affairs: The EU Approach and International and Domestic Perspectives* (Bologna University Press 2025) 559, 571–572.

[81] Romano (n. 75) 25.

However, the 'race to the bottom' risk should arguably only be the primary concern for policymakers when it has systemic repercussions. In these instances, financial regulators should prioritise coordination initiatives with a global scope of application. Correspondingly, the prevailing areas of focus for the 'convergence' efforts by the above-mentioned transnational fora have traditionally been systemic risk[82] and other cross-border externalities.[83] These were also the primary objectives behind the previously mentioned IOSCO and Basel Committee principles on securitisation. Systemic risk concerns also drove initiatives in the sustainable finance area, like the Task Force on Climate-Related Financial Disclosures (hereinafter TCFD), which aimed at supporting institutions in managing the financial repercussions of climate-related risks.[84] While this framework was discontinued and superseded by the previously mentioned ISSB Standards in 2024, both focus on the financial materiality of sustainability matters.[85] This suggests that systemic risk considerations have arguably represented the key driver for global coordination efforts thus far, and should constitute the proxy for the desirable level of 'regulatory competition'.

Instead, the legal framework for 'sustainable securitisation' shows different goals and drivers. It should attract more capital to the internal market and enable the EU to pursue its other general and sustainability-related objectives. Therefore, 'race to the bottom' risks should not be overstated and should not overshadow the paramount benefits of 'regulatory competition' between EU and non-EU countries.

In addition, a persuasive point of reference to understand the optimal remit of transnational coordination mechanisms is a compendium,

[82] See Armour and others, 'The Goals and Strategies of Financial Regulation' (n. 82).; on their role in preserving financial stability within the COVID-19 pandemic, see Gortsos (n. 1) 119 ff.

[83] E.g. in the context of transnational corporate law principles, environmental harms and human rights violations, which have driven the formulation of the UN Guiding Principles on Business and Human Rights, see Pargendler (n. 69) 3–4, 31.

[84] For the first report see TCFD, 'Final recommendations' (2017) https://assets.bbhub.io/company/sites/60/2021/10/FINAL-2017-TCFD-Report.pdf accessed 30 December 2025.

[85] TCFD's disclosures aim at *'enable[ing] stakeholders to understand better the concentrations of carbon-related assets in the financial sector and the financial system's exposures to climate-related risk'*, ibid., 2. See also *'The IFRS Foundation's focus is on meeting the information needs of investors. The definition of material information is aligned with that used in IFRS Accounting Standards – that is, information is material if omitting, obscuring or misstating it could be reasonably expected to influence investor decisions'*. ISSB, 'Frequently Asked questions' (n. 71).

compiled and constantly updated by the FSB. It lists what the FSB considers to be the key global financial standards that countries are expected to implement as well as articulating a series of criteria which justify their inclusion to the list.[86] One of the criteria that brings all financial standards together is their scope, as they all cover areas which should be of relevance to all jurisdictions ('universality').[87] This distinctive trait does not match the characteristics of securitisation, which is strictly dependent on the broader regulatory environment, and especially on property law, company law, and financial law rules, with considerable differences across jurisdictions. The popularity of this financing technique also widely varies across financial markets depending on the respective regulatory treatment and on broader issues, like taxation and the available corporate forms.

Therefore, the regulatory diversity and the different appetite for securitisation across all jurisdictions prevents such universality, making global coordination extremely costly and cumbersome, to the point that it may become unfeasible. Any regulatory effort drawing from such divergencies also risks being too general.[88]

According to the FSB, another criterion which should inspire regulatory coordination initiatives is the 'broad endorsement' of standards. The FSB recommends that a widely recognised international body acts as standard setter following extensive consultation processes with stakeholders.[89] Arguably, such endorsement on sustainable securitisations is more feasible at the EU level than at the global level. The multi-stage legislative process, also entailing several consultation opportunities with interested parties and Member States, would facilitate consensus.[90] Relatedly, even the process for the design of sustainable finance rules was, in principle, highly collaborative, due to the wide participation of civil society experts.[91]

[86] '*key for sound financial systems and deserving of priority implementation depending on country circumstances. These standards are broadly accepted as representing minimum requirements for good practice that countries are encouraged to meet or exceed'*, FSB (n. 68).

[87] Ibid.

[88] On the risks of global coordination when too costly, see Pargendler (n. 69) 39.

[89] As also highlighted by FSB (n. 68).

[90] See For example, the procedure under art 352 TFEU, requiring the involvement of national Parliaments.

[91] Such as the 'Platform on Sustainable Finance', which has developed the technical criteria for the Taxonomy, see EU Commission, 'Communication From The Commission To The European Parliament, The Council, The European Economic And Social Committee And The Committee Of The Regions, 'Action plan: financing Sustainable Growth' (2018) 097, Action 1.2.

Lastly, FSB highlights that global regulatory efforts are likely to focus on 'relevant' and 'critical' issues in the transnational political agenda. Even though many countries have agreed on a high-level global sustainability agenda through the UN Sustainable Development Goals[92] and the Paris Agreement[93], they have so far demonstrated various levels of commitment to it. Global coordination efforts on sustainable finance, like the UN Principles for Responsible Investment (UNPRI)[94] or the above-mentioned ISSB have so far prioritised other high-level issues, such as investment and corporate reporting. By contrast, the EU has positioned itself in recent years as a fore-runner in the global sustainable finance domain at all levels. Even though previous initiatives have drawn limited attention to securitisation specifically[95], the momentum suggests that further regulation on sustainable securitisation is realistic and achievable at the EU level. In the light of these points, a harmonised approach in the EU context, ensuring competition with non-EU legal frameworks, would therefore be the preferable course of action.

6.3 Overview of Harmonisation Initiatives on Securitisation

EU law-making initiatives targeting securitisations have significantly increased, especially in response to the weaknesses that had emerged throughout the 2007 global financial crisis,[96] giving rise to a complex and comprehensive legal framework.

The key harmonisation effort has been the Regulation (EU) 2017/2402 (Securitisation Regulation),[97] amended in 2021,[98] along with various

[92] UN Transforming our world: the 2030 Agenda for Sustainable Development (2015) RES 70/1.

[93] Paris Agreement to the United Nations Framework Convention on Climate Change, (2015) TIAS 16-1104.

[94] UNPRI, 'What Are the Principles for Responsible Investment?' (2023)https://www.unpri.org/about-us/what-are-the-principles-for-responsible-investment; See above Chap. 5, s. 5.7.

[95] See below Sects. 6.4.–6.5.

[96] For an overview of salient initiatives see Steven L Schwarcz, 'Securitisation and Post-Crisis Financial Regulation' (2017) 102 Cornell Law Review 115, 121–125.

[97] (n. 33).

[98] Regulation (EU) 2021/557 of the European Parliament and of the Council of 31 March 2021 amending Regulation (EU) 2017/2402 laying down a general framework for securitisation and creating a specific framework for simple, transparent and standardised securitisation to help the recovery from the COVID-19 crisis (2021) OJ 116/1.

implementing and delegated acts providing the related technical standards for its application.[99] The EU Commission has also proposed a comprehensive review of the framework in 2025; however, it is unlikely to result in an overhaul of the broader regime for 'sustainable securitisations'.[100]

The Regulation has the twofold objective of setting common rules for all securitisations as well as of creating a differentiated regime for the so-called 'simple, transparent and standardised' (hereinafter STS) transactions. The latter meet a series of additional requirements, transposed to EU law and adapted from the above-mentioned transnational BCBS and IOSCO principles. [101] Reunited under the Task Force on Securitisation Markets, these standard-setters have set general criteria for the identification and for the capital treatment of STS deals.[102]

The rules of the STS regime centre around disclosure and external supervision. In other words, it is only possible to award the STS label to securitisations when originating entities provide extensive information about the characteristics of the deal following standard templates. These should facilitate other market actors, especially investors, in their assessment of the underlying risks.[103] The European Securities and Markets Authority (ESMA) is in charge of awarding the label[104] and of supervising compliance with the criteria.[105] It also keeps a common register of all outstanding STS-aligned deals,[106] which should increase their visibility.

Complying with the STS regime evidently represents a voluntary, yet common,[107] choice for market actors. Alongside the reputational advantages of signalling to investors the reliability of the transaction, it also increases its marketability. More importantly, the regime provides for a

[99] See For example, Sect. 5.3.1.

[100] See below Sect. 'The (limited) effects of the review of the securitisation framework'. Federica Agostini, Bart Joosen, 'The future of green securitisations in the European Union' (2025) 2 Kapitalmaarktenrecht, paragraph 4.

[101] Incorporated under EU legislation in Ch.4 (artt 18 –28) Securitisation Regulation.

[102] BCBS, IOSCO (n. 70); BCBS, 'Basel III framework' (n. 70) 10.

[103] For example, under Art 4 Securitisation Regulation.

[104] Art 27 (1) ibid.

[105] Art 10(1), 14, 15 ibid. Some have argued that 'a microsystem of sanctions' following from non-compliance would be an additional goal of the regime. Pierre De Gioia Carabellese, Camilla Della Giustina, *The law of securitisations: from crisis to techno-sustainability* (Routledge 2023) 68.

[106] Art 27 (3) ibid.

[107] For some data on the compliance with STS regime in 2022 and 2023, see below (n. 165).

more advantageous capital treatment for STS over non-STS transactions[108] since the compliance with the criteria should testify to the use of less risky assets.[109] The choice to link a lower capital surcharge to a voluntary standard is meant to create lower interference over parties' freedom.[110]

While the scope of the securitisation framework is broader, some provisions also target sustainability-related aspects of these deals. At the same time, various initiatives of the ambitious and comprehensive EU agenda on sustainable finance have indirectly affected and will affect securitisations. The present work will predominantly focus on the links between the Green Bond Standard *(hereinafter* EUGBS), included in the EU Green Bond Regulation (*hereinafter* EUGBR) and the various components of the Taxonomy Regulation.[111]

The EUGBR[112] has introduced an EU Green Bond (EUGB) label, which market actors may *voluntarily* choose to adopt to market debt securities, including ABSs, as 'green'. The idea behind the Standard resonates with some of the soft law principles analysed in Chap. 4, such as the ICMA's Green Bond Principles,[113] yet it will considerably differ from them due to the link to statutory requirements ('hard law'). For the bonds that do comply with the EUGBS, it will be mandatory to abide by the various requirements concerning reporting, external reviews, and tracking of expenses in their entirety.[114]

[108] Art 243, 260, 262 Regulation 2017/2401. Commission Delegated Regulation (EU) 2019/1851 of 28 May 2019 supplementing Regulation (EU) 2017/2402 of the European Parliament and of the Council with regard to regulatory technical standards on the homogeneity of the underlying exposures in securitisation.

[109] (4) Regulation 2017/2401; As also recommended by BCBS, IOSCO (n. 70) 10. See For example, Art 13 (1) – (2) Commission Delegated Regulation (EU) 2015/61 of 10 October 2014 to supplement Regulation (EU) No 575/2013 of the European Parliament and the Council with regard to liquidity coverage requirement for Credit Institutions. De Gioia Carabellese, Della Giustina (n. 105) 69.

[110] Vincenzo Bavoso, 'Good Securitisation, Bad Securitisation and the Quest for Sustainable EU Capital Markets' (2015) 4 *Butterworth Journal of International Banking and Financial Law* 221.

[111] Especially in Sect. 6.4.2.–6.5.2.

[112] The EUGBR entered into force in 2023 and is applicable since 21 December 2024. Art 72 (2) Regulation (EU) 2023/2631 on European green bonds and optional disclosures for bonds marketed as environmentally sustainable and for sustainability-linked bonds (2023) PE/27/2023/REV/1.

[113] See Chap. 4, Sect. 5.2.3.–5.4.2.1.

[114] See further in Sect.s. 6.4.2–6.5.2.–6.8.2.

The basis to label securitisations as 'sustainable' is the Taxonomy, which encompasses the Regulation (EU) 2020/8521[115] as well as a series of delegated acts.[116] They aim at introducing a classification system to qualify economic activities and financial instruments as 'environmentally sustainable' under EU law.[117] While the former Regulation lists the general principles to comply with the Taxonomy,[118] the delegated acts provide very detailed 'technical screening criteria' for a wide array of sectors, ranging from forestry, energy, transport,[119] to the much debated gas and nuclear-related activities.[120] For the time being, the Taxonomy only focuses on six environmental objectives (with different technical criteria developed for climate-related and other environmental goals[121]), even though there were initial investigations on the merits of a parallel classification system for

[115] Regulation (EU) 2020/852 of the European Parliament and of the Council of 18 June 2020 on the establishment of a framework to facilitate sustainable investment, and amending Regulation (EU) 2019/2088 (2020) OJ L 198/13.

[116] Commission Delegated Regulation (EU) 2021/2139 of 4 June 2021 supplementing Regulation (EU) 2020/852 of the European Parliament and of the Council by establishing the technical screening criteria for determining the conditions under which an economic activity qualifies as contributing Sect.tantially to climate change mitigation or climate change adaptation and for determining whether that economic activity causes no significant harm to any of the other environmental objectives (2021) OJ L 442/1; Commission Delegated Regulation (EU) 2022/1214 of 9 March 2022 amending Delegated Regulation (EU) 2021/2139 as regards economic activities in certain energy sectors and Delegated Regulation (EU) 2021/2178 as regards specific public disclosures for those economic activities (2022) OJ L 188/1.

[117] It distinguishes between 'environmentally sustainable', 'transitional' and 'enabling' activities, Art 3, 10(2), 16 Regulation 2020/852.

[118] Inter alia, Art 3 ibid.

[119] Commission Delegated Regulation (EU) 2021/2139.

[120] Commission Delegated Regulation (EU) 2022/1214.

[121] On climate change mitigation and adaptation, see Commission Delegated Regulation (EU) 2021/2139 and ibid.; for other environmental matters see Commission Delegated Regulation (EU) 2023/2486 of 27 June 2023 supplementing Regulation (EU) 2020/852 of the European Parliament and of the Council by establishing the technical screening criteria for determining the conditions under which an economic activity qualifies as contributing Sect.tantially to the sustainable use and protection of water and marine resources, to the transition to a circular economy, to pollution prevention and control, or to the protection and restoration of biodiversity and ecosystems and for determining whether that economic activity causes no significant harm to any of the other environmental objectives and amending Commission Delegated Regulation (EU) 2021/2178 as regards specific public disclosures for those economic activities (2023) C/2023/3851.

socially beneficial activities ('Social Taxonomy'[122]). It is also worth noting that the technical criteria draw boundaries to 'green' activities by providing 'do no significant harm' (DNSH) criteria that taxonomy-aligned assets may not breach: for instance, the Appendix C lists the types of activities that may cause pollution due to the use of certain chemicals.[123]

In the following sections, I will examine how these initiatives affect the 'focal points' of securitisation law. As discussed in Chap. 3,[124] regulating these areas is strategic since they directly shape sustainability-related aspects of securitisations. Law can also act on the incentives, encouraging more market actors to put these transactions into motion.

For ease of reference, I will differentiate between the initiatives for any type of securitisation and those specifically targeting 'sustainable' elements. Later, I will provide normative considerations on potential reforms which could have a *transformative impact* on securitisations.

6.4 (Limited) Harmonisation on the Composition of Portfolios

6.4.1 Sustainability-Related Elements of 'Traditional' Securitisations

The regime for traditional securitisations has very few rules requiring financial institutions to consider sustainability-related factors when originating credit claims or putting together securitisation portfolios. This also resonates with the findings of a survey by UNPRI across investors, which has identified the lack of data on the ESG profile of Securitisation portfolios as one of the largest hurdles to investments in 'sustainable' securitised products.[125]

6.4.1.1 Salience of Sustainability Factors in the Origination Process

In the context of the present analysis, it must be recalled that the Securitisation Regulation provides assignors with the option to either

[122] Platform on sustainable finance, 'Report on social taxonomy' (2022) https://finance.ec.europa.eu/system/files/2022-08/220228-sustainable-finance-platform-finance-report-social-taxonomy_en.pdf accessed 10 December 2025.

[123] Commission Delegated Regulation (EU) 2021/2139.

[124] S. 3.4.

[125] UNPRI, 'ESG Incorporation in Securitised Products: the challenges ahead' (2021) 24https://gsfo.org/resource/esg-incorporation-securitised-products-challenges-ahead accessed 20 December 2025.

create the obligations and bundle them together for the purposes of a securitisation deal, or to purchase them from a third party: in both circumstances they will act as 'originators'.[126] Under EU law there are currently no general rules on the environmental and social impact of lending activities, which would require such actors to take sustainability risks into account before the origination of loans or after the provision of finance.

Similarly, securitisation originators are also not required to assess whether the credit claims included within portfolios provide finance for environmental and social goals. Admittedly, the Corporate Sustainability Due Diligence Directive (CSDDD) is expected to require (larger-sized) companies to conduct 'due diligence' assessments to prevent, identify and address sustainability-related risks (like environmental risks or adverse impacts on human rights across supply chains) on their activities.[127] Yet, it only covers 'upstream' activities for financial institutions, like the supply of goods and provision of services, but not lending and investment activities.[128] This means that originators of securitisations would not need to assess and report about the risks related to their securitisation portfolios. At the same time, as the following sections will show, the CSDDD leaves limited room for SPVs to perform such assessments.[129]

[126] Art 2(3) – (5) Securitisation Regulation.

[127] (26); Art 3-15, 16, Directive (EU) 2024/1760 of the European Parliament and of the Council of 13 June 2024 on corporate sustainability due diligence (2024) OJ L 2024/1760 (*hereinafter*CSDDD). Following the 'stop-the-clock' initiative in 2025, the Directive is expected to only start applying from 26 July 2028 or 2029, Art 2 (1) Directive (EU) 2025/794 of the European Parliament and of the Council of 14 April 2025 amending Directives (EU) 2022/2464 and (EU) 2024/1760 as regards the dates from which Member States are to apply certain corporate sustainability reporting and due diligence requirements 2025/794.

[128] This was the result of a lengthy and tortious legislative process, culminating in the exclusion of the financial sector, originally included in the first text of the Proposal. For an overview of the legislative process, see Sara Addamo, 'Il dovere di diligenza ai fini della sostenibilità nel settore bancario' (2024) 3 *Rivista di Diritto Bancario*, 585–590. Some have argued for an expansive interpretation of the Directive, which would also cover some financing activities. Litwin, Daniel and Savourey, Elsa, Human Rights in EU Sustainable Finance (2024) 11 Journal of Financial Regulation 1 41, 53. However, this interpretation is arguably in contrast with the recitals and literal wording of the provisions of the Directive, which refer to 'companies' and not to 'financial undertakings'. A review provision of the Directive also outlines a plan to reconsider the inclusion of 'financial services and investment activities' at a later stage, implying the exclusion from the current text. See art 36 Directive (EU) 2024/1760.

[129] See Sect. 5.5.

One unresolved question relates to the value of soft law principles, like the OECD Guidelines. These Guidelines, while not securitisation-specific, effectively conceptualise the meaning of 'due diligence' for underwriting and lending activities.[130] The non-binding nature of the instrument for EU Member States places them outside the scope of the present chapter (as 'soft law' instruments). As the OECD represents a transgovernmental and not a market-led organisation,[131] its Guidelines do not fall within the perimeter of 'private law', as discussed in Chap. 5, either. Yet, OECD Guidelines could certainly fill the gaps of the CSDDD and operationalise the concept of 'due diligence', as also suggested in the recitals.[132] The added value of complying therewith in the securitisation context is for originators to demonstrate a minimum level of consideration for environmental and social risks, even when the securitised assets are not 'sustainable'.[133]

Additionally, when originating entities are financial institutions, the composition of the portfolio is also indirectly relevant for the purposes of their treatment in the context of central bank operations, as I shall discuss in Sect. 6.4.3.

6.4.1.2 *Disclosure of Sustainability-Related Attributes of Portfolios*

One cornerstone of EU securitisation rules, especially following the 2007 global financial crisis, relates to disclosure requirements for securitisation originators, which also relate to the duty to communicate certain characteristics of securitisation portfolios. The main rationale for these rules relates to the need to address the information asymmetry between issuers and capital market investors and to prevent the assignment of flawed

[130] OECD, 'Due Diligence for Responsible Corporate Lending and Securities Underwriting: Key Considerations for Banks Implementing the OECD Guidelines for Multinational Enterprises' (2019) https://www.oecd.org/en/publications/due-diligence-for-responsible-corporate-lending-and-securities-underwriting_71fd2895-en.html accessed 20 December 2025.

[131] On the OECD as a formal regulator in the international financial arena, see Jordana J, 'Transgovernmental Networks as Regulatory Intermediaries: Horizontal Collaboration and the Realities of Soft Power' in *Regulatory Intermediaries in the Age of Governance* (2017) 670 *The Annals of the American Academy of Political and Social Science* 245, 258–259.

[132] These emphasise the interpretive value of the OECD Guidelines and the duty of financial undertakings to exercise 'leverage' to influence their clients. (51) Directive (EU) 2024/1760.

[133] On the possibility to use 'non-sustainable' assets in 'sustainable' securitisations, see Chap. 4, Sect. 4.4.

assets.[134] However, the wide variety of information that market actors currently disclose provides very limited insight into the sustainability risks and impact of the underlying credit claims.

Originators have to report on several aspects of the exposures in the prospectus (or in a transaction summary, if they are not required to draft the former[135]) as well as in separate quarterly reports to current investors (and prospective ones, upon request).[136] Two Delegated Regulations provide a very detailed list of the mandatory items[137] and a series of standardised templates for reporting.[138] In recent consultations on the EU securitisation framework, the majority of respondent market actors have highlighted the disproportionate nature of such requirements, to the point that they may constitute a barrier to engaging in securitisation transactions.[139]

The STS label requires the compliance with even stricter disclosure requirements surrounding the composition of the portfolio. The origination of the assets must have occurred 'in the ordinary course of the

[134] See, among others, BCBS, IOSCO (n. 70) 3; Regulation (EU) 2021/557 amending Regulation 2017/2402; on the evolution of the securitisation legal framework after the Crisis, see Steven L Schwarcz, 'The Future of Securitisation' (2009) 41 *Connecticut Law Review* 1317–1320; 1322–1323; ibid., 'Securitisation and Post-Crisis Financial Regulation' (n. 98) 117–119, 124.

[135] See Sect. 6.5.1. on exemptions from prospectus requirements.

[136] Art 7 (1)(a) – 7(1)(c)(ii) – 7 (1)(c)(ii) – 7 (1)(e) (i) Securitisation Regulation. Such information is also made available to competent authorities. Art 7(1) ibid.

[137] For example, info on the debtor, maturity date, balance, see Annex II-IX commission Delegated Regulation (EU) 2020/1224 of 16 October 2019 supplementing Regulation (EU) 2017/2402 of the European Parliament and of the Council with regard to regulatory technical standards specifying the information and the details of a securitisation to be made available by the originator, sponsor and SSPE (2020) OJ L 289\1.

[138] Commission Implementing Regulation (EU) 2020/1225 of 29 October 2019 laying down implementing technical standards with regard to the format and standardised templates for making available the information and details of a securitisation by the originator, 2020 OJ 289\17. ibid.

[139] European Commission, 'Summary report – –Targeted consultation on the functioning of the EU securitisation framework' (2021) 10 https://finance.ec.europa.eu/system/files/2022-09/2021-eu-securitisation-framework-summary-of-responses_en.pdf accessed 10 December 2025; on a similar point in relation to disclosure requirements in general, see EU Commission, 'Factual summary report of the targeted consultation on the functioning of the EU securitisation framework' (2025) 6 https://finance.ec.europa.eu/document/download/35846dc8-8ebc-4bc9-967d-de77ef5d9234_en?filename=2024-eu-securitisation-framework-summary-of-responses_en.pdf For an overview of industry concerns in relation to disclosure requirements and proposals, see Thomas Papadogiannis Varouchakis, 'The European Securitisation Market: Effects of an Uneven Regulatory Playing Field, (2024) 10 *Journal of Financial Regulation* 2, 143, 152–153, 154–155.

business' of the original lender or of the originator.[140] In order to facilitate investors' assessment of risks, the exposures must be 'simple' and comparable.[141] One indicator of 'simplicity' is the clear specification of the criteria that make assets eligible for inclusion in the underlying portfolio in the transactional documentation. As discussed in Chap. 3, the document articulating such criteria will generally be the loan purchase agreement, through which the assets are assigned to SPVs.[142] The STS label also prevents all 'cherry-picking' exercises involving a discretionary choice of the assets. As a result, while it is possible across *traditional* securitisation operations to delegate the selection of the exposures to be securitised to a third party ('sponsor')[143], this is not possible for STS deals.[144]

Besides, a transaction is 'simple' if the underlying portfolio is 'homogenous', as also suggested by the BCBS recommendations.[145] 'Homogeneity' follows from the combination of different factors, including the adoption of similar underwriting standards for all securitised receivables. The coexistence of these factors should indicate that the originator has applied a consistent approach to the assessment of credit risks.[146] A corollary of the 'homogeneity' requirement is also the need to only use one type of assets.[147] A Delegated Regulation [148] sets out all these criteria and provides

140 Art 20 (10) Securitisation Regulation.

141 Hill CA, 'Securitisation: A Low-Cost Sweetener' (n. 58) 1074.

142 See Chap. 3; Phillip Wood, 'Transfer and Administration Agreement', in Wood, *Project Finance, Securitisations and Subordinated Debt, para 7-003.*

143 Art 2 (5)(b), Art 20 (7) ibid.

144 Art 20 (7) Securitisation Regulation; Criterion 5 BCBS, IOSCO (n. 71).

145 (27); Art 20 (7)-(8) ibid.; Criterion 1 ibid. (27);

146 (3) – (4) ibid. EBA, Consultation Paper – –Draft Regulatory Technical Standards – –On the homogeneity of the underlying exposures in securitisation under Art. 20(14) and 24(21) of [Regulation (EU) XXX/201X ... laying down a general framework for securitisation and creating a specific framework for simple, transparent and standardised securitisation] (2019) 28 https://www.eba.europa.eu/documents/10180/2298183/1ecb2150-fd3d-4aef-ac8f-393dc314deea/Draft%20RTS%20on%20homogeneity%20of%20underlying%20exposures%20in%20securitisation%20(EBA-RTS-2018-02%20).pdfaccessed 3 June 2026..

147 Art 20(8) ibid.; Art 1 (a)(b) Commission Delegated Regulation 2019/1851; art 26b (8) Regulation (EU) 2021/557; EBA, ibid.

148 Commission Delegated Regulation (EU) 2024/584 of 7 November 2023 amending the regulatory technical standards laid down in Delegated Regulation (EU) 2019/1851 as regards the homogeneity of the underlying exposures in simple, transparent and standardised securitisations; amending ibid. It builds on EBA's work, EBA, 'Final Draft Regulatory Technical Standards On the Homogeneity of the Underlying Exposures in Securitisation under Articles 20(14) and 24(21) of Regulation (EU) No 2017/2402 Laying down a General Framework for Securitisation and Creating a Specific Framework' (2018).

a non-exhaustive list of the eligible asset types, including residential and commercial loans when secured or guaranteed, consumer credit, commercial credit, auto loans and leases, credit card receivables, trade receivables, as well as other exposures deemed as distinct on the basis of internal methodologies and procedures.[149] For assets other than consumer credit and credit card receivables, the Regulation also requires originators to take additional homogeneity 'factors' into account, such as the jurisdiction, the ranking of security rights, as well as the type of obligors.[150]

Originating entities must disclose how the assets meet these homogeneity criteria as part of their notification to ESMA, which oversees the compliance with the STS requirements.[151]

The most striking feature of the requirements for traditional and for STS transactions is their granularity. However, mandatory reporting items and templates currently warrant very little attention to the environmental and social impact or risks of the underlying exposures.

Originators of 'traditional' securitisation operations, not falling under the STS regime, should only report on the 'energy performance' of the underlying assets, as well as providing the related certification by a third-party provider, when the exposures include automobile, residential real estate or consumer receivables.[152] However, the Delegated Regulation also leaves issuers the option not to report on the matter if they indicate that the information is unavailable for justifiable reasons ('no data' (ND)), for which they may have to provide an explanation to competent authorities upon request. [153] Such circumstances may include the cases where they did not collect such information because they purchased the portfolio from third parties. As a result, there is empirical evidence of the very low reporting levels on energy performance certificates, with significant divergencies across market actors.[154] In 2023, ESMA and the other supervisory authorities over

[149] Art. 1 (a) Commission Delegated Regulation 2019/1851 of 28 May 2019 supplementing Regulation (EU) 2017/2402 of the European Parliament and of the Council with regard to regulatory technical standards on the homogeneity of the underlying exposures in securitisation; EBA, ibid.

[150] Art 2 (6) ibid.; EBA (n. 146) 11–16.

[151] See below Sect. 6.5.

[152] AUTL57-58; RREC10-RREC11; CMRL68-69 Commission Delegated Regulation (EU) 2020/1224; Commission Delegated Regulation (EU) 2020/1225.

[153] (13) Commission Delegated Regulation (EU) 2020/1224

[154] As noted by the ECB, 'ECB response to the ESAs' joint consultation on sustainability-related disclosures for STS securitisations' (2022) 3 https://www.ecb.europa.eu/pub/pdf/other/ecb.response_ESAs_consultation_sustainability_related_disclosures_STS_securitisations~ee2f3fb909.cs.pdf?0c0d2818008690c368bdfa8593b9afa8 accessed 10 December 2025.

financial markets have announced that they were considering the 'opportunity of introducing new, proportionate and targeted climate change-related metrics' in their review of these standardised templates.[155] This initiative, which ultimately did not move forward, could have considerably strengthened issuers' transparency on the environmental impact of securitisation portfolios. However, as will be discussed later, the prevailing view was that strengthening disclosure requirements would have added to the reporting hurdles that market participants were already perceiving as too burdensome, disproportionate and of limited use to investors.[156]

As regards STS transactions, the Securitisation Regulation requires originators to report on the 'environmental performance' of the exposures if these are residential or auto mortgages or leases. However, an important caveat to the application of such requirement applies: Disclosure is only required if such information is available and, if not, its absence does not prevent the award of the STS label.[157] As an alternative, the 2021 amendments to the Securitisation Regulation allow originating entities to report on the 'principal adverse impact' (PAI) of these exposures 'on sustainability factors'.[158] Such wording resonates with a parallel initiative of the EU's sustainable finance debate, i.e. the Sustainable Finance Disclosure Regulation (SFDR).[159] Building on this approach, the European supervisory authorities (EBA, ESMA, EIOPA) have introduced regulatory technical standards (RTSs)[160], based on a previous

[155] EBA, ESMA, European Insurance and Occupational Pensions Authority (EIOPA), ECB, 'Joint ESAs-ECB Statement on disclosure on climate change for structured finance products' (2023) 3 https://www.esma.europa.eu/sites/default/files/library/ESAs_ECB%20Joint%20Statement%20on%20disclosures%20for%20securitisations_FINAL_6%20March%202023_0.pdf accessed 10 December 2025.

[156] Olivia Hauet, 'Reviving securitisation in the EU: A critical analysis of the reporting requirements' (2025) 7 Journal of Financial Compliance 2, 5 https://www.true-sale-international.de/fileadmin/tsi-gmbh/tsi_downloads/aktuelles/ECB_Olivia_Hauet_A_critical_analysis_of_the_securitisation_reporting_2023-12.pdf accessed 10 December 2025; Thomas Papadogiannis Varouchakis (n. 139); EU Commission (n. 139).

[157] EBA, 'Final report – –on Guidelines on the STS criteria for non-ABCP securitisation' (2018) 20, 82 https://www.bde.es/f/webbde/INF/MenuHorizontal/Normativa/guias/EBA-GL-2018-09-EN.pdf

[158] Art 22(4) – 26d(4) Regulation (EU) 2021/557 amending Regulation 2017/2402. The provision is only applicable to ABSs and not to ABCP, see artt 24-25-26 ibid.

[159] Regulation (EU) 2019/2088 of the European Parliament and of the Council of 27 November 2019 on sustainability-related disclosures in the financial services sector (2019).

[160] Commission Delegated Regulation (EU) 2024/1700 of 5 March 2024 supplementing Regulation (EU) 2017/2402 of the European Parliament and of the Council with regard

consultation[161], that bring these reporting requirements in line with the ones under SFDR.[162] Originating companies choosing to disclose about the PAI of their activity will have to follow the standardised template under the RTSs, consisting of a series of indicators and metrics.[163]

The RTSs represent a significant step ahead, which could address the current inconsistencies when it comes to measuring and assessing the environmental performance of assets, which is particularly evident in relation to auto loans.[164]

Nevertheless, the scope of the RTSs is too limited to increase the transparency around the sustainability risks and impact of all securitisation portfolios. The exemptions from the application of the requirements linked to the lack of data provide market participants with an easy justification to avoid gathering and disseminating such information, as demonstrated by the low levels of reporting. Additionally, disclosure concerning 'environmental performance' and 'principal adverse impacts' both relate to STS-compliant transactions. While the STS regime is likely to remain the preferred option for market actors, there is still in principle room for

to regulatory technical standards specifying, for simple, transparent and standardised non-ABCP traditional securitisation, and for simple, transparent and standardised on-balance-sheet securitisation, the content, methodologies and presentation of information related to the principal adverse impacts of the assets financed by the underlying exposures on sustainability factors (2024) C/2024/1344;

[161] See EBA, ESMA, EIOPA, 'Joint Consultation Paper – –STS securitisations-related sustainability disclosures' (2022) https://www.eba.europa.eu/activities/single-rulebook/regulatory-activities/securitisation-and-covered-bonds/joint-regulatory-technical-standards-sts-securitisations-related-sustainability-disclosures accessed 3 June 2026;, as well as ESMA, Final Report on draft Regulatory Technical Standards With regard to the content, methodologies and presentation of disclosures in respect of the sustainability indicators in relation to adverse impacts of the assets financed by the underlying exposures for STS securitisations on the climate and other environmental, social, and governance-related adverse impacts pursuant to Article 22(6) and 26d(6) of Regulation (EU) 2017/2402 (2023) https://www.esma.europa.eu/sites/default/files/2023-05/JC_2023_13_-_Final_report_on_ESG_disclosure_for_STS_securitisations.pdf accessed 10 December 2025.

[162] Ibid., 5–6.

[163] They have clarified that these templates are mandatory for those choosing to report on the PAI. Ibid., 11.

[164] Asset Management and Investor Council (AMIC) ICMA, 'Discussion Paper on ESG KPIs for Auto-Loans/Leases ABS' (2021) 7 6 https://www.icmagroup.org/assets/documents/Regulatory/AMIC/AMIC-discussion-paper-ESG-auto-loan-ABS-FINAL-170521.pdf; accessed 10 December 2025.

non-STS-compliant transactions[165], which would be exempt from the disclosure requirements under the RTSs.

6.4.1.3 The (Limited) Effects of the Ongoing Review of the Securitisation Framework

Other initiatives that warrant attention are those related to the broader and ongoing review of the securitisation framework.[166] In particular, a 2023 consultation by ESMA has sought stakeholders' input on potential revisions to securitisation reporting templates. Among other options, it considered requiring originators to provide further information about the climate risks linked to the portfolio, such as GHG emissions data and more details on location and physical characteristics of real estate assets. These would support investors' assessment of climate-related risks. At the same time, ESMA itself acknowledged the challenges linked to limited data availability and to inconsistent regulatory requirements across EU jurisdictions, which would prevent a meaningful assessment by investors of such information, even if supplied. The consultation paper also highlighted the widespread concern around potential compliance burdens. [167]

Thus, it comes as no surprise that the following step in the review process left this option off the table. Following wide criticism from market actors about the reporting hurdles,[168] a 2025 consultation paper proposed

[165] See, among others, AFME's statistics on the use of the STS regime in the most recent years. It highlighted how 35% of EU issuances in 2023 was in line with STS, AFME, 'European Structured Finance – Securitisation Report, Q4 2023 & 2023 FY' (2023) 5 https://www.afme.eu/Portals/0/DispatchFeaturedImages/Securitisation%20Data%20Report%20Q4%202023%20and%202023%20Full%20Year.pdf ESMA highlights that public STS-compliant transactions accounted for around 60% of the public securitisation market in 2022, ESMA, 'TRV Risk Analysis – The EU securitisation market – an overview' (2023) 10 https://www.esma.europa.eu/sites/default/files/2023-09/ESMA50-524821-2908_TRV_risk_analysis_-_EU_securitisation_markets_overview.pdf.

[166] Spearheaded by a 2022 report by the EU Commission on the functioning of the framework. EU Commission, 'Report from the Commission to the European Parliament and the Council on the functioning of the Securitisation Regulation' (2022) 517 https://eur-lex.europa.eu/legal-content/EN/TXT/PDF/?uri=CELEX:52022DC0517.

[167] ESMA, 'Consultation Paper On the securitisation disclosure templates under Article 7 of the
Securitisation Regulation' (2023) 38–41 https://www.esma.europa.eu/sites/default/files/2023-12/ESMA12-2121844265-3053_-_Consultation_Paper_on_the_Securitisation_Disclosure_Templates.pdf accessed 15 December 2025.

[168] EU Commission (2025) (n. 166).

concentrating on a review of the disclosure templates for private securitisations only.[169]

Such developments underscore the broader policy challenge underpinning the design of disclosure-oriented rules on sustainability matters, i.e. striking the right balance between enhancing originators' accountability while ensuring that reporting obligations are feasible for originators, and of effective use to market participants.

6.4.2 *A Stronger, Yet Fragmented, EU Regime for 'Sustainable' Portfolios*

While the rules for *traditional* securitisations have so far devoted limited focus to the sustainability-related qualities of portfolios, recent harmonisation initiatives are expected to strengthen the requirements for the transactions labelled as 'sustainable'. Yet, the intersection between the various initiatives remains unclear.

6.4.2.1 *Lights and Shadows of the EU Green Bond Regulation*

One of the thorny questions in the EU legislative debate was whether securitisation should be 'sustainable' only on the basis of the final allocation of investors' proceeds (what I have previously called 'sustainable proceeds securitisations') or also on the basis of the portfolio ('sustainable portfolio securitisations'[170]). To this end, a consultation on the revisions to the securitisation framework also sought market participants' feedback on the meaning previously given to 'sustainable' securitisations.[171] As part of the responses, the ECB has confirmed that the securitised products that the Eurosystem purchased did not always have 'sustainable' portfolios.[172]

[169] ESMA, 'Consultation Paper on the revision of the disclosure framework for private securitisation under Article 7 of the Securitisation Regulation' (2025) https://www.esma.europa.eu/sites/default/files/2025-02/ESMA12-2121844265-4462_-_Consultation_Paper_on_Private_Securitisation.pdf. Accessed 10 December 2025.

[170] See Chap. 3, Sect. 3.3.

[171] European Commission, 'Consultation Document – –Targeted Consultation on the Functioning of the EU Securitisation Framework' (2021) question 6.6 https://finance.ec.europa.eu/regulation-and-supervision/consultations/2021-eu-securitisation-framework_en accessed 10 December 2025.

[172] European Central Bank, Eurosystem, 'Eurosystem Contribution to the European Commission's Targeted Consultation on the Functioning of the EU Securitisation Framework' (2020) 7. https://www.ecb.europa.eu/pub/pdf/other/ecb.eurosystemcontributiontargetedconsultationEUsecuritisationframework2109~cade37d248.en.pdf accessed 10 December 2025.

Conversely, the preliminary reports for the design of an EU green bond standard[173] and the wording of the consultation[174] suggested that the EU Commission initially intended to qualify securitisations as 'sustainable' only when backed by 'sustainable' assets, also following the ICMA's Green Bond Principles.[175]

The key instrument that brought clarity was the EU Green Bond Standard (hereinafterEUGBS or 'the Standard'), which subjects 'securitisation bonds' to the same legal treatment as 'green' bonds.[176] As previously mentioned, compliance with the Standard entails a series of additional mandatory disclosure and review requirements.[177] On the contrary, market actors using securitisations as a way to achieve certain sustainability-related targets (through the use of 'sustainability-linked' securities[178]) would have the option, but not the duty, to use standardised templates to report information ahead of and after the issuance.[179]

As a result, the distinctive criterion for eligibility for the EUGBS is the 'use of proceeds' instead of the types of assets composing the portfolio. According to the recitals to the EUGBR, this approach is more 'efficient and pragmatic' given the scarcity of 'sustainable' assets that originating entities could bundle together as part of the portfolio otherwise.[180]Indeed, empirical research has underlined the general scarcity of financial assets, including loans, that comply with the Taxonomy technical screening

[173] Using 'green real estate mortgages' as examples of eligible 'financial green assets', see Technical Expert Group on Sustainable Finance (TEG), 'Usability Guide EU Green Bond Standard' (2020) 15. https://finance.ec.europa.eu/system/files/2020-06/200309-sustainable-finance-teg-green-bond-standard-usability-guide_en.pdf accessed 15 September 2023.

[174] See *'According to the Commission proposal for a European green bond standard, a securitisation bond may qualify as EU green bond if the proceeds of the securitisation are used by the issuing special purpose vehicle to purchase the underlying portfolio of Taxonomy-aligned assets'* EU Commission, Consultation Document (n. 171) 13.

[175] See Chap. 5, Sect. 5.4.2.2, ICMA, 'GBPs' (2025), 8.

[176] See 'Explanatory memorandum' Proposal for a Regulation of the European Parliament and of the Council on European green bonds (2021) 390 final (*hereinafter*EUGBR).

[177] See above Sect. 5.3.

[178] See Chap. 3, Sect. 3.3.

[179] Art 13g EUGBR.

[180] (20b) also following EBA'S recommendations, see EBA, 'Developing a framework for Sustainable Securitisations' (2022) 36, 38. REP/2022/06 https://www.eba.europa.eu/sites/default/files/document_library/Publications/Reports/2022/1027593/EBA%20report%20on%20sustainable%20securitisation.pdf accessed 3 June 2026.

criteria within the EU market.[181] In other words, it is possible for securitisation originators to comply with the EUGBS whilst using portfolio assets not in line with the Taxonomy.

At the same time, the Regulation also introduces constraints on the composition of the portfolio, which should also provide investors with some information on the environmental impact and risks of the securitised assets. Exposures, such as loans and leases, financing fossil fuel-related activities, cannot back EUGBS-compliant securitisations.[182] Exposures must be excluded on the basis of their predominant purpose[183]: For instance, originators may include them if the link with fossil fuel activities is marginal or incidental (e.g. a commercial building with gas storage tank).[184] Originators will have to indicate how securitisations abide by these requirements in a pre-issuance document (the 'Factsheet').[185]

More importantly, originators have to report about sustainability-related risks, and about the impact, of the portfolio assets on an aggregate basis. Before the transaction, they have to disclose the share of assets 'eligible' for the EU taxonomy, the share of assets 'aligned' with its technical screening criteria and the share of assets considered as harmful (breaching the so-called 'do no significant harm' (DNSH) criteria.[186] This information is part of both the prospectus and of the dedicated 'Factsheet'

[181] Lucia Alessi, Stefano Battiston, Ana Sofia Melo et al, 'The EU Sustainability Taxonomy: a Financial Impact Assessment' (2019) European Commission, Joint Research centre, https://publications.jrc.ec.europa.eu/repository/bitstream/JRC118663/2020.01.10_technical_report_commission_taxonomy_published.pdf accessed 10 December 2025. While the research is based on estimates ahead of the introduction of the Taxonomy's technical screening criteria, they proved correct in the following years. Lucia Alessi, 'Concluding remarks' (5th Summer School on Sustainable Finance, Joint Research Centre, 14 July 2023) https://webcast.ec.europa.eu/5th-jrc-summer-school-on-sustainable-finance-14th-of-july accessed 17 March 2025. In 2023, EBA has estimated that green loans amount to around 4.5% of the total loans granted by EU banks. EBA, 'Report in response to the call of advice from the European Commission on green loans and mortgages' (2023) EBA/REP/2023/38, 19 https://www.eba.europa.eu/sites/default/files/2023-12/e7bcc22e-7fc2-4ca9-b50d-b6e922f99513/EBA%20report%20on%20green%20loans%20and%20mortgages_0.pdf accessed 3 June 2026..

[182] Art 20b EUGBR.

[183] Specifically, securitised exposures financing electricity, co-generation or power generation of heat and cool from fossil fuels. 13c (2) ibid.

[184] Art 13c (1) ibid.

[185] Art 10 ibid.

[186] Art 13d(2) ibid. on DNSH criteria, see above Sect. 6.3.

which originators have to publish ahead of the issuance[187] following a standardised template.[188]

The expansion of disclosure requirements on the portfolio under the EUGBR certainly represents a step forward in reducing the ambiguity and inconsistencies among current transactions, where market actors have labelled securitisations as a whole as 'sustainable', while the characteristics of the underlying assets have not always been clear.[189]

However, the EUGBR still leaves room for issuers not to be fully transparent around the composition of the portfolio. They have to report on the compliance with the taxonomy criteria 'on a best effort basis', 'to the best of the originator's ability' and 'based on available data'.[190] Certainly some data on the environmental impact of securitisation exposures may be still unavailable at the moment, which would justify less stringent reporting requirements to some extent. However, giving issuers flexibility to determine when they do not report could still leave room for opportunistic behaviours. It may favour under-reporting, which may in practice give rise to omissions of relevant information, resulting in 'greenwashing' practices.[191]

The other risk arising from the limitations of the current drafting of the disclosure requirements is the multiplication of costs for prospective investors to conduct 'due diligence' assessments.[192] In other words, incomplete reporting on environmental and social risks may require investors to make additional research, especially when they may need to demonstrate that they are making 'sustainable investments' for regulatory purposes.[193] Therefore, it may also constitute a barrier to investments.

In addition, the most striking barrier to transparency is the voluntary scope of the EUGBS.[194] In other words, market actors may decide to comply with the Standard, and would thereby be subject to the more robust disclosure requirements, but they may also prefer to independently

[187] Art 13d(1)(4) ibid.

[188] Annex I ibid.

[189] See Chap. 4, s. 4.6.

[190] 13d(3) EUGBR.

[191] On the definition of 'greenwashing', see Chap. 3, Sect. 3.4.2.; for some examples see Chap. 4, Sect. 4.6.2.–4.7.2.

[192] See above Chap. 3, Sect. 3.4.4.

[193] See below s. 6.7.

[194] See 'This Regulation: (a) lays down uniform requirements for issuers of bonds *who wish* to use the designation 'European Green Bond' or 'EuGB' for their bonds that are made available to investors in the Union', Art 1 EUGBR.

label ABSs as 'sustainable' or to comply with less stringent guidelines, like the ICMA's GBP. [195] While the compliance with the EUGBS is expected to have reputational advantages, it is hard to ascertain the uptake of the Standard among market participants at this stage. At the time of writing, it was only possible to trace seminal cases of aligned green bonds[196], and no cases of aligned green securitisations. The high compliance costs suggest, however, that self-labelled 'sustainable' securitisations, deviating from the EUGBR and thus not providing sufficient information around the composition of the portfolio, could remain a widespread practice.

At the same time, the EUGBR fails to ensure an effective 'gatekeeping' over the characteristics of portfolios. While it requires reviews over the issuance as well as empowering competent authorities with supervisory powers[197], it does not mandate external reviews over the characteristics of the portfolio. Instead, external reviewers should focus on the *'alignment of the use of proceeds of the European Green Bond (or Bonds) with the Taxonomy Regulation'*.[198] This gap sits unwell within other reporting requirements under the EUGBR, since other statements on the compliance with the Taxonomy (on the Factsheet and 'Allocation Reports'[199]) are subject to external review. [200] Besides, it also fails to address the most remarkable shortcoming of the current external reviews highlighted in Chap. 4, i.e. their tendency to overlook the composition of the portfolio, leaving room for 'greenwashing' practices.[201]

6.4.2.2 Other Policy Initiatives on 'Green Loans'

Another question that remains unanswered is the link between the EU Green Bond Regulation and other potential initiatives for 'green' asset

[195] As they currently do, see Chap. 3, Sect. 4.3.2–4.4.2.

[196] i.e. A2A and the revised ABN AMRO, A2A, 'European Green Bond Factsheet' (2025) https://content.gruppoa2a.it/sites/default/files/2025-01/a2a-european-green-bond-factsheet.pdf accessed 17 March 2025; ABN AMRO, 'European Green Bond Factsheet', 2025, https://assets.ctfassets.net/1u811bvgvthc/2mKUOfOk6jp1eSYbm9qj0P/cd6973465c36bfbb3124e6d25673e445/250214_Green_Bond_Factsheet_-vF.pdf.

[197] Art 20c-20d, Art 8, Art 36 EUGBR. For external assurance requirements over sustainability-related reports, see art 26a, 27a, 28a Directive (EU) 2022/2464 of the European Parliament and of the Council of 14 December 2022 amending Regulation (EU) No 537/2014, Directive 2004/109/EC, Directive 2006/43/EC and Directive 2013/34/EU, as regards corporate sustainability reporting.

[198] 3. Annex IV EUGBR.

[199] See below Sect. 5.8.2.

[200] See Art 8 EUGBR.

[201] See Chap. 4, Sect. 4.6.4.–4.7.3.

claims. In a 2023 report, the EBA has also reflected on the merits of standalone legal definition for 'green mortgages' and 'green loans'.[202] After noting the small size of the EU 'green loan' market and the wide discrepancies among debt instruments labelled as such, the EBA has concluded that ad hoc legal requirements would be premature. At the same time, the report provides policy recommendations in the mid- to long-term, including a definition of 'green loans', which should eventually feed into the revisions to the Mortgage Credit Directive (MCD).[203]

EBA's most significant proposal relates to the introduction of a 'green loan' label, leveraging the market-designed definition by the LMA discussed in Chap. 5[204], yet with a similar scope to the Green Bond Standard. In other words, the EBA proposes a voluntary label anchored to the compliance with the Taxonomy, where the borrowers would have to use the finance for sustainability-related projects in line with the technical screening criteria.[205] The pre-contractual term sheet should then articulate sustainability-related information, such as the energy efficiency levels of the secured buildings and any additional obligations that borrowers would need to comply with.[206]

Such policy initiative would enable market actors to label their securitisations as 'green' based on the use of the proceeds raised from the issuance of ABSs as well as on the purpose of the underlying loans. Yet, the same limitations highlighted above, inherent to the voluntary scope of such

[202] EBA, 'Report in response to the call of advice from the European Commission on green loans and mortgages' (n. 181). The report stems the 2021 EU sustainable finance strategy and from a Sect.equent call for advice by the European Commission. See Action 2 (A) EU Commission,;Communication from the Commission to the European Parliament, the Council, the European Economic and Social Committee and the Committee of the Regions, 'Strategy for Financing the Transition to a Sustainable Economy' (2021) 390; EBA, 'Call for advice to the European Banking Authority on green loans and mortgages' (2022) https://www.eba.europa.eu/sites/default/files/document_library/About%20Us/Missions%20and%20tasks/Call%20for%20Advice/2022/CfA%20on%20green%20loans%20and%20mortgages/1043881/EBA%20Call%20for%20Advice%20Green%20Loans%20and%20Mortgages_Clean.pdfaccessed 10 D 2026.

[203] Directive 2014/17/EU of the European Parliament and of the Council of 4 February 2014 on credit agreements for consumers relating to residential immovable property and amending Directives 2008/48/EC and 2013/36/EU and Regulation (EU) No 1093/2010 (2014) OJ L 60/34.

[204] See Chap. 5, Sect. 5.4.2.1.

[205] EBA, 'Report in response to the call of advice from the European Commission on green loans and mortgages' (n. 181), 46–55, 74.

[206] Ibid.,

label, would apply.[207] As a result, the main question underlying the design of further policy initiatives in this realm should relate to the real *incentives* prompting market actors to opt for the use of burdensome labels. To this end, the report also maps existing incentives to extend 'green loans' across EU countries, stemming from public guarantees and subsidies, preferential financing conditions, and tax reliefs, among others.[208] Should the reputational advantage of aligning with standards with the EUGBR not suffice to ensure a wide uptake, it will be crucial for the EU legislator to complement further initiatives on green loans with public schemes.

6.4.3 *Indirect Incentives: The Role of the ECB's Frameworks and Policies*

The inclusion of certain types of exposures within the portfolio also depend on a series of rules which can encourage or deter market actors to set up securitisations, and 'sustainable' securitisations in turn. For instance, rules in the fiscal or prudential domain can make it more costly to create and securitise certain types of assets, or they can provide an advantageous treatment to certain assets for the purposes of capital requirements.[209] Despite the wide-ranging initiatives within the EU to align the prudential framework to sustainability risks;[210] these have not gone as far as to

[207] See above 'Sect. 5.4.2.1.

[208] EBA (n. 181) 84–87.

[209] See the capital treatment of STS transactions, (8), Art 260, 262 Regulation 2017/2401. On how prudential rules shape the incentives to engage in securitisation operations, see Alessandro Scopelliti, 'Securitisation, Bank Capital and Financial Regulation: Evidence from European Banks', *ECB Forum* (2016). 18–28 https://www.ecb.europa.eu/pub/conferences/ecbforum/shared/pdf/2016/scopelliti_paper.pdf accessed 8 December 2025; See For example, Art 13 (1) – (2) Commission Delegated Regulation (EU) 2015/61 of 10 October 2014 to supplement Regulation (EU) No 575/2013 of the European Parliament and the Council with regard to liquidity coverage requirement for Credit Institutions.

[210] Among others see EBA, 'The role of Environmental Risks in the Prudential Framework' (2022) EBA Discussion Paper 2. https://www.eba.europa.eu/sites/default/files/document_library/Publications/Discussions/2022/Discussion%20paper%20on%20the%20role%20of%20environmental%20risk%20in%20the%20prudential%20framework/1031947/Discussion%20paper%20on%20role%20of%20ESG%20risks%20in%20prudential%20framework.pdf accessed 10 December 2025. For an overview see Jens-Hinrich Binder, 'Prudential Requirements Framework and Sustainability' in: K. Alexander, M. Gargantini, M. Siri (eds.), *The Cambridge Handbook of EU Sustainable Finance: Regulation, Supervision and Governance* (2025) Cambridge Law Handbooks. Cambridge University Press, 526–551.

adjust the prudential treatment of securitised exposures with this goal in mind.[211]

On the contrary, the legal treatment of assets within central bank operations has increasingly occupied the policy and scholarly debate and could most effectively act on the incentives to incorporate sustainability-related considerations before the creation of the portfolio. All central banks conduct open market operations, where they purchase securities, including ABSs, from companies, as in the case of the ECB's (currently discontinued) Asset-Backed Purchase Programme (ABSPP).[212] They also have specific 'collateral frameworks' where they set out the criteria to accept assets as collateral within their lending activities, such as liquidity-providing standing facilities.[213]

The origination of assets like credit claims also depends on their attractiveness for central banks' collateral frameworks and open market operations. As *Nyborg* argues, 'if central bank money is only available against (…) igloo-backed securities, igloos would be built'.[214] In practice, this means that financial institutions are more incentivised to extend certain types of loans and to securitise them if the related ABSs could be purchased by, or could be used for borrowing from, central banks.[215] Indeed, academic research and ECB data point to an increase in the issuance of ABSs before the 2007 global financial crisis by market actors that used them as collateral for repo transactions with central banks (the so-called 'run on repos'[216]).[217] Asset-backed securities currently represent a large

[211]For example, see Art 465 (13)(a)-(b) Regulation (EU) 2024/1623 of the European Parliament and of the Council of 31 May 2024 amending Regulation (EU) No 575/2013 as regards requirements for credit risk, credit valuation adjustment risk, operational risk, market risk and the output floor (2024) PE/80/2023/INIT; A report by the EBA, in close cooperation with ESMA, is expected by 31 December 2026, which is expected to feed into a regulatory review of the prudential treatment of securitisations. Art 506d (1) ibid.

[212]Introduced through the Decision (EU) 2015/5 of the European Central Bank of 19 November 2014 on the implementation of the asset-backed securities purchase programme, ECB/2014/45.

[213]Ulrich Bindseil et al, 'The Eurosystem Collateral Framework Explained' (2017) ECB Occasional Paper Series 13–14 https://www.ecb.europa.eu/pub/pdf/scpops/ecb.op189.en.pdf accessed 3 June2026.

[214]Nyborg, ibid., 204.

[215]Ibid.; Scopelliti (n. 209) 27–30.

[216]A repurchase agreement ('repo') is, in essence, a sale of securities with an agreement to repurchase them at a later stage. It is essential for short-term liquidity. See Armour et al, *Principles of Financial Regulation* (n. 69) 317.

[217]Gary B Gorton, Andrew Metrick, 'Securitised Banking and the Run on Repo' (2012) 104 Journal of Financial Economics 3 425, 427–428. The ECB has also documented the

proportion of the overall collateral accepted by the ECB (around 14% in 2021).[218] Thus, the Eurosystem's collateral framework,[219] which sets out the rules for transactions with the ECB, represents another hard law instrument which indirectly affects the appetite for securitisations.

Notably, market actors are incentivised to originate or purchase assets from third parties which may comply with the ECB's collateral eligibility criteria if securitised.[220] These pertain, among other things, to the composition of the portfolio, by requiring that credit claims backing ABSs are 'homogeneous'.[221] However, the assessment of 'homogeneity' in the context of ECB collateral is stricter than for the award of the STS label. The ECB exclusively accepts assets from an exhaustive list,[222] including residential mortgages, commercial real estate mortgages, loans to small and medium-sized enterprises (SMEs), auto loans, consumer finance loans, leasing receivables, and credit card receivables.[223] Therefore, assets like trade receivables or corporate loans to large enterprises are not included.[224] The prospectus is the key document which has to demonstrate compliance with such requirements.[225] At the same time, originators are incentivised to gather data and monitor their credit claims since the ECB only accepts securitised assets as collateral if originators provide loan-by-loan information at least quarterly.[226]

steady increase in the use of ABS as collateral for repos between 2004 and 2013, Nyborg (n. 209) 206.

[218] Loris André, Alice Grept, Nadia Laut et al, 'Climate Risk Measurement of Assets Eligible as Collateral for Refinancing Operations – Focus on Asset Backed Securities (ABS)' (2022) Banque de france Working Paper 3. https://publications.banque-france.fr/sites/default/files/medias/documents/wp-858.pdf accessed 10 December 2025.

[219] Enshrined under the Guideline 2015/510 of the European Central Bank of 19 December 2014 on the implementation of the Eurosystem monetary policy framework (General Documentation Guideline) 2014 (OJ L 91) 5.

[220] Bindseil et al (n. 203) 15, 21, 22–23.

[221] Artt 72–79 Guideline 2015/510.

[222] On the contrary, assets are also considered as homogenous for the purposes of STS transactions based on 'internal methodologies and parameters', see Art 1 (a)(viii), Art 2 Commission Delegated Regulation 2019/1851.

[223] 'Art 73 (1) Guideline 2015/510.

[224] Art 116 ibid.

[225] For example, art 73 ibid.; Danny Busch and others, *Prospectus Regulation and Prospectus Liability*, paras 4.70–4.73.

[226] Following ECB's decision on 16 December 2010, see ECB, 'ECB introduces ABS loan-by-loan information requirements in the Eurosystem collateral framework' (2010) https://www.ecb.europa.eu/press/pr/date/2010/html/pr101216.en.html accessed 10 December

As part of its collateral framework, the ECB also sets criteria for the 'haircut' applicable to securitised assets.[227] In other words, when financial institutions conclude repo transactions with the ECB, the value of their collateral is lower than the market value due to an initial margin, which should reflect the perceived potential fluctuations in the value of the assets.[228] The calculation is the same for all ABSs, regardless of the types of underlying assets.[229]

For the purposes of the present analysis, it is important to note that the ECB currently accepts green bonds and sustainability-linked bonds as collateral.[230] However, the eligibility criteria and haircut rules currently fail to take into account sustainability risks and the impact of the portfolio. Various empirical studies have highlighted a general lack of data on the carbon impact of the ABSs currently accepted by ECB.[231] Even the limited currently available evidence shows that only a minority of the collateral eligible for the operations with ECB complies with Paris Agreement targets.[232] Conversely, some studies have highlighted how the eligibility criteria and the rules on haircut disproportionately favour fossil fuel and carbon-intensive companies, giving rise to a 'carbon bias'.[233]

2025; for the templates, see ECB, 'Data templates' (2023) https://www.ecb.europa.eu/paym/coll/loanlevel/transmission/html/index.en.html accessed 10 December 2025.

[227] Art. 3(4)–3(5) Guideline (EU) 2016/65 of the European Central Bank of 18 November 2015 on the valuation haircuts applied in the implementation of the Eurosystem monetary policy framework.

[228] Armour et al (n. 69) 317.

[229] Table 10, Art 129 Guideline 2015/510.

[230] On sustainability-linked bonds, see Art 63(1)(c) ibid. see further below, 5.5.3.

[231] André et al (n. 218) 23–24.

[232] Antoine Oustry, Erkan Bunyamin, Romain Svartzman et al, 'Climate-Related Risks and Central Banks' Collateral Policy: A Methodological Experiment' (2021) Banque de france Working Paper, figs 19, 26 https://publications.banque-france.fr/sites/default/files/medias/documents/wp-790_0.pdf accessed 20 December 2025.

[233] Yannis Dafermos, Daniela Gabor, Maria Nikolaidi et al, 'Greening the Eurosystem Collateral Framework: How to Decarbonise the ECB's Monetary Policy' (2021) New Economics Foundation, SOAS University of London, Greenpeace et al 3–4; 14–21 https://neweconomics.org/uploads/files/Collateral-Framework.pdf accessed 10 December 2025; Studies conducted in 2024 and 2025 have highlighted that the ECB's collateral pool included assets belonging to fourty-four fossil fuel companies. The latter also have an overage lower haircuts than the average haircut of all bonds from ECB's collateral pool. According to the authors, this would result in an implicit support to carbon-intensive infrastructure projects. Economist Intelligence Units, 'ECB collateral policy threatens EU's path to net-zero' (2024) https://www.eiu.com/n/ecb-collateral-policy-threatens-eus-path-to-net-zero/ accessed 10 December 2025; Urgewald, Reclaim Finance, 'Collateral damage: ending the ECB's support to fossil fuel companies' (2024) 1824 https://reclaimfinance.org/site/wp-content/

While the ECB had announced in 2022 its intention to revise the collateral framework to incorporate climate change considerations[234], it con firmed its intention not to revise eligibility criteria and haircuts 'at [that] point'.[235] As a result, there are currently no advantages of building 'sustainable' portfolios for the purposes of central bank borrowing from the ECB. The incentives may change in the years to come, following ECB's commitment to revise the collateral framework with a view to climate-related risks in 2026.[236]

When it comes to monetary policy purchase programmes, holding 'sustainable' assets could, in principle, give a comparative advantage to companies when such programmes were still in place.[237] The ECB also announced in 2023 its intention to 'tilt' its corporate bond purchases to ensure the compliance of the issuing companies with Paris Agreement targets over time. It also designed 'climate scores' for companies depending on their past and future GHG emissions as well as on the quality of their reporting of emissions-related data. The aim of such 'tilting' was to reduce its purchases from companies with low climate scores and to favour companies demonstrating lower emissions or more ambitious decarbonisation

uploads//04/BRIEFING-2024.04.04_Reclaim-Finance_Collateral-damage-FINAL.pdf accessed 10 December 2025; Urgewald, Reclaim finance, 'Still backing fossi fuels – the costs the European Central Bank's inaction (2025) https://reclaimfinance.org/site/wp-content/uploads/2025/04/ECB_Collateral_April2025_En.pdf accessed 10 December 2025.

[234] As part of its 2022 action plan on climate change, ECB, 'Annex I – –Detailed roadmap of climate change-related actions' (2021) https://www.ecb.europa.eu/press/pr/date/2021/html/ecb.pr210708_1_annex~f84ab35968.en.pdf accessed 10 December 2025.

[235] ECB, 'ECB provides details on how it aims to decarbonise its corporate bond holdings' https://www.ecb.europa.eu/press/pr/date/2022/html/ecb.pr220919~fae53c59bd.en.html, accessed 10 December 2025; ECB, 'ECB reviews its risk control framework for credit operations' (2022) https://www.ecb.europa.eu/press/pr/date/2022/html/ecb.pr221220_1~ca6ca2cc09.mt.html accessed 10 December 2025.

[236] By introducing *'pool limits [. . .], assuming all technical conditions are in place'* and climate-related disclosure requirements. ECB, 'Climate and nature action plan 2024–2025' (2024) https://www.ecb.europa.eu/ecb/climate/our-climate-and-nature-plan/shared/pdf/ecb.climate_nature_plan_2024-2025.en.pdf accessed 10 December 2025.

[237] The last prong of the Asset Purchase Programme consisted of reinvestments of already purchased securities, and it continued until July 2023. Reinvestments under the Pandemic Emergency Purchase Programme continued until December 2024. ECB, 'FAQ on incorporating climate change considerations into corporate bond purchases' (2025) https://www.ecb.europa.eu/mopo/implement/app/html/ecb.faq_cspp_climate_change.en.html accessed 10 December 2025.

targets.[238] These changes could, in principle, indirectly incentivise companies to originate 'sustainable' assets and to securitise them in order to be eligible for monetary policy purchase programmes.

6.4.4 Normative Considerations

Overall, the current EU framework only marginally regulates sustainability-related elements of securitisation portfolios. Over time, the EU has strengthened the disclosure requirements on environmental and social risks, as well as restricting the types of credit claims which could be securitised across EUGB-compliant securitisations. The scope of these obligations is, however, limited, giving market actors excessive flexibility to deviate from the requirements. This may also leave room for underreporting. Thus, there is room for improvements to disclosure-oriented rules both for traditional securitisations and for 'sustainable' securitisation more specifically.

As regards traditional deals, the Securitisation Regulation should mandate all issuers to ensure a baseline level of transparency concerning the sustainability risks within their portfolio, regardless of whether securitisations fall under the STS regime. [239] One possibility is to build on the RTSs by requiring disclosure about the environmental and social 'PAI' of all types of securitised assets, across all securitisation programmes in the EU market.

Further research is imperative to design the optimal metrics and indicators to report on some types of assets, like commercial loans or credit card receivables, as market participants' responses to the consultation on said RTSs have highlighted.[240] This illustrates why the standardisation work by industry associations towards new sustainability-related metrics, discussed in Chap. 5, should be complementary to EU harmonisation measures.[241] In the longer run, it has the potential to lay the groundwork for the refinement of RTSs.

[238] ECB, 'Combined monetary policy decisions and Statement' (2 February 2023); https://www.ecb.europa.eu/press/pressconf/shared/pdf/ecb.ds230202~40a069a7b4.en.pdf accessed 3 June 2026; on the methodology see ECB, ibid.

[239] As also recommended by EBA, 'Developing a framework for sustainable securitisations' (n. 180) 6, 29, 51.

[240] EBA, ESMA, EIOPA, 'Joint Consultation Paper – –STS securitisations-related sustainability disclosures' (n. 161) 18–19, 24–25.

[241] See Chap. 5, Sect. 5.4.3.

Despite the limited availability of metrics, requiring the disclosure of PAIs in all circumstances would still be preferable to retaining the current option of non-reporting. One option may be to allow market actors to temporarily use their own metrics, at least for those categories where RTSs are unavailable for the time being. The added value of admitting such an alternative would be to encourage market actors to start gathering further sustainability-related data until the design of harmonised RTSs.

Arguably, the recommended strengthening of reporting requirements is realistic and in line with the EU agenda, as demonstrated by the plethora of disclosure-oriented rules at all levels of the financial sector.[242] Further evidence of the interest in such rules is a statement by EU supervisory authorities, where they have also called on regulators to enhance transparency on sustainability-related elements in all securitisations.[243]

As regards 'sustainable' securitisations, making sustainability-related portfolio information available and accessible to investors is paramount. One option could involve mandating issuers to indicate, on the first page of the prospectus, not only whether the transaction is EUGB-compliant, as the EUGBR requires,[244] but also whether the transaction involves a 'sustainable use of proceeds' and/or a 'sustainable portfolio', or if it facilitates the achievement of certain targets ('sustainability-linked').[245]

While it is, in principle, realistic to not require all securitisations to have a 'sustainable' portfolio, the EUGBR should prevent the use of purely 'brown' assets. As previously discussed, the EUGBR exclusively prohibits the use of carbon-intensive assets, and requires disclosure of *what share of portfolio assets* are considered to be harmful under the EU Taxonomy (i.e. breaching the DNSH criteria[246]). However, there is room for the EUGBR to be more restrictive, by requiring originators to provide loan-to-loan information on assets violating DNSH requirements.

Another obstacle to the transparency around sustainability-related information on the portfolio is the voluntary nature of the EUGBS. This could also encourage market actors to apply less stringent standards in the light of the lower compliance costs.[247] Thus, the way forward should

[242] On the role of disclosure across EU sustainable finance initiatives, see Chap. 2, Sect. 2.6.

[243] EBA, ESMA, EIOPA, ECB, 'Joint ESAs-ECB Statement on disclosure on climate change for structured finance products' (n. 155).

[244] Art 12 (1)(a) EUGBR.

[245] See Chap. 4, Sect. 4.4.

[246] See above Sect. 5.4.2.

[247] See above Sect. 5.4.2.

also entail making the Standard mandatory for all issuers of 'sustainable' financial products, like bonds and ABSs, in the EU market, as I will also discuss later.[248]

Alongside the potential changes to disclosure requirements, it is also essential to overcome the current limitations of reviews, which fail to provide adequate 'gatekeeping' over securitisation portfolios, as discussed in Chap. 3.[249] Therefore, the EUGBS should also require reviewers to verify whether the reported information on the alignment of securitised assets with the Taxonomy criteria is accurate.

In addition, Sect. 5.4.3. has underscored how the legal framework governing central banks' open market operations and lending activities affects the origination of credit claims and their aggregation within securitisation portfolios. Therefore, an interesting question relates to whether changes to 'eligibility criteria' and 'haircuts' within central banks' collateral frameworks can incentivise market actors to create more 'sustainable' assets and also to engage in 'sustainable' securitisations in turn.

Some studies have advocated for more stringent eligibility criteria, leading to the complete exclusion of companies engaging in carbon-intensive businesses or particularly exposed to climate-related risks from operations with the ECB.[250] However, leaving 'brown' activities out of collateral requirements may be a disproportionate measure for the regulation of 'sustainable' securitisation deals. It would cut or considerably reduce the access to central bank liquidity for certain market actors[251], potentially raising broader macro-economic and monetary policy implications.[252] It

[248] See below Sect. 5.5.4. See also Elia Garcia Cerrato, Federica Agostini (2023) 'The Green Bonds Market in the Light of European Commission's Proposal: Implications for Greenwashing Liability' in Agnieszka Smoleńska, David Ramos Muñoz, *Greening the Bond Market: A European Perspective* (Springer 2023).

[249] Chap. 3, Sect. 3.4.1.3.

[250] See For example, Dafermos et al (n. 233) 24–26; Urgewald, Reclaim Finance, 'Collateral Damage' (n. 233) 26; ClientEarth, 'The ECB's Monetary Policy Strategy Review: Corporate Sector Asset Purchases, Climate Change and the ECB's Legal Obligations' (2021) 7–8 https://www.clientearth.org/latest/documents/letter-from-clientearth-to-christine-lagarde-president-of-the-european-central-bank/ accessed 10 December 2025. The NGFS has also mentioned such option as best practice to integrate climate risk considerations in central bank frameworks. NGFS, 'Adapting Central Bank Operations to a Hotter World Reviewing Some Options' [2021] NGFS Technical document paras 36–37 https://www.ngfs.net/en/adapting-central-bank-operations-hotter-world-reviewing-some-options accessed 10 December 2025.

[251] Dafermos et al, ibid., 26–29.

[252] NGFS (n. 250) 35–36.

also raises the question of the optimal method to identify 'brown' companies.[253] This uncertainty calls for caution on the regulators' side.[254]

An alternative regulatory approach, which is out of the scope of the present analysis, could involve a reduction in the haircuts for 'green' ABSs, as others have suggested.[255] For instance, it could consist of a discount for all ABSs if the underlying assets comply with the Taxonomy criteria. It would allow companies to obtain liquidity at lower costs by pledging 'sustainable' assets[256], thereby also incentivising their creation. However, reducing haircuts only on the basis of the sustainability impact of certain assets may have adverse systemic side effects.[257] While these unavoidable risks should not inhibit the proposal altogether, they should encourage further research and reflection on the optimal calibration of the discounts.[258] Another option may be the refinement of the requirements for the ECB to accept 'green' and 'sustainability-linked' securities as collateral. I will discuss it in Sect. 6.5.4.

6.5 (More Encouraging) Signs of Harmonisation on the Terms of Issue of ABSs

6.5.1 *For Traditional ABS Issuances*

6.5.1.1 Underwriting

The Securitisation Regulation leaves space for parties' autonomy to determine the underwriting conditions, i.e. the terms for the purchase and subsequent marketing of ABSs.[259] Underwriters (as well as all other investors[260]) are required to make a comprehensive 'due diligence' assessment of several aspects of the transactions, such as the risk profile of the securities

[253] Oustry et al (n. 232) 12; NGFS, ibid., 37.

[254] Linn Anker-Sørensen, Dirk A. Zetzsche, 'Regulating Sustainable Finance in the Dark' (2022) 23 *European Business Organisation Law Review* 47, 67-71..

[255] Dafermos et al (n. 233), 24–26; NGFS (n. 250) 36–37.

[256] See Nyborg (n. 209) 204;

[257] NGFS (n. 251) 37.

[258] While raising the systemic risk objection, the NGFS also seems to open the door to a careful calibration of risks, ibid.

[259] On underwriting, see Chap. 3, Sect. 3.4.2.1.

[260] See Chap. 3, Sect. 3.4.4., Chap. 6, s. 6.7.

and of their portfolio.[261] It does not, however, cover the sustainability risks of the portfolio, or the potential risks and impact of the issuance.

6.5.1.2 *Issuance Documentation: Prospectuses and Beyond*

When offering ABSs to the public in the EU market, prospective issuers are also generally subject to rigorous disclosure requirements under the Prospectus Regulation.[262] These include the approval by a competent authority and the publication of a prospectus (including a summary) as well as a securities note. The publication of a prospectus is not always strictly required.[263] Causes of exemptions include, among others, offers of securities solely addressed to institutional investors,1,[264] which are likely to represent a significant proportion of securitisations. However, it is good market practice and generally advantageous for issuers to draft a prospectus anyway.[265] For instance, the Securitisation Regulation still requires issuers to file an offering document with current or prospective noteholders, or to supervisory authorities, upon request.[266]

Annexes to the Regulation and a more detailed Delegated Regulation provide a list and standard templates of all the items that these documents need to include, which issuers are likely to use even when outside the formal scope of the framework.[267] Issuers falling outside the prospectus requirements still need to list the main characteristics of the issuance in a more succinct 'Transaction Summary'. [268] Another common practice is

[261] Art 5(3) EUGBR.

[262] Regulation 2017/1129 of the European Parliament and of the Council of 14 June 2017 on the prospectus to be published when securities are offered to the public or admitted to trading on a regulated market, and repealing Directive 2003/71/EC 2017. (hereinafter Prospectus Regulation). For a more detailed analysis of the role of disclosure, albeit in relation to the old prospectus regime, see Joseph Tanega, *Securitisation Law & Practice: EU & US Disclosure Regulations* (3rd edn, LexisNexis 2017) paras 3.38–3.39.

[263] Tanega, para 2.1., ibid.

[264] Art 1(2)-(3)-(4) ibid.

[265] Tanega (n. 262) para 2.1; .

[266] Art 7(1)(b)(i) Securitisation Regulation.

[267] Annex I, II; III, IV Prospectus Regulation; List of Annexes, Commission Delegated Regulation (EU) 2019/980 of 14 March 2019 supplementing Regulation (EU) 2017/1129 of the European Parliament and of the Council as regards the format, content, scrutiny and approval of the prospectus to be published when securities are offered to the public or admitted to trading on a regulated market, and repealing Commission Regulation (EC) No 809/2004 (2019) OJ L 166/26. For an overview see Busch et al, *Prospectus Regulation and Prospectus Liability,* ibid.

[268] Art 7(1)(c) ibid.

to incorporate further documents, such as financial statements or articles of association, 'by reference'. The Securitisation Regulation provides an exhaustive list of the eligible documents and requires issuers to file them, along with their Prospectuses, with their supervisory authorities.[269]

The Securitisation Regulation also leads to the 'minimum harmonisation'F[270] of the enforcement avenues related to the publication of a prospectus and to the documents incorporated by reference. Both investors and supervisory authorities can act when these documents contain misstatements and omissions. Despite the considerable country-specific differences, consequences for issuers include civil liability, administrative liability, and potentially criminal liability. [271]

It is important to highlight that, regardless of whether ABSs are marketed as 'sustainable' or not, prospectuses contain 'strategic' elements that issuers could use to describe the extent to which the issuance contributes to sustainability goals. After analysing market practices ahead of recent regulatory changes, ESMA had indeed urged market actors to provide more extensive reference to sustainability goals within prospectuses, even in the absence of ad hoc requirements. Natural venues to disclose sustainability-related information included, among others, mandatory items like the 'use of proceeds', 'risk factors', 'underlying assets' and 'post issuance reporting'[272] Notably, 'risk factors' are 'material' elements affecting investors' ability to make informed decisions.[273] These could, in principle, include sustainability risks, although the choice as to what to include will be on issuers. Moreover, issuers must provide, both ahead of the issuance and on a periodic basis, 'key information' related to the deal.[274] Issuers of

[269] Art 19(2) Prospectus Regulation.

[270] Danny Busch et al, *Prospectus Regulation and Prospectus Liability* (Oxford University Press 2020).

[271] See also chapters 20–26 ibid.; Criminal liability is an option left to Member States. Artt 11, 33, 38, Prospectus Regulation.

[272] ESMA, 'Sustainability Disclosure in Prospectuses' (2023) 3–4 https://www.esma.europa.eu/sites/default/files/2023-07/ESMA32-1399193447-441_Statement_on_sustainability_disclosure_in_prospectuses.pdf accessed 10 December 2025. Regulation (EU) 2024/2809 of the European Parliament and the Council amending Regulations (EU) 2017/1129, (EU) No 596/2014 and (EU) No 600/2014 to make public capital markets in the Union more attractive for companies and to facilitate access to capital for small and medium-sized enterprises OJ L, 2024/2809;

[273] Art 7(6)(c) Prospectus Regulation.

[274] Art 7(1)(c)(i) ibid.; see also the need to disclose about terms and conditions of the deal, Annex III (IV).

sustainability-linked ABSs[275] could thus use these opportunities to inform about the variations in financial terms linked to not achieving sustainability targets,[276] which fundamentally affect investors' rights.

At the same time, I have discussed in Chap. 4[277] as well as in other work[278] the frequently ambiguous sustainability-related language of prospectuses and of other issuance documents.

While the standardised format for prospectuses and transaction summaries provides ample space to inform investors on sustainability matters, prospectuses of 'sustainable securitisations' designed thus far have rarely incorporated such themes. Even in the instances where issuers adopt 'Sustainability frameworks' to articulate their overarching commitments to sustainability goals, they generally do not incorporate these documents within prospectuses 'by reference'.[279] This practice *de facto* deprives investors of remedies if issuers fail to substantiate said sustainability objectives.

The current gaps and ambiguities of issuance documentation could partially benefit from the recent revisions to the Prospectus Regulation,[280] which, among others, have prescribed the incorporation of 'ESG-related information' within prospectuses. Prospective issuers of debt securities 'marketed as taking into account ESG factors or pursuing ESG objectives' will particularly have to provide an 'unambiguous and fact-based explanation' of the ESG features and of any sustainability claims, such as the extent to which they align with any market standards or labels[281]

[275] As defined in Chap. 4, Sect. 4.4.

[276] See Sect. 5.3.2.

[277] Sect. 3.4.3.1.

[278] Federica Agostini, 'From 'Green Bond Principles' to 'Green Bond Clauses': Mitigating Greenwashing Through Contract Law' in Andenas M, Heidemann M (eds), *Quo vadis Commercial Contract?* (Springer 2023) LCF Studies in Commercial and Financial Law, Vol. 1https://doi.org/10.1007/978-3-031-14105-8_6

[279] Sect. 3.4.3.1.

[280] Stemming from the Listing Act. Regulation (EU) 2024/2809 of the European Parliament and of the Council of 23 October 2024 amending Regulations (EU) 2017/1129, (EU) No 596/2014 and (EU) No 600/2014 to make public capital markets in the Union more attractive for companies and to facilitate access to capital for small and medium-sized enterprises. The Regulation also amends Commission Delegated Regulation (EU) 2019/980 supplementing Regulation (EU) 2017/1129 of the European Parliament and of the Council as regards the format, content, scrutiny and approval of the prospectus to be published when securities are offered to the public or admitted to trading on a regulated market, and repealing Commission Regulation (EC) No 809/2004 [2019] OJ L166/26.

[281] This applies to the prospectus summary, the securities note, and the EU Growth Issuance Prospectus (a tailored prospectus for small and medium enterprises), see Annex I IX; Annex

The broad formulation of these provisions captures all transactions labelled as 'sustainable', whether they are designed as 'use of proceeds' or 'sustainability-linked', and regardless of whether they comply with the EUGBS or not.[282]

In turn, these changes are also expected to strengthen investors' 'voice' rights, by allowing them to take action in case of misstatements in these claims. While these regulatory changes will not address the inherent flaws of prospectuses' liability mechanisms, like the procedural challenges of filing a claim for omissions or the disproportionate burden of proof requirements, they certainly represent a step forward in broadening issuers' accountability for 'greenwashing' practices. 'Disclosure' requirements are even stronger for transactions that are labelled as EUGB-compliant, as it will be discussed below.

6.5.2 *Lights and Shadows of the EU Green Bond Regulation as Applied to the Offer of ABSs*

The EUGBS, which currently represents the key regulatory instrument for 'sustainable' securitisations, does not subject the underwriting of ABSs to any requirements.

Thus, prospective underwriters willing to undertake 'due diligence' assessments ahead of their investments may voluntarily apply the abovementioned 'soft' OECD Guidelines, which also cover underwriting activities. In so doing, they evaluate how securities issuers identify and manage sustainability-related risks in their operations. [283]

III VI; Annex VIII XI ibid. within the securities note, such information is provided together with information on the use of proceeds. Annex I, IV.C. ibid.

[282] As also recommended by the ESMA, which concluded that a narrower definition would have conflicted with market practice. ESMA, 'Final Report on the Technical advice concerning the Prospectus Regulation and the RTS updating the CDR on metadata' (2025) ESMA32-117195963-1417, 41–43 https://www.esma.europa.eu/sites/default/files/2025-06/ESMA32-117195963-1417_Final_Report_Listing_Act_Advice_on_Prospectus.pdf accessed 20 December 2025. On the distinction between 'use of proceeds' and 'sustainability-linked' securitisations see also Chap. 4, Sect. 4.4. For a more exhaustive analysis of the additional disclosure requirements for debt securities stemming from the Listing Act, see also Kostantinos Serdaris, 'The Regulatory Framework of ESG Disclosures in EU Prospectuses' in Gonzalo Larrea, Federica Agostini and Kostantinos Serdaris (eds), *Critical perspectives on EU corporate sustainability law* (forthcoming, *Palgrave MacMillan,* 2026).

[283] Outside the scope of the present analysis, see above Sect. 6.3.

As anticipated in Sect. 6.4.2, the 'use of proceeds' approach underlying the EUGBS means that ABSs are 'sustainable' on the basis of the projects and activities implemented through the issuance.[284] Originators willing to comply with the (voluntary) Standard have to devote investors' funds to capital and operating expenditures, households' assets and expenditures, fixed and financial assets (or portfolios thereof)[285] which fully meet the Taxonomy's 'technical screening criteria'.[286] The Regulation also provides for very limited exceptions, such as the possibility to allocate up to 15% of the proceeds to other activities for which Taxonomy standards are still unavailable, or the possibility for issuers to gradually align with such criteria, as long as they publish a dedicated plan and obtain external review over it.[287]

They have to include information about the 'Use of proceeds' in the issuance documentation, and specifically in the related clause within their Prospectuses,[288] as well as in the above-mentioned green bond-specific document (the 'Factsheet'[289]). The latter also needs to specifically describe how the final allocation of the proceeds complies with the Taxonomy criteria following a standardised template.[290] The EUGBS also considers all elements of the Factsheet to be an integral component of the Prospectus.[291] In principle, such 'incorporation by reference' extends the prospectus liability regime to the statements contained in the Factsheet: as previously mentioned, this should entitle investors and supervisory authorities to take action in case of misstatements or omissions.[292] However, the lack of a dedicated provision on the liability regime makes the avenues and the circumstances for investors' 'voice' remedies too vague. Notably, it remains unclear whether and to what extent investors may take action in

[284] Art 13A EUGBR.

[285] See Sect. 6.4.2. as long as they are not created later than five years after the issuance of the EU Green Bond.

[286] Art 4 EUGBR. On technical screening criteria, see above 6.3.

[287] Art 4a, 6 EUGBR.

[288] Securities are only eligible for the EUGBS upon the publication of a Prospectus, Art 12 (1) EUBGR.

[289] See above Sect. 6.4.2.

[290] Art 8, Annex I EUGBR.

[291] It constitutes 'regulated information' for the purposes of the Prospectus Regulation, Art 12 (2) EUGBR. Information not included within the Factsheet must be covered in the Prospectus, ESMA, 'Final Report on the Technical advice concerning the Prospectus Regulation' (n. 282) 56.

[292] See above Sect. 6.4.1.

case of 'greenwashing' practices, which may consist of false or misleading statements on the use of proceeds, but also of defective information. The example of the traditional liability regime under the Prospectus Regulation, whose transposition and application have been widely inconsistent across Member States, [293] illustrates that the risks of ambiguity are high.

The EUGBR also introduces 'gatekeeping' mechanisms, by requiring third-party service providers to review Factsheets and confirm the alignment with the Taxonomy criteria.[294] The underlying logic is to overcome market practices, where (voluntary) reviews rely on inconsistent methodologies, through a mandatory and standardised review. There is also centralised oversight over the process as only market actors accredited by ESMA can provide such a service. F[295] The focus of ESMA's supervisory powers will relate to ensuring the soundness of reviewers' organisations, including on matters like skills and resources.[296]

These regulatory changes, while welcome, may not suffice to reassure investors around the reliability of 'sustainable' ABSs given the inherent methodological flaws of this process. It relies on an issuer-paid model, making negative reviews, concluding that the transaction does not comply with the Taxonomy criteria, unlikely. Instead, issuers are likely to hire third parties who are more prone to provide a positive evaluation of their securities, or to turn to a different provider in the course of the relationship when they may issue a negative review. Reviewers also have a commercial interest in maintaining the relationship with issuers for future securitisation programmes, which may incentivise them to make biased

[293] Exhaustively discussed in Busch and others (n. 267) chs 20–26.

[294] Artt 8.3, 9.3, 14, 15, 43 EUGBR.

[295] Art 14 ibid. A transitional period applies until 21 June 2026, during which reviewers can operate without a registration, yet they should use their 'best efforts' to comply with the EUGBR. Art 69 (1)-(3) ibid.

[296] On the basis of a series of Regulatory technical standards and implementing technical standards designed by ESMA, both introduced in 2025. ESMA, 'Final Report – –Technical Standards on the European Green Bonds Regulation' (2025) ESMA84-858037815-195 https://www.esma.europa.eu/sites/default/files/2025-02/ESMA84-858037815-195_Final_Report_on_Technical_Standards_under_EuGB.pdf; ESMA, 'Final Report – –Technical Standards on the European Green Bonds Regulation' (2025) ESMA84-858037815-684 https://www.esma.europa.eu/sites/default/files/2025-10/ESMA84-858037815-684_Final_Report_on_Technical_Standards_under_EuGB.pdf accessed 10 December 2025. They build on an earlier consultation paper, ESMA, Consultation Paper – –Technical Standards on the European Green Bond Regulation (2025) ESMA84-858027815-176 https://www.esma.europa.eu/sites/default/files/2025-04/ESMA84-858037815-176_Consultation_Paper_on_European_Green_Bond_Regulation.pdf. accessed 10 December 2025.

statements.[297] Despite a provision preventing to link reviewers' fees to the outcome of the review,[298] this risk is likely to persist. Moreover, while the Annexes to the Proposal provide general indication of the content of external reviews,[299] they still leave several questions open about the procedure that reviews should follow. Further ESMA technical standards[300] should address this ambiguity.

Alongside 'gatekeeping' mechanisms, the EUGBS also provides national supervisory authorities as well as ESMA with additional enforcement powers.[301] The coordination between these various enforcement layers poses a series of problems beyond the scope of this analysis.[302]

Lastly, the narrow scope of the EUGBR is problematic. It leaves out of the requirements all securitisations labelled as 'sustainable' but voluntarily not complying with the Standard.[303] As the focus is on the Taxonomy-compliant activities, which currently comprise those supporting environmental goals,[304] it is not applicable to 'social' ABSs.

In addition, the EUGBR has also introduced a lighter and differentiated regime for 'sustainability-linked' transactions. For these transactions, which link the non-achievement of sustainability targets to the worsening of bond financial conditions for issuers, there are optional disclosure and review recommendations. In other words, not only complying with the EUGB label is voluntary, like for 'green' transactions, but even the market actors that will label their securitisation as aligned with the EUGB, will not *have to* follow any disclosure and external review rules. If they wish, they may follow optional disclosure templates.[305] The use of such

[297] Nikolai Badenhoop, 'Green Bonds – –An Assessment of the Proposed EU Green Bond Standard and Its Potential to Prevent Greenwashing' Study requested by the ECON Committee (2022) 66 https://www.europarl.europa.eu/thinktank/en/document/IPOL_STU(2022)703359 accessed 10 December 2025.

[298] Art 27 (2) EUGBR.

[299] Annex IV ibid.

[300] See art 20 (3) but also 18-19-21-22-23(3) ibid.

[301] Artt 37, 41, 42, 51, 52 (2) ibid.; See above Sect. 6.3.

[302] See Garcia Cerrato, Agostini (n. 248).

[303] for illustrative examples of recent issuances where this has occurred, see Chap. 4, Sect. 4.4.

[304] See above Sect. 6.3.

[305] A consultation, open until 28 January 2025, sought stakeholders' input on a draft of optional disclosure principles and optional disclosure templates. EU Commission, 'Commission Delegated Regulation supplementing Regulation (EU) 2023/2631 of the European Parliament and of the Council by establishing the content, methodologies, and presentation of the information to be voluntarily disclosed by issuers of bonds marketed as environmentally sustainable or of sustainability-linked bonds in the templates for periodic post-issuance

templates will also open up a communication channel with supervisory authorities, which will monitor compliance with the templates.[306]

The disparity of the EUGB regime for 'green' and 'sustainability-linked' transactions may incentivise market actors willing to issue a 'sustainable' securitisation programme to opt for the latter arrangement, due to the lower compliance costs. Moreover, the limited scope of application of EUGB rules falls short of harmonising the terms of issue of 'sustainable' ABSs in the EU market.

6.5.3 Indirect Incentives

Other EU legal provisions may also affect the incentives to issue 'sustainable' ABSs. For instance, large EU companies, EU medium-sized and small companies, when listed, as well as third-country companies with substantial activities in the EU,[307] have to disclose about the percentage of their economic activities in line with the Taxonomy criteria. The calculation of the percentage considers their capex, opex and turnover ('Taxonomy Ratio'). A similar obligation is also in place for financial institutions, which calculate such alignment ratios on the basis of assets and securities held and issued ('green asset ratio'[308]).

disclosures' (2024); Annex to the Commission Delegated Regulation (EU) supplementing Regulation (EU) 2023/2631 of the European Parliament and of the Council by establishing the content, methodologies, and presentation of the information to be voluntarily disclosed by issuers of bonds marketed as environmentally sustainable or of sustainability-linked bonds in the templates for periodic post-issuance disclosures, https://ec.europa.eu/info/law/better-regulation/have-your-say/initiatives/14196-Green-bonds-and-sustainability-linked-bonds-templates-for-voluntary-post-issuance-disclosures_en accessed 10 December 2025. Moreover, as part of the amendments to the Listing Act, ESMA has indicated its intention to adopt additional RTSs for the incorporation by reference of such optional templates within prospectuses. ESMA, 'Final Report on the Technical advice concerning the Prospectus Regulation' (n. 282) 59.

[306] Art 44 (1) (b) EUGBR.

[307] As defined under Art 2, 3, 19a(1), 29a(1) Directive 2013/34/EU of the European Parliament and of the Council of 26 June 2013 on the annual financial statements, consolidated financial statements and related reports of certain types of undertakings, amending Directive 2006/43/EC of the European Parliament and of the Council and repealing Council Directives 78/660/EEC and 83/349/EEC (2013) OJ L 182/19, as amended by the recent Corporate Sustainability Reporting Directive.

[308] Art 8, 8(2) Regulation 2020/852; Commission Draft Delegated Regulation (EU) 2021/2178 of 6.7.2021 supplementing Regulation (EU) 2020/852 of the European Parliament and of the Council by specifying the content and presentation of information to be disclosed by undertakings subject to Articles 19a or 29a of Directive 2013/34/EU concerning

As previously mentioned, all EUGB-compliant ABSs need to meet the Taxonomy criteria; as a result, holding such assets would increase these ratios.[309] This could be especially relevant in the case of synthetic securitisations, in which the securitised assets remain on originators' balance sheet. Thus, taxonomy-alignment requirements, while centered on disclosure, could, in turn, incentivise market actors to issue 'sustainable' ABSs instead of 'traditional' ones to improve their alignment ratios.

Even the ECB's framework encourages the issuance of 'green' and 'sustainability-linked' bonds and ABSs by accepting these securities as collateral. Sustainability-linked securities should also have an external review.[310] However, it does not strictly require the alignment with the Taxonomy or the EUGBS. While the eligibility for central banks' operations may incentivise market actors to issue 'sustainable' over traditional securities, prospective issuers may also decide to apply other non-binding green frameworks instead of the EUGBS due to the lower compliance costs. Therefore, the current ECB's collateral framework does not necessarily push market actors to implement the thorough disclosure and review practices which should make ABSs labelled as 'sustainable' reliable.

6.5.4 Normative Considerations

In principle, both the Prospectus Regulation (for 'traditional' securitisation issuances) and the EUGBR offer room for market actors to incorporate sustainability considerations in the clauses of the documentation for the issuance of 'sustainable' ABSs. Nevertheless, it is essential to strengthen investors' rights to have their say over the content of the documentation ('voice' rights[311]) and to safeguard them against 'greenwashing' risks.

Regardless of the composition of the portfolio, which may not necessarily be 'sustainable', prospective underwriters should always have a duty to assess the sustainability risks related to the underlying assets. In practice, this could be achieved by amending the existing due diligence requirements ahead of investments in securitised products, which I will analyse later in this chapter.[312]

environmentally sustainable economic activities, and specifying the methodology to comply with that disclosure obligation.

[309] 1.2.3. ibid.

[310] Art 63(1)(c) Guideline 2015/510.

[311] See Chap. 2, Sect. 2.6.

[312] See s. 6.7.

Furthermore, there is merit in being more explicit about the 'voice' mechanisms for holders of 'sustainable' ABSs, specifically on 'greenwashing' grounds. Admittedly the link between the EUGBS and the Prospectus Regulation should give investors the same remedies arising from prospectuses. However, another option to ensure issuers' accountability would be for the EUGBR to provide an indicative list of the circumstances where investors can take action, like the failure to include information about material sustainability risks, false or misleading statements in factsheets and 'use of proceeds' clauses.[313]

The EUGBR should also empower investors to take action vis-à-vis external reviewers when their assessment results in a damage[314], like the cases where their formally positive review statements are incorrect or misleading.

Although the EUGBR is a good starting point to 'frame' the conditions for the issuance[315], the link to the Taxonomy, leaving out 'social' and 'sustainability-linked' ABSs, is weak. The EUGBS should not be dependent on a voluntary decision, but it should rather become applicable to all securities traded in the EU market and qualified as 'sustainable', i.e. 'green', 'social' and 'sustainability-linked'. In other words, it would be appropriate to require all issuers to comply with the EUGBR disclosure and review requirements as well as to clearly indicate the 'sustainable' label in the first page of their prospectuses.

Furthermore, it is also crucial to incentivise the issuance of 'sustainable' ABSs by adjusting the collateral eligibility criteria for transactions with ECB. While the current framework only generally admits sustainability-linked bonds, there is merit in clarifying that ECB may only accept 'green', 'social' and 'sustainability-linked' ABSs as collateral when these comply with the EUGBS.

One may question the compatibility of this recommendation with other principles underpinning ECB's operations. One of them is 'market neutrality', which several statements by ECB's officials and official ECB documents have invoked as one of the overarching 'principles' guiding monetary policy operations like asset purchase programmes. Even though the notion has been used and interpreted inconsistently, one recurrent aspect thereof is the need for ECB to ensure the equal

[313] See art 41(2) EUGBR.

[314] *Badenhoop* also recommends strengthening civil liability remedies, Badenhoop (n. 297) 77.

[315] See Chap. 2, Sect. 2.5.

treatment of industries and market actors throughout its programmes and purchases in order to minimise the 'unintended consequences' of its operations, like distortions to price formation or other market mechanisms.[316] This principle legally represents an integral component of the Treaty principle of *'an open market economy with free competition, favouring an efficient allocation of resources'*.[317] Thus, it gives rise to an obligation for the ECB under EU primary law.[318] Therefore, one could argue that the application of the 'market neutrality' principle is incompatible with eligibility criteria that favour certain securities based on their alignment with the EUGBR.

However, the proposed recommendation would not necessarily be at odds with a market-neutral application of asset purchases, since it would not discriminate certain market actors altogether.[319] Instead, it would only exclude certain securities based on their misalignment, while allowing the same issuers to pledge other 'greener' ABSs as collateral. Moreover, there are other 'programmatic' provisions of the Treaties that should equally guide the actions and policies of the ECB alongside market neutrality. As discussed above, the Treaties also bind the ECB, as an EU institution, to Art 11 TFEU (on 'environmental protection') and to Art. 3 TEU (on 'sustainable development').[320] These concurrent principles equally justify measures that effectively prioritise certain 'greener' securities.

[316] For a typology of the legal meanings of market neutrality see Javier Solana, Marco Goldoni, 'The legal nature of market neutrality in the euro area's monetary policy' (2024) 3 *European Law Open* 1 7, 17–24. On this specific meaning, see ibid., 14–17. Consistently, see Benoît Cœuré, Member of the Executive Board of the ECB, 'Embarking on public sector asset purchases' International Conference on Sovereign Bond Markets, 10 March 2015 https://www.ecb.europa.eu/press/key/date/2015/html/sp150310_1.en.html accessed 20 December 2025; For a political economy analysis of the principle, Jens Van 't Klooster, Clément Fontan, 'The Myth of Market Neutrality: A Comparative Study of the European Central Bank's and the Swiss National Bank's Corporate Security Purchases' (2020) 25 *New Political Economy* 865.

[317] Art 127(1) TFEU.

[318] Solana, Goldoni (n. 316); for a different conclusion, see Zilioli, Ioannidis (n. 63) 291 ff.

[319] Zilioli, Ioannidis, conclude that it would breach the principle, ibid., 392. However, they interpret the market neutrality principle in a narrow sense as the need to avoid distortions to efficiency and to the price discovery mechanisms. In the present work I interpret it as a broader notion, in line with Solana, Goldoni (n. 316) 28.

[320] See Sect. 6.2.4.

6.6 The Challenges of a 'Sustainable' Governance of SPVs

6.6.1 Shortcomings of Traditional Harmonisation Initiatives

SPVs, which the Securitisation Regulation names special purpose entities (SSPEs)[321], represent another crucial player. The Regulation contemplates different organisational structures for these entities, like companies, trusts or other entities, the only condition being the distinctness of their structure and of their obligations from the originators'.[322] However, very few EU provisions have harmonised national rules surrounding the governance, i.e. the purpose and decision-making dynamics, of these entities.

The Securitisation Regulation identifies SPVs' purpose as 'carrying out one or more securitisations', limiting the scope of their activities to those appropriate to accomplishing that objective.[323] Correspondingly, Chap. 4 has already highlighted the common market practice among SPVs to undertake through a covenant in their articles of association not to engage in any parallel business activity.[324]

At the same time, no EU provisions regulate the duties of the respective decision-making bodies. Several studies have highlighted the similarities between directors' duties across EU jurisdictions. Most countries provide, at a minimum, a duty to act with care, i.e. with skills and experience when making business decisions, as well as with loyalty, i.e. acting in the best interest of their company.[325] There is also often an explicit or implicit expectation on directors to act *intra vires,* i.e. for the purposes for which their powers were conferred.[326] However, some commentators have objected to the application of the traditional regime on directors' duties to SPVs.[327] As discussed in Chap. 2, the general assumption is that SPVs'

[321] Art 2 Securitisation Regulation.

[322] Ibid. On the different types of entities that may act as SPVs, see Chap. 3, Sect. 3.3.2.

[323] Art 2 (2) Securitisation Regulation.

[324] See Chap. 3.

[325] Carsten Gerner-Beuerle, Philipp Paech and Edmund-Philipp Schuster, 'Study on Directors' Duties and Liability in the EU' (2013) LSE Enterprise, prepared for the EU Commission DG Markt, IX-X–XI.

[326] This wording is recurrent in some national provisions (like Slovakia, Malta) or in national case law (in countries like Cyprus). It has been described as a 'behavioural expectation' rather than a standalone duty, see Gerner-Beuerle, Paech, Schuster, ibid.

[327] David Ramos Muñoz, 'In Praise of Small Things: Securitisation and Governance Structure' (2010) 5 Capital Markets Law Journal 363, 397; Vincenzo Bavoso, 'Filling the

directors only perform residual decision-making powers, while originators would be effectively in charge of the operation since they put it into motion.[328] As investors are generally not shareholders, they also have limited rights to participate in SPVs' decisions by law. Correspondingly, there have been few cases where they have taken action vis-à-vis SPVs in the past and where courts have held them accountable.[329]

Therefore, the general securitisation regime provides for very limited rules concerning the governance of SPVs. In turn, it does not regulate the extent to which sustainability goals can inform their day-to-day business decisions.

6.6.2 *Towards 'Sustainable' Governance of 'Sustainable' Securitisations?*

Recent EU sustainable finance initiatives, including the EUGBR, have also failed to regulate the role of SPVs and the extent to which they should promote sustainability objectives.

The EBA has emphasised the need to place liability for all obligations under the EUGBS (like reporting and external review duties) on originators, while recommending not to impose any sanctions on SPVs.[330] It argued that SPVs' liability may sit unwell with their nature of passive entities and with their limited capitalisation as well as complicating the enforcement of sanctions. The final text of the EUGBR has incorporated EBA's recommendations, placing on 'originators' the responsibility for the completion of the factsheet and for reporting on the allocation of proceeds and their impact.[331]

Moreover, supervisory authorities have to monitor both originators and issuers.[332] Thus, the EUGBR postulates an active role for SPVs in the compliance with disclosure and review requirements.

Accountability Gap in Structured Finance Transactions: The Case for a Broader Fiduciary Obligation' [2014] 23 *Columbia Journal of European Law* 2 370, 384.

[328] See Chap. 3, Sect. 3.4.3.

[329] For example, *Roberts v Frohlich* [2011] EWHC 257 (Ch) (UK). More generally, see Bavoso (n. 327) 14.

[330] EBA, 'Developing a framework' (n. 180) 30. ibid.

[331] Art 13a(1) EUGBR. On the contrary, it must be noted that the responsibility for compliance with other disclosure requirements, like those under the Prospectus Regulation, rests on the originator, thus creating an inconsistent and practically inconvenient 'bifurcated regime'. Agostini, Joosen (n. 100) para 5.3.

[332] Art. 36(1)–36(3) ibid.

Other EU provisions also depict SPVs as more than 'empty shells'. For instance, the Corporate Sustainability Due Diligence Directive (CSDDD) also applies to SPVs.[333] This translates into a duty to publish reports where they disclose how they have identified, prevented and addressed environmental and social risks across their activities.[334] The Directive effectively compels SPVs to assess the sustainability profile of the securitisation portfolios that are assigned to them for their various securitisation programmes. It also requires them to take measures in case of risks and to disclose about their assessment. Under the CSDDD regime, investors may also, in principle, be able to take action based on misrepresentation or on general remedies for misstatements in corporate reporting if there are inaccuracies or omissions across 'due diligence' reports. However, the general challenges of holding SPVs accountable[335] may discourage investors from bringing these types of proceedings in the context of securitisation. This could leave room for SPVs not to adequately assess the assigned portfolio assets or provide accurate and exhaustive information across their reports, which may result in 'greenwashing'.

Moreover, there are constraints that may *de facto* deprive the provision of any practical relevance. As the CSDDD only applies to companies with over 1000 employees,[336] it may hardly apply to SPVs, who tend to only have directors, but not a stable worker base, unless the threshold requirements are assessed at the originator's level.

All in all, rules within the emerging EU corporate sustainability framework, albeit with a limited scope of application, expect SPVs to be proactive in their compliance with reporting activities and assessments on sustainability matters. At the same time, they provide for very limited enforcement opportunities, both by supervisory authorities and by investors. Thus, the latter have very few means and avenues to raise their 'voice' on greenwashing matters, as they could not directly challenge decisions on sustainability-related matters for breach of directors' duties or react to misstatements throughout due diligence reports.

[333] It is also applicable to a 'securitisation special purpose entity, as defined in Article 2, point (2), of Regulation (EU) 2017/2402 of the European Parliament and of the Council', see art 3(1)(a)(iii) CSDDD.

[334] Art 4(1) ibid.

[335] See Chap. 3, Sect. 3.4.3. and above, 6.6.1.

[336] Art. 2(1)(a) CSDDD.

6.6.3 *Normative Considerations*

The previous subsections have highlighted how the EU securitisation regime has failed to regulate how sustainability goals should permeate the administration and day-to-day decisions of SPVs. While governance is inherently discretionary, the absence of rules and the inconsistency among recent sustainability-oriented provisions leaves room for uncertainty.

Instead, dedicated provisions clarifying the purpose of SPVs within 'sustainable' securitisations should arguably provide a clearer context and justification for sustainability-oriented decisions. The most appropriate setting for such rules appears to be the EUGBR. Desirable amendments to the Regulation should notably clarify that the SPVs of securitisations under the Standard must exclusively promote sustainability goals and act in the best interests of ABS holders. In other words, they should exclusively issue 'sustainable' ABSs and manage the circulation of cash flows between the actors of the transaction. This entails clarifying that they must exclusively use their patrimony to repay investors and to fulfil the sustainability goals of the transaction.

This recommendation also fits squarely within the burgeoning policy and academic discussions around the possible fine-tuning of provisions surrounding companies' purpose with sustainability objectives.[337]

Regulating SPVs' purpose would serve an 'enabling' function, since it may shape a different understanding of the duties of their directors. Notably, it may have an impact on directors' duty to act *intra vires*, since achieving sustainability targets would become an integral component of the purpose of their powers. In turn, decisions like misallocating the funds of the operation for non-sustainability-related purposes, not complying with the Taxonomy criteria, or failing to adequately report on the composition

[337]For example, the British academy as part of the 'Future of the Corporation' project, see British Academy, 'Principles for purposeful business' (2019) Executive Summary, https://www.thebritishacademy.ac.uk/documents/223/future-of-the-corporation-principles-purposeful-business-executive-summary.pdf accessed 10 December 2025; David Kershaw, Edmund-Philipp Schuster, 'The Purposive Transformation of Company Law' (2021) 69 The American Journal of Comparative Law 3 478, 490. Others have concluded that legal provisions surrounding corporate purpose would be less effective than other types of legal changes, see Iain MacNeil, Irene-Marie Esser, 'Corporate Purpose as a Conduit for Sustainability in Corporate Governance' in MS Silva, A Nicolussi et al (eds) *Private Law and Sustainability* (Routledge 2024) 541–542. While some of the objections are legitimate, the normative recommendations in the present Sect.ection are intended to be *complementary* to other legal reforms targeting other strategic aspects of securitisations, proposed in Sect. 6.4.4.–6.5.4–6.7.4–6.8.3.

of the portfolio, may signal a breach of their duties. Therefore, the clarification of SPVs' purpose, coupled with the above-mentioned statutory duty to perform due diligence assessments over the securitisation portfolio under CSDDD, could considerably strengthen investors' 'voice' powers.

6.7 The Elephant in the Room: Investors' Representation Mechanisms

Another strategic aspect heavily overlooked by recent initiatives on securitisation is the representation of investors. Several specialised market actors, including pension funds, insurers and hedge funds, invest in ABSs. On the contrary, EU legislation generally restricts the sale to retail investors, unless the seller has performed an additional 'suitability' test and informed the client about it.[338] Investments in securitised products, and in 'sustainable' securitised products in turn, present several peculiarities, calling for further initiatives to strengthen the representation of such market actors.

6.7.1 The Drivers (I): Challenges to Investments in ABSs

Several practical, economic and legal constraints curb the possibility for institutional investors to take sustainability risks into account when making decisions about whether they want to invest in securitised products and which products they would like to purchase.

As previously mentioned with regard to underwriters,[339] all institutional investors must undertake a thorough due diligence assessment before investing in 'traditional' securitised products. Additionally, the securitisation regime requires them to put in place ad hoc onerous procedures when holding ABSs in order to regularly monitor the transaction as well as to perform regular stress tests.[340] Several market actors have described such due diligence requirements as particularly cumbersome and costly, to the point that they would discourage investments in ABSs altogether.[341]

When it comes to assessing 'sustainability' features such as the risks and impact of the securities and of the underlying portfolio, due diligence

[338] Art 3 Securitisation Regulation.

[339] Art 5(3)(a)-(b) ibid.

[340] Art 5(4)(a)-(b) ibid.

[341] See various responses to the consultation on securitisation (n. 154), For example, responses 3-4-5-12, EU Commission, 'Summary report – Targeted consultation on the Functioning of the EU securitisation framework' (n. 139).

costs may increase even further. ESMA has clarified in a recent Q&A that it is up to investors to assess the sustainability profile of investments.[342] Such assessments may require in-house expertise and a dedicated ESG team, which many investors that apply PRI Principles have declared not to have in a recent survey.[343] An additional hurdle is the small size of investments in ABSs within investors' portfolio: For instance, one survey of the signatories of the Principles for Responsible Investment (PRI) has reported that they only represented roughly 4% of the global assets under management in 2021.[344] This may disincentivise investors from dedicating extensive resources to analysis for these products.

Furthermore, broader legal constraints may deter investors from widely integrating sustainability factors within their portfolio-management activities. EU law allows, but does not require, pension funds to take environmental, social, and governance factors into account.[345] The above-mentioned SFDR[346] only requires various types of investors to report *if* they have considered sustainability factors (specifically on the 'principal adverse impact' of their activities on ESG issues) but also allows them to provide an explanation about why they have not.[347] Under the general framework for institutional investors, they would only ***have to*** use

[342] ESMA, 'Answers to questions on the interpretation of Regulation (EU) 2019/2088, submitted by the

European Supervisory Authorities on 9 September 2022' (2023) 2 https://www.esma.europa.eu/sites/default/files/2023-04/Answers_to_questions_on_the_interpretation_of_Regulation_%28EU%29_20192088.PDF accessed 10 December 2025.

[343] UNPRI (n. 125) 17–24.

[344] Ibid., 14.

[345] Art. 19(1)(b) Directive (EU) 2016/2341 of the European Parliament and of the Council of 14 December 2016 on the activities and supervision of institutions for occupational retirement provision. It may be different in some EU jurisdictions, where pension funds *must* take such factors into account and disclose about it, like France after the « Loi Pacte ». See Art 72 (6) Loi du 22 mai 2019 relative à la croissance et la transformation des entreprises.

[346] See Sect. 6.4.1.

[347] On their website and to end investors when concluding a contract, Art 4(1)-6(1) Regulation 2019/2088 of the European Parliament and of the Council of 27 November 2019 on sustainability-related disclosures in the financial services sector (2019) OJ 337\1. (3) Commission Delegated Regulation (EU) amending Delegated Regulation (EU) 2017/565 as regards the integration of sustainability factors, risks and preferences into certain organisational requirements and operating conditions for investment firms (2021) 2616 amending Article 23 of the MiFID II Delegated Regulation; Art 2 (1) Commission Delegated Directive (EU) amending Directive 2010/43/EU as regards the sustainability risks and sustainability factors to be taken into account for Undertakings for Collective Investment in Transferable Securities (UCITS) OJ L 277/141.

sustainability considerations as a guiding criterion when such factors affect financial risks.[348] Similarly, the UK Law Commission invited pension trustees to only take non-financial factors into account when in line with their clients' views and if the decision would not cause significant financial detriment.[349] In all other instances, it would not be incompatible with investors' mandates to purchase assets backed by 'brown' portfolios.

Even among the investors that do consider sustainability risks, some may only *exclude* some types of companies or assets due to their higher (climate-related) financial risks ('screening' approach), as the vast majority of investors have reported doing in a recent survey, rather than choosing them because of their positive environmental impact. [350] An example could be investors not purchasing any ABSs when originators are oil companies, rather than choosing the 'greenest' issuers.

As a result of these costs and legal constraints, investors are only likely to apply sustainability considerations to investments in securitised products under two circumstances: Either to discharge their statutory duties or when the terms for their appointment so require. Arguably, these two peculiar categories of investors have a special interest in the 'sustainable' qualities of their investment and want additional reassurance against 'greenwashing' risks.

6.7.2 The Drivers (II): Challenges to Investments in 'Sustainable' ABSs

Investments in 'sustainable' ABSs present even further challenges, making them hardly accessible to smaller institutional investors.[351]

[348] Because of the duty on all institutional investors to manage long-term risks, Art 25 Directive 2016/2341; UNPRI (n. 125) 13, 200, 287. In relation to pension funds see Lloyd Brown, 'Towards 'Green' Trusteeship: New Statutory Amendments for Occupational Pension Trustees' (2019) 25 Trusts & Trustees 10 978.

[349] Law Commission, 'Pension Funds and Social Investment' (2017) Law Com 374, 40–43 https://assets.publishing.service.gov.uk/media/5a750d09ed915d5c5446524c/6.3496_LC_DRAFT_190617_TT_20_06_17__WEB_-_final_.pdf accessed 10 December 2025.

[350] As also reported by UNPRI (n. 125) 17.

[351] As the responses to the UNPRI survey seem to suggest, UNPRI (n. 125) 17. David Nelson, Brendan Pierpont, 'The Challenge of Institutional Investment in Renewable Energy' (2013) Climate Policy Initiative 16 https://www.climatepolicyinitiative.org/wp-content/uploads/2013/03/The-Challenge-of-Institutional-Investment-in-Renewable-Energy.pdf accessed 10 December 2025.

Investing in these products may sometimes come with a 'Greenium', i.e. a premium that should reward companies for their sustainability performance. For investors, this entails additional costs and lower returns. There is anecdotal evidence of investors refraining from choosing these securities, under the assumption that it may entail a kill to the financial returns of beneficiaries.[352] Due diligence costs are also likely to be higher than those for 'traditional' securities, due to the preliminary state of the 'sustainable securitisation' market: Correspondingly, it has been estimated that only larger investors, managing large portfolios of assets, are likely to be able to make direct investments in renewable energy projects.[353]

The legal environment, as well as the interconnected litigation risks, may also discourage investors from opting for 'sustainable' ABSs.[354] There are several general principles surrounding institutional investors' duties that have been interpreted as implying a duty to prioritise financial over sustainability considerations. For instance, one of the key EU principles is their need to pursue the 'best interests' of their beneficiaries.[355] Long-standing academic scholarship and judicial precedents have linked the duty to the Modern Portfolio Theory, which provides a mathematical and microeconomic model for the construction of portfolios. A corollary of the theory would be the need for investors to prioritise financial products that maximise beneficiaries' risk-adjusted returns over those that advance sustainability-oriented goals.[356] Another traditional assumption often associated with the need to pursue 'best interests' is the need for investors to balance the interests of all beneficiaries.[357] One could argue that not all

[352] According to estimates, only investors owning over $50 bln may directly invest in renewable energy projects, UNPRI, ibid., 17.

[353] The argument referred to bonds, but it seems reasonable to transpose it to all 'sustainable' ABSs. Tommy Stubbington, Squeeze on 'greenium' as ESG bond investors demand more value (FT 11th October 2021) https://www.ft.com/content/ecbed322-1709-4ed6-9f7f-d974f6e181da accessed 10 December 2025.

[354] It has been estimated that only investors owning over $50 bln may directly invest in renewable energy projects, Nelson, Pierpont (n. 332) 16, 47. On transaction costs, see also 3.5.

[355] UNPRI (n. 331) 17.

[356] Ibid.; on the relationship between investments driven by sustainability considerations and liability standards see Tom Gosling, Iain Macneil, 'Can Investors Save the Planet? – –NZAMI and Fiduciary Duty' (2023) 18 Capital Markets Law 2 172.

[357] For example, see art 18(1)(a) Directive 2003/41/Ec of The European Parliament and of the Council of 3 June 2003 on the activities and supervision of institutions for occupational retirement provision; For example, 41 (1)(a) Regulation (EU) 2019/1238 of the European Parliament and of the Council of 20 June 2019 on a pan-European Personal Pension Product (2019) OJ 198/1; Art 22 Directive 2010/43/EU of 1 July 2010 implementing

end beneficiaries have an interest in sustainability goals. The EU sectoral legislation on pension funds[358] also sets out the duty for pension funds to act in line with the 'prudent person' standard[359]: this would involve, among other things, the need to diversify the assets within portfolios.[360] Liquidity should also be one of the principles guiding investments.[361] As a result of these general principles, institutional investors may intentionally decide not to invest in 'sustainable' securitised products under the perception that it may contravene concurrent statutory duties, exposing them to liability risks.[362]

Instead, they are likely to opt for these types of securities when pursuing sustainability objectives *clearly* falls within their mandate. Acting in accordance with the terms of contractual agreements or trust deeds represents one of the overarching principles guiding contractual and trust-type relationships,[363] like those between pension funds and their beneficiaries, also acknowledged under EU law on pension funds.[364]

Directive 2009/65/EC of the European Parliament and of the Council as regards organisational requirements, conflicts of interest, conduct of business, risk management and content of the agreement between a depositary and a management company (2010) OJ 176\42.

358 For the old case law, *Cowan v Scargill* (UK); on the link with the portfolio theory, see Jeffrey N Gordon, 'Systematic Stewardship' (2022) 47 J Corp L 627. More recent commentary has progressively discarded this old-standing interpretation. See UNPRI, 'A legal framework for Impact: sustainability impact in investor decision-making' (2021) 449 https://www.unpri.org/download?ac=13902 accessed 20 December 2025.

359 See *Cowan v Scargill* (UK); See For example, UNPRI, ibid., 377, 74, 92.

360 Comprising the IORP II Directive, applicable to any *'institution, irrespective of its legal form, operating on a funded basis, established separately from any sponsoring undertaking or trade for the purpose of providing retirement benefits in the context of an occupational activity on the basis of an agreement or a contract'*, and the PEPP Regulation, applicable to *'pan-European pension products (PEPPs) that complement public or private pension schemes on a voluntary basis and on the basis of a contract'*, see Art 6 Directive (EU) 2016/2341, Regulation (EU) 2019/1238 of the European Parliament and of the Council of 20 June 2019 on a pan-European Personal Pension Product (2019) OJ 198/1.

361 Art 19(1) Directive 2016/2341; Art 41(1) Regulation 2019/1238. For an overview of the framework, see UNPRI (n. 125), 277–278.

362 Art 19(1)(f) ibid.; art 41(1)(f) Regulation 2019/1238.

363 Art 19(1)(c) ibid.; art 41(1)(c) Regulation 2019/1238. On the concept of liquidity, see Chap. 2.

364 See also *'(...) the proper purpose and best interests duties implicit in the duty of loyalty mean that an asset manager could be on shaky ground if they pursued climate goals that were not explicitly mandated by the asset owner even if pursuit of those goals was in parallel to, and not in opposition to, delivery of the financial return'* Gosling, MacNeil (n. 354) 176.

Therefore, institutional investors may choose 'sustainable' ABSs when clauses within their agreements with beneficiaries[365] clearly refer to sustainability considerations as their key investment objective. Chap. 4 has acknowledged the lack of publicly available data on the frequency of such clauses within securitisation documentation.[366] However, the number of EU funds with an explicit mandate to promote ESG objectives has significantly grown in recent years and, while the majority invests in equity, a growing proportion is also active in the fixed income market.[367] Such funds are likely to choose securitised products too, as these investments may help them comply with the now stricter reporting obligations on their environmental impact: under the SFDR, they must periodically disclose the percentage of their holdings complying with the technical criteria of the Taxonomy.[368]

All in all, various financial and legal considerations may hinder investments in securitised products even when they are clearly labelled as 'sustainable'. Those market actors who do choose these products are likely to have an additional interest in the 'sustainable' profile of the securities and in their reliability.

6.7.3 *The Drivers (III): Limited 'Voice' Rights*

Market actors are only likely to incur the additional costs and to bear the potential legal risks of considering sustainability factors when investing in securitised products in order to comply with other statutory obligations or with the terms of their appointment. However, it will arguably be of utmost importance for such investors to monitor and ensure that the ABSs or the underlying assets indeed contribute to sustainability goals. Several provisions nevertheless restrict the ability of ABS holders to *individually* raise their 'voice' vis-à-vis investee companies.[369]

[365] UNPRI (n. 125) 77–79. Luke Broadway, 'Trustee Duties versus Equity's Moral Conscience: Trust Law ' s Inadequate Facilitation of Socially Responsible Investment' (2021) Trusts & Trustees, 1.

[366] Art 6(b) Directive 2003/41.

[367] See Chap. 3, Sect. 3.2., 3.4.4.

[368] See Chap. 4, Sect. 4.6.1.

[369] AFME, 'ESG Finance Report – –Q2 2021' (n. 12) https://www.afme.eu/publications/data-research/details/esg-finance-q2-2021---european-sustainable-finance accessed 10 December 2025. More recent data confirms the growth of such funds in spite of the general contraction of the supply of 'sustainable' financial instruments, AFME, 'European sustainable finance – ESG finance report – Q3 2025' (2025) 29 https://www.afme.eu/media/mhgayv3c/esg-finance-report-3q25.pdf accessed 20 December 2025.

As discussed in Chap. 2, ABS holders have very limited rights in the governance of SPVs, since securitisation generally leads to the issuance of debt securities. Traditional shareholders' rights, such as participating in general meetings, voting, and holding directors and companies judicially accountable for their actions, are also largely unavailable to investors in ABSs. [370] While recent EU harmonisation initiatives have further stimulated and enhanced shareholders' rights to participate and vote in their respective companies[371], they have neglected investors in debt securities.

Such investments also inhibit 'collaboration' activities, which are more recurrent among shareholders, leading to coalitions for the exercise of voting rights or to bring a resolution. [372] Instead, ABS holders almost exclusively engage on a bilateral basis with issuers, while only rarely collaborating with other investors for renegotiation or debt restructuring purposes.[373]

Additional coordination costs are also inherent in the characteristics of debt instruments, since bond issuances may attract different investors with different time horizons and objectives.[374] They are even more prominent across securitisation operations, where investors can choose between different 'tranches' of securities with different priority rights and returns.[375] Their different preferences may prevent the reaching of a consensus and exacerbate the lack of collaboration, especially on ESG matters, where preferences widely vary.

Facing the challenges to coordination among ABS holders, Chap. 3 has discussed the merits of the appointment of a representative to act on their behalf.[376] While the EU securitisation regime does not require it, it envisages their involvement. For example, it acknowledges documents such as 'trust deeds' as well as 'administration and management agreements' as part of the documentation to circulate with investors.[377] Indeed, trustees

[370] Artt 5, 6 Regulation 2020/852.

[371] On the definition of 'voice' rights, see Chap. 2, Sect. 2.6.

[372] See Chap. 2, Sect. 2.3.4.

[373] See Art 3c Directive (EU) 2017/828 of the European Parliament and of the Council of 17 May 2017 amending Directive 2007/36/EC as regards the encouragement of long-term shareholder engagement (2017) OJ L 132\1.

[374] Iain MacNeil, 'Activism and Collaboration among Shareholders in UK Listed Companies' (2010) 5 Capital Markets Law Journal 419, 424, 430–431; UNPRI, 'ESG Engagement for fixed-income investors: managing risks, enhancing returns' 9 https://www.unpri.org/download?ac=4449 accessed 10 December 2025.

[375] UNPRI, ibid., 27, 33, 37.

[376] Ibid., 27.

[377] See Chap. 3, Sect. 3.2.–3.4.2.

are one of the possible legal forms for these representatives. Under EU law, it is also compulsory to clearly incorporate mechanisms for 'the timely resolution of conflicts between different classes of investors' for the award of the STS label.[378]

In the absence of ad hoc EU rules, Member States and the contractual documentation have the crucial role of articulating the powers of such representatives and the extent to which they can engage with issuers or originating entities. For instance, a precondition for the award of the STS label is to clearly specify in the documentation 'contractual obligations, duties and responsibilities of trustees and other entities owing fiduciary duties to investors'.[379] This provision grants securitisation parties the flexibility to set out representatives' powers on a deal-by-deal basis. At the same time, some EU countries, like Italy and Luxembourg, have their own legislation. [380] In other cases, general principles such as fiduciary principles or the clauses of the contractual documentation will come into play.[381] Law or the contractual documentation can, in principle, give such market actors enforcement powers, which may include the power to bring action vis-à-vis SPVs. They may be able to convene ad hoc decision-making fora among ABS holders. [382] Accordingly, the legal framework for investors' representatives has the crucial potential to facilitate their powers to raise their 'voice' on sustainability matters.

6.7.4 Normative Considerations

The challenges to applying sustainability considerations to investments in securitised products, as well as the limited 'voice' rights of ABS holders, call for the involvement of additional intermediaries. EU law provisions are crucial in order to ensure uniform 'voice' powers on sustainability matters.

The due diligence requirements under the Securitisation Regulation[383] should also come to encompass the assessment of sustainability risks. They should identify sustainability-related risks in the securitisation portfolio

[378] See Chap. 3, Sect. 3.4.4.

[379] Art 7(1)(v)-(vi) Securitisation Regulation.

[380] Art 21(10) ibid. The provision is not applicable to ABCP deals as it is not replicated in Artt 24 (19) and in Art 26(7) ibid., providing for the requirements for the documentation of ABCP STS deals.

[381] Art 21(7)(a) ibid.

[382] See For example, In Luxembourg, see Art 67–68 Law of 22 March 2004 on securitisation; in Italy, see Art 7 (8) L 130\1999 (IT). see Chap. 2, Sect. 2.3.4.

[383] David Ramos Muñoz, *The Law of Transnational Securitisation*, 154; Gibson (n. 242).

and in all the underlying positions before holding ABSs. They should also be required to put in place risk-management techniques and monitoring procedures for the environmental and social impact of the securities and of the securitised asset pool *ex post*. As the current due diligence requirements also encompass regular stress tests,[384] a possible risk management technique could be climate-related stress tests. However, as these are a relatively recent development, only marginally used by private market actors,[385] legislation should give market actors flexibility to design their own techniques.

A broader question is whether EU law should also *require*, and not merely allow, institutional investors to apply sustainability considerations to their portfolio-management activities. The possibility of regulating the requirement for investors to engage with companies and to achieve a sustainability impact was initially on the table under the EU sustainable finance strategy[386], but later abandoned. While such reforms may also indirectly affect the incentives for market actors to purchase 'sustainable' ABSs, they remain outwith the scope of the present analysis.

Furthermore, ad hoc provisions should also facilitate investors' 'voice' rights. Some commentators have already advocated for an expansion of trustees' monitoring and enforcement powers within traditional bond issuances ('super-trustees').[387] Others have also suggested rethinking the role of trustees in securitisation transactions on similar grounds.[388] Arguably, a welcome development could be the amendment of the EUGBR to mandate the appointment of a 'Lead investor' across all 'sustainable' securitisation issuances. The EUGBR should clarify the role of such representatives, as well as introduce a statutory duty for them to act on behalf of all investors in 'sustainable' ABSs.

Among others, the Regulation should specify that Lead Investors should meet enhanced professionalism requirements, such as prior expertise in sustainability matters. Such a requirement would align with national provisions on the 'representatives of bondholders' in some EU countries, such as Luxembourg and Italy.[389] Even though investors should have wide

[384] For example, in Germany, Estonia, Croatia, Gerner-Beuerle et al (n. 325) 64–65, 186, 187.
[385] Art 5 Securitisation Regulation, see Sect. 6.6.1.
[386] Art 5 (4)(2) ibid.
[387] NGFS, 'Adapting Central Bank Operations' (n. 238) 28.
[388] Communication from the Commission, Financing the transition (n. 196) 14.
[389] Yakov Amihud, Kenneth Gerbade, Marcel Kahan, 'A New Governance Structure for Corporate Bonds' (1999) 51 Stanford Law Review 3 447, 472–479. Park has recommend introducing 'supertrustees' for 'green' bondholders, Stephen Kim Park, 'Green Bonds and

freedom in their choice of representative, one option could be to create a list of accredited entities at EU level, like that for bonds' external reviewers.[390] Promising market actors to serve this role could be, among others, multilateral development banks, which *Park* suggests appointing as representatives of green bond investors, due to their wide experience in the sector and in the development of best practices.[391] The appointment of 'Lead investors' should take place ahead of the issuance of ABSs in order to give them a say over the transaction documentation and in the 'underwriting' process more generally.[392] Their early involvement in the structuring of the deal may foster constructive dialogue,[393] leading to the inclusion of sustainability-oriented terms in the transactional documentation.

EU law should also clarify that Lead investors should serve as intermediaries between the interests of ABS holders and originators. For example, they should be in charge of calling and convening a meeting among all investors[394] in order to collect their views on sustainability matters. They should also actively monitor originators' sustainability performance. To do so, they should have the statutory right to access and inspect the transaction documentation, which is sometimes given to bondholders on a deal-by-deal basis through contractual provisions.[395] Notably, they should be able to view the loan purchase agreements or other documentation related to the securitised assets, which is often not publicly available. They should also have a right to ask questions of the directors of SPVs and originating companies on sustainability matters, in line with existing shareholders' rights at general meetings.[396]

Lastly, ad hoc EU statutory provisions should empower such representatives to bring judicial action against SPVs and originating companies on sustainability matters. Such a proposal does not seem unrealistic in the face of the national corporate law systems within the EU. Traditionally, shareholders have been the category in charge of bringing enforcement

beyond: Debt Financing as a Sustainability Driver', *Cambridge Handbook of Corporate Law, Corporate Governance and Sustainability* (2020) 608.

[390] Gibson (n. 242).

[391] Art 7.1 (8) Law 300\1999 (IT); Art 80 (3)-(4) Loi du 22 mars 2004 relative à la titrisation (Luxembourg).

[392] See 6.5.2.

[393] Park (n. 1315) 608.

[394] See above Sect.... 5.5.1.

[395] As advocated by UNPRI for all bond investors in 'sustainable' products, see 'ESG Engagement for fixed-income investors' (n. 372) 33.

[396] Like the representative of bondholders in Italy, see Art 2415 cc (IT).

powers vis-à-vis companies since restricting directors' liability vis-à-vis other constituencies was believed to encourage directors to take on more risks, maximising the value of the company, without the fear of their liability risks vis-à-vis other actors.[397] Nevertheless, the rationale behind sustainability-oriented legal initiatives is very different: it is not just about the management of risks, but also about the trust and credibility of 'sustainable' investments.[398] This different objective justifies the expansion of enforcement powers to ABS holders, albeit not shareholders, for circumstances like originating companies or SPVs failing to comply with the sustainability targets of the operation.

This proposal should be read in conjunction with the above-mentioned recommendation concerning the clarification of SPVs' purpose.[399] A broader understanding of SPVs' purpose is complementary to the expansion of investors' enforcement remedies and to the appointment of a representative to protect their interests.

All in all, appointing a Lead investor should mitigate the high due diligence costs of assessing the sustainability features of the transaction, as well as the coordination costs of finding and pursuing a common position on sustainability matters among ABS holders. It should enable investors to adequately monitor issuers' performance and to exercise their 'voice' rights through various means, such as direct involvement in the negotiation of the terms of the issue or the enforcement of breaches of sustainability-related clauses.

6.8 On Monitoring Processes

6.8.1 Fragmentation of the Traditional EU Securitisation Regime

As discussed in Chap. 3, various processes allow market actors to ensure the insulation of the funds and their allocation for sustainability purposes, a concept referred to as 'shielding'.[400] These include the appointment of third parties to oversee the circulation of cash flows ('servicers' or 'management companies') or mechanisms to ring-fence bank accounts.[401]

[397] See above.

[398] See art 9 Directive 2007/36/EC of the European Parliament and of the Council of 11 July 2007 on the exercise of certain rights of shareholders in listed companies (2007) OJ 184\17.

[399] Ramos Muñoz, In praise of small things (n. 327), 377–378.

[400] See explanatory memorandum, EUGBR.

[401] See Chap. 4, Sect. 6.6.2.

EU harmonisation efforts have only indirectly affected the mechanisms to monitor expenses in securitisation operations. The Securitisation Regulation postulates the involvement of third parties by including 'servicing, back-up servicing, administration and cash management agreements' within the documentation to be made available to investors.[402] It also explicitly permits the delegation of the servicing tasks to a regulated asset manager.[403] The STS regime provides for additional rules for 'servicing' concerning the expertise of servicers and the disclosure of their responsibilities in the documentation.[404]

The Securitisation Regulation also contemplates 'bank account agreements' as part of the documentation to circulate with investors[405], but provides no further requirements on their content. Chap. 3 has examined how these agreements designate one bank account where all the cash flows for the operation are deposited, whilst highlighting their differences in forms and provisions.[406] There have also been few cases of national legislation in the EU, like Luxembourg, where legislation have *mandated* the compartmentalisation of funds for all securitisation operations.[407]

Overall, due to the flexibility of the EU and of most national securitisation regimes, market actors have tended to implement mechanisms to monitor securitisation funds on a deal-by-deal basis. However, as noted in Chaps. 4 and 5, parties' arrangements do not always provide for the segregation of funds from external creditors or for the 'earmarking' of funds for a specific purpose. In fact, because these types of provisions are restrictive to issuers' activities, they are most likely to be rare.[408]

6.8.2 The Limited Impact of the EU Green Bond Standard

Another aspect which the EUGBR seeks to regulate is how originators and SPVs monitor sustainability-related and non-sustainability-related expenses following the issuance of ABSs.

[402] On 'shielding', see Chap. 2, Sect. 2.6.
[403] See Chap. 2, Sect. 2.6.; Chap. 3, Sect. 3.4.5.
[404] Art 7(b)(4) Securitisation Regulation.
[405] (7) ibid.
[406] See recital (4), Art 21 Commission Delegated Regulation 2019/1851. following Basel Recommendations, see Criteria 14 BCBS (n?) 12.
[407] Art 7(b)(4) Securitisation Regulation.
[408] Chap. 3, Sect. 3.4.5.

As previously discussed, market actors aiming to comply with the Standard will need to publish a prior factsheet on how they plan to meet the Taxonomy criteria. After the issuance of ABSs, they will have to publish yearly 'allocation reports', also following a standardised template.[409] These reports will, however, primarily provide high-level information on the congruence of the expenses with the information on the factsheet. In essence, originators will disclose the share of the proceeds allocated towards activities in line with the Taxonomy criteria and the share which deviates from them. They will have to explain why they have not met the Taxonomy requirements. The template also allows restriction of the amount of information in the light of *'confidentiality agreements, competitive considerations'* or if several projects are concurrently involved.[410] The style of the template, however, inevitably poses the risk of the allocation reports becoming a 'tick-the-box' exercise.

Similar 'gatekeeping' mechanisms to the ones over factsheets will also be mandatory for allocation reports. Third-party reviewers, accredited by ESMA, will have to confirm in a statement whether the expenses over investors' funds as described in these documents indeed comply with the Taxonomy criteria.[411]

Another means for investors to monitor issuers' commitment to sustainability goals after the issuance will be so-called 'impact reports', to be drawn at least once during the lifetime of securitisation bonds and after the full allocation of proceeds (i.e. most likely after the full implementation of 'sustainable' projects). These will provide more general information linking the issuance to the broader environmental strategy of issuers.[412]

Despite the welcome steps to increase disclosure and to introduce some 'gatekeeping' processes over allocation and impact reports, these measures are expected to have a partial effect. As previously discussed, it would appear from the wording of the proposal that only factsheets, but not allocation and impact reports, would be part of the prospectus.[413] External reviews over impact reports should remain optional.[414] Besides, what investors could do if such reports contain misstatements remains unclear.

[409] Like French and Luxembourg law, see Chap. 3, Sect. 3.4.5.
[410] See Chap. 4, Sect. 4.10; Chap. 5, Sect. 5.8.
[411] See 4., Annex I., and Annexes II-III; see also above Sect. 6.4.2., 6.5.2.
[412] Annex II
[413] Art 13(1)(d), 3. Annex IV. EUGBR.
[414] Art 10 (1), Annex III.

Moreover, Chap. 5 has discussed the best practices enshrined in the soft law principles for green bonds. ICMA's Principles recommend more accurate tracking of the expenses over the proceeds through 'a sub-account, sub-portfolio' or 'in another appropriate manner'.[415] The Climate Bonds Standard Scheme also requires 'earmarking' of funds and restricts the possible uses of the funds where issuers decide to re-allocate some of the proceeds for other purposes on a temporary basis.[416] These mechanisms would more effectively ensure the appropriate allocation of investors' funds towards 'sustainable' projects. However, the non-binding nature of such guidelines arguably does not currently provide adequate incentives for market actors to set up such mechanisms.

The EUGBS, which has failed to reflect these principles, is unlikely to incentivise market actors to adopt such practices. As a result, the Standard will not ensure the effective 'shielding' of the assets for sustainability purposes.

6.8.3 Normative Considerations

The reference to 'ring-fencing' mechanisms throughout the soft law principles for green bonds suggests their feasibility as well as their suitability for financial transactions that promote sustainability goals. These mechanisms could effectively prevent originating companies from 'greenwashing', which also involves misusing ABS proceeds for non-sustainability-related purposes. Therefore, another welcome development for the EUGBS should be the requirement to disclose the 'tracking' mechanisms over the expenses (following ICMA's terminology[417]) in the factsheet.

However, such a harmonisation initiative should consider jurisdiction-specific differences, particularly concerning proprietary and financial rules for fund segregation. As a result, the provision should be formulated in open-ended wording, allowing all EU Member States to develop their own solutions in the implementation phase.[418]

Moreover, there is room to strengthen the means for investors and external reviewers to have a say over allocation and impact reports. The EUGBR should specifically empower ABS holders to take action in case of

[415] See above Sect. 6.5.2.
[416] Art 10 (2b), Annex IV.
[417] ICMA, GBPs, 8, SBPs, 4–5.
[418] CBI, 'Climate Bonds Standard 4.0' (2023) Clauses 3.1.1.–3.1.3.; 7.3.1.–7.3.3.; see also Chap. 5, Sect. 5.8.2.

misstatements or omissions in these documents ('voice') as well as make external reviews of impact reports mandatory ('gatekeeping').

6.9 Conclusions

The chapter has examined possible mechanisms to coordinate sustainability-related aspects of securitisation through EU law, which can be broadly characterised as examples of 'harmonisation'.

EU law offers intriguing insights into the potential of binding legislation, designed at the supranational level ('public hard law'), for regulating the conditions under which market actors can promote sustainability goals through securitisation. The EU also represents a peculiar actor since, unlike other examples of transnational law-making bodies (such as IOSCO or the BCBS), it can introduce rules which EU Member States are *required* to transpose, as well as create new rules through harmonisation. These distinctive features are particularly strategic for the regulation of sustainable securitisations as they may address the enforcement limitations and accountability concerns raised by market-designed mechanisms discussed in Chap. 4.43[419] Further harmonisation initiatives at EU level, as opposed to globally coordinated standards, also appear more realistic and in line with the EU's broader policy agenda.

Against this background, the chapter has provided an in-depth analysis of the implications of salient EU harmonisation initiatives on sustainability-related elements of securitisations.

The general regime for securitisations, while introducing various complex disclosure requirements, draws very limited attention to sustainability-related aspects. For instance, the general duty to provide information about the environmental and social impact of the portfolio only applies to the transactions under the STS securitisation regime (which, in principle, do not necessarily cover the whole spectrum of EU securitisations), as well as providing several exceptions. As a result of such provisions and of limited reporting, investors and other market actors can currently gain very limited insight into environmental and social risks of the underlying assets or into the final impact of the transaction.

Despite the EU's leadership position in the global sustainable finance policy discussion, recent measures have tended to regulate securitisations only tangentially. An example is the EUGBR, providing for similar legal

[419] See Chap. 4, Sect. 4.7.2.

treatment for green bonds and securitisation, except for a few ad hoc provisions. This approach has given rise to a fragmented framework that fails to fully take the peculiarities of these deals into account. Furthermore, the EUGBR fails to fully address the question of the legal definition of 'sustainable securitisation'. In other words, its voluntary scope still allows market actors to qualify securitisations as 'sustainable' even when they do not align with the requirements of the Standard. Its scope is also narrower than the broader universe of 'sustainable' securitisations in the EU market discussed in Chap. 4. For instance, as it focuses on 'green' bonds, and only provides non-binding rules for sustainability-linked ones, it would not apply to 'social' securitisations. [420]

EU rules predominantly revolve around the dissemination of further sustainability-related elements about securitisation deals ('disclosure'). For instance, the EUGBR will require issuers to be more transparent around how they contribute to environmental goals through investors' proceeds. However, as the proposal is centred on green bonds, it imposes less stringent reporting requirements in relation to the composition of the portfolio. Besides, the persisting exemptions from reporting are likely to render these regulatory efforts inadequate, leaving room for ambiguity and under-reporting.

The analysis of the 'gatekeeping' mechanisms likewise reveals the shortcomings of directly transposing the green bond regime to securitisation. For instance, while the EUGBR mandates external reviews of various green bond-specific documents, these processes will not necessarily need to cover the qualities of the portfolio. It also remains to be seen whether EU authorities will further harmonise the procedures for external reviews (e.g. through ESMA's regulatory technical standards), thus addressing the inconsistencies highlighted in Chap. 4.[421]

By considering 'sustainable' ABSs as an example of green bonds, recent EU initiatives, including the EUGBR, have also largely neglected other strategic aspects distinctive to securitisation. For instance, they have only minimally regulated the 'shielding' of securitisation by requiring the publication of reports on the allocation of proceeds and their impact. At the same time, the EUGBR fails to require the implementation of mechanisms

[420] On the flexibility on EU Member States in the transposition of EU harmonisation initiatives, especially directives, see above s. 6.2. On some examples of country-specific differences, see further Chap. 3, Sect. 3.3.1.

[421] See Sect. 4.8.

to ring-fence the funds of the operation, such as segregated funds, for exclusive use for sustainability purposes.

The other aspect unique to securitisations, marking the difference from other examples of green bond issuances, is the involvement of SPVs. The EUGBS and Securitisation Regulation contain no provisions on the purpose of these entities and the extent to which they have to promote sustainability goals. Relatedly, investors may have no enforcement mechanisms to raise their 'voice' when SPVs' directors make decisions and incur expenses from SPVs' patrimony that deviate from sustainability goals. More generally, the EUGBR leaves limited room for investors' 'voice' by failing to clarify when and under what circumstances investors may take action for 'greenwashing' based on the statements within prospectuses, allocation and impact reports, or directors' actions. It also falls short of providing any remedies for investors damaged by incorrect or untruthful external review statements.

Lastly, EU law fails to provide sufficient incentives for market actors to consider sustainability factors when building securitisation portfolios or to issue 'sustainable' instead of traditional securities. A starting point could be amendments to ECB's collateral frameworks, making only 'sustainable' portfolio assets or EUGB-compliant ABSs eligible as collateral.

The shortcomings of the current law for 'sustainable' securitisations call for additional ad hoc provisions, going beyond the assimilation between securitisation and green bonds. Further, more targeted reforms have the potential to yield a twofold effect, which would also resonate with the EU's agenda and with the goals of the Capital Markets Union. First, they are likely to attract more investments to the EU market, mobilising further capital towards sustainability goals. Second, they are likely to incentivise the creation of further 'sustainable' assets.[422] Therefore, they have the potential to create the EU's 'global gold standard' for sustainable securitisations, inspiring other regulators to engage in similar initiatives.

References

Addamo S, 'Il dovere di diligenza ai fini della sostenibilità nel settore bancario' (2024) 3 *Rivista di Diritto Bancario* 563.

AFME, 'European Structured Finance – Securitisation Report, Q4 2023 & 2023 FY' (2023) https://www.afme.eu/Portals/0/DispatchFeaturedImages/Sec

[422] See Chap. 4, Sect. 4.4.

uritisation%20Data%20Report%20Q4%202023%20and%202023%20Full%20Year.pdf
AFME, 'European Sustainable Finance – ESG Finance Report – Q2 2021' (2021) https://www.afme.eu/publications/data-research/details/esg-finance-q2-2021---european-sustainable-finance
AFME, 'European Sustainable Finance – ESG finance report – Q3 2025' (2025) https://www.afme.eu/media/mhgayv3c/esg-finance-report-3q25.pdf
Agostini F, 'From'Green Bond Principles' to 'Green Bond Clauses': Mitigating Greenwashing Through Contract Law' in Andenas M, Heidemann M (eds), *Quo Vadis Commercial Contract?* (Springer2023) LCF Studies in Commercial and Financial Law, vol 1 https://doi.org/10.1007/978-3-031-14105-8_6
Agostini F, 'Gold' standards for 'Greening' Financial Flows? The 'Brussels Effect' Of The Eu Sustainable Finance Agenda' In E.Baroncini, P. Manzini, & G. Tellarini (Eds.), *Sustainable Development as Fundamental Pillar of Economic Governance and Public Affairs: The EU Approach and International and Domestic Perspectives* (Bologna University Press 2025)
Agostini F, Joosen B, 'The future of green securitisations in the European Union' (2025) 2 *Kapitalmaarktenrecht*,https://portal.denhollander.info/articles/41610
Alessi L, Battiston S, Melo AS et al, 'The EU Sustainability Taxonomy: a Financial Impact Assessment' (2019) European Commission, Joint Research centre, https://publications.jrc.ec.europa.eu/repository/bitstream/JRC118663/2020.01.10_technical_report_commission_taxonomy_published.pdf
Alessi L, 'Concluding remarks' (5th Summer School on Sustainable Finance, Joint Research Centre, 14 July 2023) https://webcast.ec.europa.eu/5th-jrc-summer-school-on-sustainable-finance-14th-of-july
Amihud Y, Garbade K, Kahan M, 'A New Governance Structure for Corporate Bonds' (1999) 51 *Stanford Law Review* 3 447.
André L, Grept A, Laut Net al, 'Climate Risk Measurement of Assets Eligible as Collateral for Refinancing Operations – Focus on Asset Backed Securities (ABS)' (2022) Banque de france Working Paper 3. https://publications.banque-france.fr/sites/default/files/medias/documents/wp-858.pdf
Anker-Sørensen L, Zetzsche DA, 'Regulating Sustainable Finance in the Dark' (2022) 23 *European Business Organisation Law Review* 47.
Armour J, Awrey D, Davies P and others, *Principles of Financial Regulation* (OUP2016)
Asset Management and Investor Council (AMIC) ICMA, 'Discussion Paper on ESG KPIs for Auto-Loans/Leases ABS' (2021) https://www.icmagroup.org/assets/documents/Regulatory/AMIC/AMIC-discussion-paper-ESG-auto-loan-ABS-FINAL-170521.pdf
Baasch Andersen C, 'Applied Uniformity of a Uniform Commercial Law: Ensuring Functional Harmonisation of Uniform Texts through a Global Jurisconsultorium of the CISG' in MadsAndenas and Camilla BaaschAndersen, *Theory and Practice of Harmonisation* (Edward Elgar2012)

Badenhoop N, 'Green Bonds – An Assessment of the Proposed EU Green Bond Standard and Its Potential to Prevent Greenwashing' Study requested by the ECON Committee (2022) https://www.europarl.europa.eu/thinktank/en/document/IPOL_STU(2022)703359

Bavoso V, 'Good Securitisation, Bad Securitisation and the Quest for Sustainable EU Capital Markets' (2015) 4 *Butterworth Journal of International Banking and Financial Law* 221.

Bavoso V, 'Filling the Accountability Gap in Structured Finance Transactions: The Case for a Broader Fiduciary Obligation' [2014] 23 *Columbia Journal of European Law* 2 370, 384.

BCBS, Board of IOSCO, 'Criteria for Identifying Simple, Transparent and Comparable Securitisations' (2015) https://www.iosco.org/library/pubdocs/pdf/IOSCOPD494.pdf

BCBS, Basel III framework (2023) https://www.bis.org/basel_framework/chapter/CRE/40.htm?inforce=20230101&published=20201126

Binder JH, 'Prudential Requirements Framework and Sustainability' in Alexander K, Gargantini M, Siri M (eds.), *The Cambridge Handbook of EU Sustainable Finance: Regulation, Supervision and Governance* (Cambridge Law Handbooks, Cambridge University Press, 2025).

Bindseil Uet al, 'The Eurosystem Collateral Framework Explained' (2017) ECB Occasional Paper Series https://www.ecb.europa.eu/pub/pdf/scpops/ecb.op189.en.pdf

British Academy, 'Principles for purposeful business' (2019) https://www.thebritishacademy.ac.uk/documents/223/future-of-the-corporation-principles-purposeful-business-executive-summary.pdf

Broadway L, 'Trustee Duties versus Equity's Moral Conscience: Trust Law's Inadequate Facilitation of Socially Responsible Investment' (2021) 27 *Trusts & Trustees* 8 809.

Busch D and others, *Prospectus Regulation and Prospectus Liability* (Oxford University Press2020)

ChristopherBrummer, *Soft Law and the Global Financial System: Rule Making in the twenty-first century* (2nd ed., Cambridge University Press, 2020).

ClientEarth, 'The ECB's Monetary Policy Strategy Review: Corporate Sector Asset Purchases, Climate Change and the ECB's Legal Obligations' (2021) https://www.clientearth.org/latest/documents/letter-from-clientearth-to-christine-lagarde-president-of-the-european-central-bank/

Cœuré B, 'Embarking on public sector asset purchases' (10 March 2015) International Conference on Sovereign Bond Markets, https://www.ecb.europa.eu/press/key/date/2015/html/sp150310_1.en.html

Coffee JC, 'The Future as History: The Prospects for Global Convergence in Corporate Governance and Its Implications' (1999) 93 *Northwestern University Law Review* 641;

Committee on economic and monetary Affairs, 'Report on Further Development of the Capital Markets Union (CMU): Improving Access to Capital Market

Finance, in Particular by SMEs, and Further Enabling Retail Investor Participation' (2020) 2020/2036(INI))
Craig P, De Búrca G, *EU Law: Text, Cases and Materials* (7th edn, Oxford University Press, 2020)
Dafermos Y, Gabor D, Nikolaidi Met al, 'Greening the Eurosystem Collateral Framework: How to Decarbonise the ECB's Monetary Policy' (2021) New Economics Foundation, SOAS University of London, Greenpeace et al, https://neweconomics.org/uploads/files/Collateral-Framework.pdf
EBA, 'Final report – on Guidelines on the STS criteria for non-ABCP securitisation' (2018) https://www.bde.es/f/webbde/INF/MenuHorizontal/Normativa/guias/EBA-GL-2018-09-EN.pdf
EBA, 'Developing a framework for Sustainable Securitisations' (2022) REP/2022/06 https://www.eba.europa.eu/sites/default/files/document_library/Publications/Reports/2022/1027593/EBA%20report%20on%20sustainable%20securitisation.pdf
EBA, 'Report in response to the call of advice from the European Commission on green loans and mortgages' (2023) EBA/REP/2023/38 https://www.eba.europa.eu/sites/default/files/2023-12/e7bcc22e-7fc2-4ca9-b50d-b6e922f99513/EBA%20report%20on%20green%20loans%20and%20mortgages_0.pdf
EBA, 'The role of Environmental Risks in the Prudential Framework' (2022) EBA Discussion Paper 2 https://www.eba.europa.eu/activities/single-rulebook/regulatory-activities/sustainable-finance/discussion-paper-role-environmental-risks-prudential-framework
EBA, ESMA, EIOPA, 'Joint Consultation Paper – STS securitisations-related sustainability disclosures' (2022) https://www.eba.europa.eu/activities/single-rulebook/regulatory-activities/securitisation-and-covered-bonds/joint-regulatory-technical-standards-sts-securitisations-related-sustainability-disclosures
EBA, ESMA, European Insurance and Occupational Pensions Authority (EIOPA), ECB, 'Joint ESAs-ECB Statement on disclosure on climate change for structured finance products' (2023) https://www.esma.europa.eu/sites/default/files/library/ESAs_ECB%20Joint%20Statement%20on%20disclosures%20for%20securitisations_FINAL_6%20March%202023_0.pdf
EBA, ESMA, EIOPA, ECB, 'Joint ESAs-ECB Statement on disclosure on climate change for structured finance products' (2023) https://www.esma.europa.eu/sites/default/files/library/ESAs_ECB%20Joint%20Statement%20on%20disclosures%20for%20securitisations_FINAL_6%20March%202023_0.pdf
Economist Intelligence Units, 'ECB collateral policy threatens EU's path to net-zero' (2024) https://www.eiu.com/n/ecb-collateral-policy-threatens-eus-path-to-net-zero/
ECB, 'Annex I – Detailed roadmap of climate change-related actions' (2021) https://www.ecb.europa.eu/press/pr/date/2021/html/ecb.pr210708_1_annex~f84ab35968.en.pdf

ECB, 'Climate and nature action plan 2024-2025' (2024) https://www.ecb.europa.eu/ecb/climate/our-climate-and-nature-plan/shared/pdf/ecb.climate_nature_plan_2024-2025.en.pdf

ECB, 'Combined monetary policy decisions and Statement' (2 February 2023); https://www.ecb.europa.eu/press/pressconf/shared/pdf/ecb.ds230202~40a069a7b4.en.pdf

ECB, 'Data templates' (2023) https://www.ecb.europa.eu/paym/coll/loanlevel/transmission/html/index.en.html

ECB, 'FAQ on incorporating climate change considerations into corporate bond purchases' (2025) https://www.ecb.europa.eu/mopo/implement/app/html/ecb.faq_cspp_climate_change.en.html

ECB, 'ECB introduces ABS loan-by-loan information requirements in the Eurosystem collateral framework' (2010) https://www.ecb.europa.eu/press/pr/date/2010/html/pr101216.en.html

ECB, 'ECB provides details on how it aims to decarbonise its corporate bond holdings'https://www.ecb.europa.eu/press/pr/date/2022/html/ecb.pr220919~fae53c59bd.en.html

ECB, 'ECB response to the ESAs' joint consultation on sustainability-related disclosures for STS securitisations' (2022) https://www.ecb.europa.eu/pub/pdf/other/ecb.response_ESAs_consultation_sustainability_related_disclosures_STS_securitisations~ee2f3fb909.cs.pdf?0c0d2818008690c368bdfa8593b9afa8

ECB, 'ECB reviews its risk control framework for credit operations' (2022) https://www.ecb.europa.eu/press/pr/date/2022/html/ecb.pr221220_1~ca6ca2cc09.mt.html

ECB, Eurosystem, 'Eurosystem Contribution to the European Commission's Targeted Consultation on the Functioning of the EU Securitisation Framework' (2020) https://www.ecb.europa.eu/pub/pdf/other/ecb.eurosystemcontributiontargetedconsultationEUsecuritisationframework2109~cade37d248.en.pdf

ESMA, 'Answers to questions on the interpretation of Regulation (EU) 2019/2088, submitted by the European Supervisory Authorities on 9 September 2022' (2023) https://www.esma.europa.eu/sites/default/files/2023-04/Answers_to_questions_on_the_interpretation_of_Regulation_%28EU%29_20192088.PDF

ESMA, 'Consultation Paper on the revision of the disclosure framework for private securitisation under Article 7 of the Securitisation Regulation' (2025) https://www.esma.europa.eu/sites/default/files/2025-02/ESMA12-2121844265-4462_-_Consultation_Paper_on_Private_Securitisation.pdf

ESMA, 'Consultation Paper On the securitisation disclosure templates under Article 7 of the Securitisation Regulation' (2023) https://www.esma.europa.eu/sites/default/files/2023-12/ESMA12-2121844265-3053_-_Consultation_Paper_on_the_Securitisation_Disclosure_Templates.pdf

ESMA, 'Consultation Paper – Technical Standards on the European Green Bond Regulation' (2025) ESMA84-858027815-176 https://www.esma.europa.eu/sites/default/files/2025-04/ESMA84-858037815-176_Consultation_Paper_on_European_Green_Bond_Regulation.pdf

ESMA, 'Final Report on draft Regulatory Technical Standards With regard to the content, methodologies and presentation of disclosures in respect of the sustainability indicators in relation to adverse impacts of the assets financed by the underlying exposures for STS securitisations on the climate and other environmental, social, and governance-related adverse impacts pursuant to Article 22(6) and 26d(6) of Regulation (EU) 2017/2402' (2023) https://www.esma.europa.eu/sites/default/files/2023-05/JC_2023_13_-_Final_report_on_ESG_disclosure_for_STS_securitisations.pdf

ESMA, 'Final Report –Technical Standards on the European Green Bonds Regulation' (2025) ESMA84-858037815-195 https://www.esma.europa.eu/sites/default/files/2025-02/ESMA84-858037815-195_Final_Report_on_Technical_Standards_under_EuGB.pdf

ESMA, 'Final Report –Technical Standards on the European Green Bonds Regulation' (2025) ESMA84-858037815-684 https://www.esma.europa.eu/sites/default/files/2025-10/ESMA84-858037815-684_Final_Report_on_Technical_Standards_under_EuGB.pdf

ESMA, 'Final Report –Technical advice concerning the Prospectus Regulation and the RTS updating the CDR on metadata' (2025) ESMA32-117195963-1417, https://www.esma.europa.eu/sites/default/files/2025-06/ESMA32-117195963-1417_Final_Report_Listing_Act_Advice_on_Prospectus.pdf

ESMA, 'Sustainability Disclosure in Prospectuses' (2023) https://www.esma.europa.eu/sites/default/files/2023-07/ESMA32-1399193447-441_Statement_on_sustainability_disclosure_in_prospectuses.pdf

ESMA, 'TRV Risk Analysis – The EU securitisation market – an overview' (2023) https://www.esma.europa.eu/sites/default/files/2023-09/ESMA50-524821-2908_TRV_risk_analysis_-_EU_securitisation_markets_overview.pdf

EU Commission, 'Commission Draft Delegated Regulation supplementing Regulation (EU) 2023/2631 of the European Parliament and of the Council by establishing the content, methodologies, and presentation of the information to be voluntarily disclosed by issuers of bonds marketed as environmentally sustainable or of sustainability-linked bonds in the templates for periodic post-issuance disclosures' (2024); Annex to the Commission Delegated Regulation (EU) supplementing Regulation (EU) 2023/2631 of the European Parliament and of the Council by establishing the content, methodologies, and presentation of the information to be voluntarily disclosed by issuers of bonds marketed as environmentally sustainable or of sustainability-linked bonds in the templates for periodic post-issuance disclosures' (2025) https://ec.europa.eu/info/law/better-regulation/have-your-say/initiati

ves/14196-Green-bonds-and-sustainability-linked-bonds-templates-for-voluntary-post-issuance-disclosures_en

EU Commission, 'Communication From The Commission To The European Parliament, The Council, The European Economic And Social Committee And The Committee Of The Regions A Capital Markets Union for people and businesses-new action plan' (2020) 590

EU Commission, 'Communication From The Commission To The European Parliament, The Council, The European Economic And Social Committee And The Committee Of The Regions, 'Action plan: financing Sustainable Growth' (2018) 097

EU Commission, 'Communication From The Commission To The European And The European Economic And Social Committee and the Committee of the Regions, Parliament, The Council, 'Strategy for Financing the Transition to a sustainable economy' (2021) 390.

EU Commission, 'Communication From The Commission To The European And The European Economic And Social Committee and the Committee of the Regions, Parliament, The Council, 'The European Green Deal 2019' (2019) 640;

EU Commission, 'Consultation Document –Targeted Consultation on the Functioning of the EU Securitisation Framework' (2021) https://finance.ec.europa.eu/regulation-and-supervision/consultations/2021-eu-securitisation-framework_en

EU Commission, 'Factual summary report of the targeted consultation on the functioning of the EU securitisation framework' (2025) https://finance.ec.europa.eu/document/download/35846dc8-8ebc-4bc9-967d-de77ef5d9234_en?filename=2024-eu-securitisation-framework-summary-of-responses_en.pdf

EU Commission, 'Report from the Commission to the European Parliament and the Council on the functioning of the Securitisation Regulation' (2022) 517 https://eur-lex.europa.eu/legal-content/EN/TXT/PDF/?uri=CELEX:52022DC0517

EU Commission, 'Summary report –Targeted consultation on the functioning of the EU securitisation framework' (2021) https://finance.ec.europa.eu/system/files/2022-09/2021-eu-securitisation-framework-summary-of-responses_en.pdf

De Gioia Carabellese P, Della Giustina C, *The law of securitisations: from crisis to techno-sustainability* (Routledge2023)

Garcia Cerrato E, Agostini F'The Green Bonds Market in the Light of European Commission's Proposal: Implications for Greenwashing Liability' in SmoleńskaA, Ramos MuñozD, *Greening the Bond Market: A European Perspective* (Springer2023).

Gerner-Beuerle C, Paech P, EP Schuster, 'Study on Directors' Duties and Liability in the EU' (2013) LSE Enterprise, prepared for the EU Commission DG Markt.

Gorton GB, Metrick A, 'Securitised Banking and the Run on Repo' (2012) 104 *Journal of Financial Economics* 3 425

Gortsos C, *The Evolution of European (EU) Banking Law under the Influence of (Public) International Banking Law: A Comprehensive Overview* (4th edn 2022) https://ssrn.com/abstract=3334493

Dougan M, 'Minimum Harmonisation and the Internal Market' (2009) 46 *Common Market Law Review* 3 854.

FSB, 'Key Standards for Sound Financial Systems' (2001) https://www.fsb.org/work-of-the-fsb/about-the-compendium-of-standards/key_standards/

Gosling T, MacNeil I, 'Can Investors Save the Planet? –NZAMI and Fiduciary Duty' (2023) 18 *Capital Markets Law Journal* 2 172.

Gordon JN, 'Systematic Stewardship' (2022) 47 *Journal of Corporation Law* 627.

Hauet O, 'Reviving securitisation in the EU: A critical analysis of the reporting requirements' (2025) 7 *Journal of Financial Compliance*2, https://www.true-sale-international.de/fileadmin/tsi-gmbh/tsi_downloads/aktuelles/ECB_Olivia_Hauet_A_critical_analysis_of_the_securitisation_reporting_2023-12.pdf

Hill C, 'Securitisation: a low-cost sweetener' for Lemons' (1997) 10 *Journal of Applied Corporate Finance* 64.

IFRS, 'Who we are' (2023) https://www.ifrs.org/about-us/who-we-are/

ISSB, 'Frequently Asked questions' (2025) https://www.ifrs.org/groups/international-sustainability-standards-board/issb-frequently-asked-questions/

Jackson HE, 'Centralisation, Competition, and Privatisation in Financial Regulation' (2001) 2 *Theoretical Inquiries in Law* 4.

Jordana J, 'Transgovernmental Networks as Regulatory Intermediaries: Horizontal Collaboration and the Realities of Soft Power' in *Regulatory Intermediaries in the Age of Governance* (2017) 670 *The Annals of the American Academy of Political and Social Science* 245.

Kershaw D, Schuster EP, 'The Purposive Transformation of Company Law' (2021) 69 *The American Journal of Comparative Law* 3 478.

Lastra RM, Bodellini M, 'Background paper on soft law and sovereign debt' (2018) https://unctad.org/system/files/non-official-document/SOFT%20LAW%20AND%20SOVEREIGN%20DEBT.pdf

Lehmann A, 'The EU green bond standard: sensible implementation could define a new asset class' (*Brugel*, 13 July 2021) https://www.bruegel.org/blog-post/eu-green-bond-standard-sensible-implementation-could-define-new-asset-class

Lenschow A, *Environmental Policy Integration: Greening Sectoral Policies In Europe* (Earthscan 2002),

Litwin D, Savourey E, 'Human Rights in EU Sustainable Finance' (2024) 11 *Journal of Financial Regulation* 1 41

Lohse EJ, 'The meaning of harmonisation in the context of European union law – a process in need of definition' in Andenas M, Baasch Andersen C, *Theory and Practice of Harmonisation* (Edward Elgar2012)

Kahan M, Kamar E, 'The Myth of State Competition in Corporate Law' (2002) 55 *Stanford Law Review* 3
Klamert M, 'What We Talk about When We Talk about Harmonisation' (2015) 17 *Cambridge Yearbook of European Legal Studies*360.
Law Commission, 'Pension Funds and Social Investment' (2017) Law Com 374, https://assets.publishing.service.gov.uk/media/5a750d09ed915d5c5446524c/6.3496_LC_DRAFT_190617_TT_20_06_17__WEB_-_final_.pdf
MacNeil I, 'Activism and Collaboration among Shareholders in UK Listed Companies' (2010) 5 *Capital Markets Law Journal*419;
MacNeil I, Esser IM, 'Corporate Purpose as a Conduit for Sustainability in Corporate Governance' in Silva MS, Nicolussi A et al (eds) *Private Law and Sustainability* (Routledge 2024)
Moloney N, 'Capital Markets Union: 'ever closer union' for the EU financial system' (2016) 41 *European Law Review* 3 307.
Nelson D, Pierpont B, 'The Challenge of Institutional Investment in Renewable Energy' (2013) Climate Policy Initiative16https://www.climatepolicyinitiative.org/wp-content/uploads/2013/03/The-Challenge-of-Institutional-Investment-in-Renewable-Energy.pdf
NGFS, 'Adapting Central Bank Operations to a Hotter World Reviewing Some Options' (2021) NGFS Technical document https://www.ngfs.net/en/adapting-central-bank-operations-hotter-world-reviewing-some-options
Nowag J, 'The sky is the limit' in Sjåfjell B, Wiesbrock A (eds) *The Greening Of European Business Under EU Law: Taking Art 11 TFEU Seriously* (1st ed, Routledge2015).
Nyborg KG, 'Central Bank Collateral Frameworks' (2017) 76 *Journal of Banking and Finance* 198;
OECD, 'Due Diligence for Responsible Corporate Lending and Securities Underwriting: Key Considerations for Banks Implementing the OECD Guidelines for Multinational Enterprises' (2019) https://www.oecd.org/en/publications/due-diligence-for-responsible-corporate-lending-and-securities-underwriting_71fd2895-en.html
Oustry A, Bunyamin E, Svartzman R et al, 'Climate-Related Risks and Central Banks' Collateral Policy: A Methodological Experiment' (2021) Banque de france Working Paper, https://publications.banque-france.fr/sites/default/files/medias/documents/wp-790_0.pdf
Papadogiannis Varouchakis T, 'The European Securitisation Market: Effects of an Uneven Regulatory Playing Field', (2024) 10 *Journal of Financial Regulation*2.
Mariana Pargendler, 'The Rise of International Corporate Law' (2021) 98 *Washington University Law Review*1765.
Park SK, 'Green Bonds and beyond: Debt Financing as a Sustainability Driver', in Sjåfjell B, Bruner CM (eds), *Cambridge Handbook of Corporate Law, Corporate Governance and Sustainability* (Cambridge University Press2020).

Pistor K, 'A Legal Theory of Finance' (2013) 41 *Journal of Comparative Economics* 315.

Platform on sustainable finance, 'Report on social taxonomy' (2022) https://finance.ec.europa.eu/system/files/2022-08/220228-sustainable-finance-platform-finance-report-social-taxonomy_en.pdf

Ramos Muñoz D, 'In Praise of Small Things: Securitisation and Governance Structure' (2010) 5 *Capital Markets Law Journal* 363

Ringe WG, 'Regulatory Competition in Global Financial Markets – The Case for a Special Resolution Regime' (2016) 1 *Annals of Corporate Governance* 175.

Romano R, 'The Need for Competition in International Securities Regulation' (2009) 2 *Theoretical Inquiries in Law 2* 389;

Serdaris K, 'The Regulatory Framework of ESG Disclosures in EU Prospectuses' in Larrea K, Agostini F and Serdaris K (eds), *Critical perspectives on EU corporate sustainability law* (forthcoming, Palgrave MacMillan, 2026).

Schwarcz SL, 'Securitisation and Post-Crisis Financial Regulation' (2017) 102 *Cornell Law Review 115*,

Scopelliti A, 'Securitisation, Bank Capital and Financial Regulation: Evidence from European Banks', (2016) *ECB Forum*https://www.ecb.europa.eu/pub/conferences/ecbforum/shared/pdf/2016/scopelliti_paper.pdf

Solana J, 'The Roles of Law in Finance' (2022) on file with the author.

Solana J, 'Market Efficiency as a Directive Principle of EU Monetary Policy' in Anna Chadwick, Eleonora Lozano-Rodríguez, Andrés Palacios-Lleras, Javier Solana (eds.), *Markets, Constitutions, and Inequality* (Routledge2022).

Solana J, 'The Power of the Eurosystem to Promote Environmental Protection' (2019) 30 *European Business Law Review* 4 547.

Solana J, Goldoni M, 'The legal nature of market neutrality in the euro area's monetary policy' (2024) 3 *European Law Open* 1 7.

Stubbington T, Squeeze on 'greenium' as ESG bond investors demand more value (FT 11th October 2021)https://www.ft.com/content/ecbed322-1709-4ed6-9f7f-d974f6e181da

Tanega J, *Securitisation Law & Practice: EU & US Disclosure Regulations* (3rd edn, LexisNexis2017)

Technical Expert Group on Sustainable Finance (TEG), 'Usability Guide EU Green Bond Standard' (2020) https://finance.ec.europa.eu/system/files/2020-06/200309-sustainable-finance-teg-green-bond-standard-usability-guide_en.pdf

TCFD, 'Final recommendations' (2017) https://assets.bbhub.io/company/sites/60/2021/10/FINAL-2017-TCFD-Report.pdf

TietjeC in GrabitzEet al (eds), *Das Recht der Europäischen Union: EUV/AEUV* (56th ed, Beck2015)

Urgewald, Reclaim Finance, 'Collateral damage: ending the ECB's support to fossil fuel companies' (2024) https://reclaimfinance.org/site/wp-content/uplo

ads/2024/04/BRIEFING-2024.04.04_Reclaim-Finance_Collateral-damage-FINAL.pdf accessed 10

Urgewald, Reclaim finance, 'Still backing fossil fuels – the costs the European Central Bank's inaction' (2025) https://reclaimfinance.org/site/wp-content/uploads/2025/04/ECB_Collateral_April2025_En.pdf

UNPRI, 'A legal framework for Impact: sustainability impact in investor decision-making' (2021) https://public.unpri.org/policy/a-legal-framework-for-impact

UNPRI, 'ESG Engagement for fixed-income investors: managing risks, enhancing returns' 9 https://www.unpri.org/download?ac=4449 accessed 10 December 2025.

UNPRI, 'ESG Incorporation in Securitised Products: The challenges ahead' (2021) https://gsfo.org/resource/esg-incorporation-securitised-products-challenges-ahead

UNPRI, 'What Are the Principles for Responsible Investment?' (2023) https://www.unpri.org/about-us/what-are-the-principles-for-responsible-investment

Van 't Klooster J, Fontan C, 'The Myth of Market Neutrality: A Comparative Study of the European Central Bank's and the Swiss National Bank's Corporate Security Purchases' (2020) 25 *New Political Economy* 865.

Wood P, *Project finance, securitisations and subordinated debt* (3rd edn, Sweet & Maxwell2019)

Zilioli C, Ioannidis M, 'Climate change and the mandate of the ECB: Potential and limits of monetary contribution to European green policies' (2022) 59 *Common Market Law Review* 2 363.

CHAPTER 7

Concluding Remarks

7.1 The Case for Sustainable Securitisations at a Critical Historical Juncture

David Barmes from the NGO Positive Money wrote in 2019, 'securitisation is back, and green finance must stray far away'. He went on to describe the risks of using securitisations to promote sustainability goals, since they would foster systemic risks, allow market actors to bypass capital requirements and reinforce the funding of 'brown' assets rather than transition. He concluded that there would be '*no role for securitisation in green finance*'.[1]

The analysis in the book aimed to move away from the long-standing assumption, also widely propagated in academic scholarship and policy discourse, that securitisation is inherently risky and problematic. Instead, its purpose was to highlight the overlooked added value of securitisation in promoting environmental, social, and governance objectives. It also aimed to demonstrate that 'sustainable securitisations' are legally constructed; thus, the growth and development of the sector, as well as the management of the risks, heavily depend on the underlying legal framework.

[1] David Barmes, *Securitisation is back, and green finance must stray far away* (Positive Money, September 2019) https://positivemoney.org/2019/09/securitisation-is-back-and-green-finance-must-stray-far-away/ accessed 10 December 2025.

F. Agostini, *The Law of Sustainable Securitisations*, EBI Studies in Banking and Capital Markets Law,
https://doi.org/10.1007/978-3-032-25046-9_7

The book's inquiry is also situated at a critical historical juncture. On the one hand, EU institutions and regulators globally have drawn increasing attention to the potentials of securitisation, increasingly perceived as an indispensable tool for revitalising capital markets.

On the other hand, policy ambitions regarding sustainability goals, and the broader agenda to promote sustainable finance, appear to be weakening. This is evidenced, among other things, by the EU policy debate on the so-called 'Omnibus packages',[2] which aim to simplify core statutory instruments and alleviate regulatory burdens for companies. Within this ever-evolving context, one may wonder what added value securitisations may provide, if any.

The following sections will summarise the main contributions of the previous chapters and the normative considerations. They will identify potential directions for future research and reflect on the sector's future.

7.2 The Book's Main Contributions

Chap. 2 has laid out the theoretical framework. By situating the analysis within the scholarly insights into the inextricable link between law and finance, it has shown how law can operate to advance environmental, social, and governance (ESG) goals in securitisations.

One of its central contributions has been the identification of four 'legal strategies' which appear particularly promising for promoting sustainability-related practices in securitisations, while being common to other areas of company and financial law. Law can require the dissemination of sustainability-related information ('disclosure') or the external review of sustainability-related aspects ('gatekeeping') of these transactions. It can insulate the funds raised from investors and from the portfolio assets in order to ensure their exclusive allocation for sustainability purposes ('shielding'). It can provide investors in securitised products with rights and procedures to engage with originators and take action on sustainability matters ('voice').

[2] See EU Commission, 'Omnibus Package' (1 April 2025) https://finance.ec.europa.eu/news/omnibus-package-2025-04-01_en accessed 10 December 2025; Apostholos Tomadakis, 'The EU is walking a fine line between simplification and deregulation' (2025) CEPS, https://www.ceps.eu/the-eu-is-walking-the-fine-line-between-simplification-and-deregulation/ accessed 10 December 2025.

Chap. 3 has contributed to the scholarly literature on the function and legal architecture of securitisation by identifying five promising aspects of securitisation law which could serve as a starting point for sustainability-oriented legal initiatives ('focal points'). It has shown that these areas should become the primary target of any legal reform since they could most effectively prompt sustainability-related practices in securitisations. These include the composition of the portfolio, the terms of issue of ABSs, the governance of SPVs, the mechanisms for the representation of investors and the processes to monitor the allocation of the funds.

The aim of the analysis was also to show that analysing securitisation through the lens of sustainability goals requires a shift in the narrative. The predominant academic scholarship depicts securitisation as a risk-management, capital-releasing and liquidity-providing technique revolving around the 'refinancing' of *existing* assets assigned to SPVs. Instead, the present work sought to highlight how securitisation could ultimately contribute to *further* goals thanks to the final allocation of its funds, i.e. the advances on the underlying loans and the proceeds from investors.

Chap. 4 aimed to map all the circumstances where market actors have qualified securitisations as 'sustainable' so far in order to conceptualise the notion. Building on previous findings,[3] it has provided a typology of 'sustainable securitisations', differentiating transactions where the underlying credit claims finance projects positively contributing to ESG goals ('sustainable portfolio securitisations') or where issuers seek to advance such goals through the financial flows raised from investors ('sustainable proceeds securitisations').[4] At the same time, it added further complexity to the demarcation, by identifying transactions where the 'sustainable' label derives from the types of *activities* backing the operation, or developed through the issuance ('green' or 'social' securitisations), or from the *targets* which issuers intend to achieve ('sustainability-linked' securitisations).

[3] Christy Petit, Pierre Schlosser, 'Rationale, Potential and Pitfalls of Green Securitisation' (2020) RSCAS Working Paper 35 https://cadmus.eui.eu/bitstream/handle/1814/67018/RSCAS%202020_35.pdf?sequence=1; Kathryn James, Isabel Parker, 'New Growth in the Green Securitisation Market' (2019) Simmons & Simmons, 1. www.simmons-simmons.com/-/media/files/articles/2019/capital%20markets/new%20growth%20in%20the%20green%20securitisation%20market%20january%202019.pdf accessed 5 December 2025.

[4] Such typology also builds, but also expands on, conclusions previously reached in Federica Agostini, 'One Step Forward, One Step Back': 'Sustainable Securitisations' Before and After the EU Green Bond Standard' in Mario La Torre and Sabrina Leo (eds.), *Contemporary Issues in Sustainable Finance* (2025) Palgrave Studies in Impact Finance, 221–227.

It has also shown how sustainability considerations have permeated other types of transactions, like ‘synthetic’ securitisations and collateralised loan obligations (CLOs).

The other aim of Chap. 4 was to show that the current legal framework surrounding sustainability-related elements in securitisations represents an obstacle to the development of this market. In principle, following *Pistor*’s insights into the rule-making powers of market actors[5], securitisation parties have the capacity to independently articulate the sustainability goals of their own transactions. They can do so by incorporating binding clauses and introducing enforcement mechanisms within their transactional documentation (what I have described as ‘private hard law’). However, the comparison of several examples of documentation from previous deals marketed as ‘sustainable’ has demonstrated that this is largely not the case. Instead, the analysis has revealed a general lack of clauses clearly articulating how and to what extent the assets within securitisation portfolios contributed to environmental and social goals. Market actors have also shown a tendency not to incorporate robust sustainability-related commitments in the documentation for the issuance of ABSs, like prospectuses, due to vaguely worded provisions and disclaimers. Moreover, they tend to provide limited to no information to prospective investors about how SPVs are run or about the processes to monitor the allocation of the funds. Such practices are not only problematic from the point of view of the lack of information available to prospective investors. They also result in a contraction of the remedies available to investors (‘voice’) if issuers fail to attain the goals announced before the issuance.

Overall, the current *status quo* requires prospective issuers to incur high transaction costs, disincentivising the inclusion of robust sustainability-related clauses. It also imposes high ‘due diligence’ costs while posing legal uncertainty and ‘greenwashing’ risks. Facing the inherent limitations of the powers of market actors, I argue that refining the legal regime governing securitisations is crucial. The analysis has identified two possible solutions to coordinate the law, i.e. standardisation and harmonisation, which rest on the rule-making powers of financial industry associations and EU institutions, respectively.

[5] See, among others, Katharina Pistor, ‘A Legal Theory of Finance’ (2013) 41 Journal of Comparative Economics 315. see further Chap. 3, Sect. 3.1.

Chap. 5 has delved into the prospects for standardisation of 'sustainable' securitisations at market level ('private soft law'). The analysis contributed to the predominant legal and law and economics scholarship, which assumes that state-designed regulation represents the only solution to the 'failures of the market'. Instead, it pointed to the role of industry associations, like the LMA and ICMA, which act as 'intermediaries' between market actors and state authorities. While overlooked in the academic discourse on sustainable finance, they have a crucial role in the design and enforcement of the law for 'sustainable' financial transactions. In particular, they could introduce guidelines, metrics, and standard provisions which market actors could directly apply throughout their securitisation documentation. These have the potential to cut costs on investors' and originators' side as well as to foster the circulation of consistent market practices throughout the securitisation market. Chap. 4 also examined the previous contribution of financial industry associations towards the standardisation of sustainability-related elements of securitisations. For instance, they have formulated a number of guidelines and a few model provisions for 'sustainable' financial instruments, like ICMA's Green Bond principles and LMA's Green Loan Principles. Their high-level nature, however, often gives limited guidance on how market actors should draft the clauses of the documentation in order to create legally binding obligations on sustainability-related matters. In the few cases where such associations have indeed formulated model provisions, like LMA's model clauses for sustainability-linked loans, they were designed with other financial instruments in mind. Therefore, they inevitably failed to take into account several strategic aspects of securitisations identified in Chap. 2, like the governance of SPVs or monitoring processes over the cash flows.

The chapter also highlighted the boundaries of the standardisation tools introduced by financial industry associations. While market actors are likely to replicate them widely across their securitisation documentation, they may also inevitably 'cherry-pick' the clauses that might appear as the least restrictive and leave out the others. Financial industry associations also have no robust enforcement procedures in place to ensure the uniform application of their legal tools. Lastly, there is a concern that they may prioritise the interests of the members of the market segment they represent, rather than the general interests of the public. This may also result in suboptimal legal tools, which may not meet the ambitions of the current sustainability agenda and exacerbate current 'greenwashing' risks. Therefore, 'soft law' necessarily needs to operate in tandem with state-designed initiatives, introducing minimum requirements to qualify securitisations as 'sustainable' as well as minimum enforcement mechanisms.

Chap. 6 contributed to the vivid academic scholarship and policy debate around optimal regulatory initiatives to mobilise financial flows towards sustainability goals. By taking the unexplored angle of securitisation, it shed light on the gaps of the previous and ongoing agenda, especially at EU level. Yet, it highlighted how the EU is in a unique position to regulate this subject matter compared to other transnational bodies. Various ongoing EU initiatives, including the Taxonomy Regulation and EU Green Bond Standard (EUGBS), have indirectly affected, or are expected to indirectly affect market actors willing to issue securitised products in the EU and qualify them as 'green'.

At the same time, the chapter has concluded that their main driver and rationale has often been to regulate other segments of the sustainable finance market, like green bonds or corporate reporting. Such a different focus is often not fit for the purpose of regulating all the complexities of securitisations. For instance, the currently discussed Proposal is expected to provide for the same legal treatment for green bonds and 'securitisation bonds'. Such assimilation is, however, problematic since it overlooks some of the strategic 'focal points' of securitisation law, like the role of SPVs.

The chapter also identified a strong focus on disclosure, which resonates with a recurrent trend in financial regulation and in sustainable finance policies. Simultaneously, mechanisms allowing prospective and current investors to raise their 'voice' during the negotiation of the terms of the transaction, or to hold issuers accountable for not complying with sustainability targets, remain weak.

Lastly, the plethora of initiatives has only partially addressed the very question behind the present study, namely the definition of 'sustainable securitisations'. While the EUGBS is expected to impose minimum requirements in relation to the final allocation of proceeds and to the underlying portfolio, it will not apply to the whole spectrum of transactions with 'sustainable' elements identified in Chap. 4. Its voluntary scope and the lack of binding requirements for 'sustainability-linked' transactions represent missed opportunities, which leave significant leeway for market actors to independently label their securitisations as 'sustainable' whilst deviating from the other requirements. The shortcomings of the currently discussed EU framework for 'sustainable securitisations' were the starting point for a series of normative considerations.

7.3 Normative Considerations: A Synthesis

The book also aimed to contribute to the lingering question around the optimal policy measures for achieving sustainability goals and mitigating climate change, as well as the financing mechanisms to fulfil these goals.

Chaps. 5 and 6 have culminated in a series of recommendations on potential legal initiatives at market level (in the form of guidelines, metrics, and standard clauses) and at EU level (in the form of legislative acts). Following the demarcation presented in Chap. 2,[6] these have the potential to advance sustainability goals in securitisations by serving three different functions.

First, law should set prerequisites for market actors to label and market securitisation operations as 'sustainable' ('enabling' function). One promising option should be an amendment to the EUGBR making the Standard mandatory for all 'sustainable' securitisations. EU rules could also require issuers to disclose in the first page of the prospectus whether their securitised products have a 'sustainable portfolio', a 'sustainable use of proceeds', or whether they are 'sustainability-linked'. The consequence for market actors not complying with such requirements would be the inability to trade ABSs with these labels in the EU market.

Second, further legal initiatives should lead to the formulation of high-level environmental, social, and governance (ESG) principles, which should serve as guides for the design of further policies or internal decisions ('programmatic' function). For instance, industry associations should develop model articles for SPVs, delineating their object and purpose in 'sustainable' transactions.

Third, EU institutions and financial industry associations should further engage in defining the boundaries of these transactions ('framing' function), with a view to mitigating risks like 'greenwashing' practices. For instance, the EUGBR is expected to prevent market actors from using loans towards carbon-intensive activities as part of the portfolio.[7] However, further reforms should go beyond and prohibit the inclusion of all assets posing considerable environmental and social risks (e.g. those breaching the 'Do no significant harm' criteria of the Taxonomy Regulation).[8]

The following table (Table 7.1) summarises the main proposals. For easier reference, the categorisation follows the *substance* and types of legal

[6] And building on Javier Solana, 'The Roles of Law in Finance' (2022) on file with author.
[7] Art 13c EUGBR.
[8] See Chap. 5, Sect. 5.5.4.

initiatives, i.e. the four legal strategies mentioned throughout the chapters, rather than their 'function', as outlined in Chap. 2, Sect. 2.6.

7.4 Further Areas for Research

Due to the focus on the most strategic aspects of securitisation law, the present work could not consider all possible indirect incentives to the issuance of securitised products. For instance, other angles worth pursuing and only briefly mentioned in the present work include potential legal reforms to the fiscal treatment or capital requirements. Tax incentives or lower capital surcharges (like the so-called 'green supporting factors' that some central banks,[9] as well as the EBA,[10] have introduced or explored) may also indirectly encourage market actors to contribute to the sustainability agenda through securitisation. Chap. 6 has also briefly pointed to the need for further research on the potential recalibration of central banks' haircut criteria for 'sustainable' ABSs in order to incentivise their creation.[11] Relatedly, the present work also could not consider guarantees or other forms of 'credit enhancement' to support originators by multilateral development banks (MDBs) or other public entities.[12]

Furthermore, the analysis of the legal bases for harmonisation of 'sustainable securitisations' led to the conclusion that EU law-making authorities would be in the best position to introduce further measures to coordinate these transactions. However, as global initiatives on sustainability

[9] See For example, MNB, the Hungarian national central bank, which has introduced such factor in 2021, MNB, 'Banks will soon have a wider variety of green loans to offer to companies and local governments' (2021)
https://www.mnb.hu/letoltes/preferential-capital-requirements-for-green-corporate-and-municipal-financing-summary.pdf accessed 10 December 2025.

[10] EBA's reflection did not culminate in the introduction of such factor. EBA, 'The role of Environmental Risks in the Prudential Framework' (2022) EBA Discussion Paper 2. https://www.eba.europa.eu/sites/default/files/document_library/Publications/Discussions/2022/Discussion%20paper%20on%20the%20role%20of%20environmental%20risk%20in%20the%20prudential%20framework/1031947/Discussion%20paper%20on%20role%20of%20ESG%20risks%20in%20prudential%20framework.pdf

[11] See Sect. 5.4.4.

[12] Partially covered by Daniela Gabor, 'Securitisation for sustainability: does it help achieve the Sustainable development goals?' (2019) Heinrich Böll Stiftung, 7–9, 16–17, https://us.boell.org/sites/default/files/gabor_finalised.pdf accessed 10 December 2025; Petit, Schlosser (n. 3) 15–17.

Table 7.1 An Overview of Normative Considerations

Legal Strategies		*'Standardisation' Proposals*	*'Harmonisation' Proposals*
Disclosure	Composition of the Portfolio	Standard clauses for portfolio assignment: 'Eligibility criteria' clause, representations on sustainability risks	• Mandatory disclosure of sustainability risks (PAI) • No assets breaching DNSH criteria
	Terms of Issue of ABSs	Standard clauses for prospectus: 'Use of proceeds' and 'variations to interest rates' clauses, sustainability-related covenants, conditions precedent, sustainability-related events of default and declassification events	• Mandatory disclosure on the first page of prospectus ('sustainable portfolio', 'sustainable use of proceeds', 'sustainability-linked') • Mandatory reporting requirements for social and sustainability-linked ABSs • Mandatory compliance with EUGBS collateral eligibility criteria
	Monitoring Processes		• Mandatory disclosure of 'tracking' mechanisms
Gatekeeping	Composition of the Portfolio		• Mandatory review of the portfolio
	Terms of Issue of ABSs	Guidelines and model clauses on underwriters' responsibilities	• Mandatory review of impact reports
Shielding	Monitoring Processes	Model bank account agreement Model servicing agreement	Mandatory tracking mechanism
Voice	Governance of SPVs	Model articles of association	• Purpose of SPVs and directors' duties
	Terms of Issue of ABSs	Guidelines and model clauses on underwriters' responsibilities	• Mandatory sustainability-related due diligence requirements for underwriters • Remedies for investors for misstatements in EUGB-compliant factsheets/prospectuses • Remedies for investors for damages arising from the review process
	Investors' Representation Mechanisms	Model intermediation agreement on appointment and duties of lead investor	• Mandatory appointment of a lead investor
	Monitoring Processes		• Enforcement remedies for misstatements in allocation/impact reports

reporting demonstrate,[13] there may be scope for other transnational fora to introduce 'public soft law' for these transactions further down the line, should the global policy priorities change and the momentum for securitisation increase.

Another interesting question, which goes beyond the scope of an inquiry into securitisation, relates to the extent to which investments in 'sustainable' ABSs and other 'sustainable' financial products are compatible with investors' mandates.

Lastly, the quantification of the effective impact of these operations on sustainability goals ('additionality') is also a critical question that goes beyond the purview of a doctrinal study of the law. Further empirical research is recommended in order to identify optimal metrics.

7.5 Future Market and Policy Outlook: Initial Reflections

The future trajectory of the EU sustainable securitisation market remains hard to predict. According to some estimates by rating agencies, the unprecedented macroeconomic conditions experienced in 2022 and 2023, characterised by high inflation and high interest rates, led to a contraction of ABS issuances.[14] It also affected lending volumes, the performance of loans, and default risks.[15] Conversely, 2024 and early predictions for 2025 highlighted growing consumer confidence and improved conditions in the lending market, as well as a subsequent growth in EU issuances, which reached post-global-financial-crisis records.[16]

[13] IFRS, 'Sustainability Disclosure Standard – –Climate-related disclosures (2023) https://www.ifrs.org/content/dam/ifrs/publications/pdf-standards-issb/english/2023/issued/part-a/issb-2023-a-ifrs-s2-climate-related-disclosures.pdf?bypass=on accessed 10 December 2025

[14] According to S&P Ratings, it reached a 20-year low in 2022. S&P Ratings, 'European ABS Outlook 2023 – –Under Pressure' (2023) 20 https://spratings.spglobal.com/ratings/en/regulatory/delegate/getPDF?articleId=2935594&type=COMMENTS&defaultFormat=PDF accessed 10 December 2025. Yet, supervisory data highlighted a limited upward trend between 2020 and 2022. EBA, ESMA, EIOPA, Joint Committee Report on the implementation and functioning of the Securitisation Regulation (Article 44) (2025) 14 https://www.esma.europa.eu/sites/default/files/2025-03/JC_2025_14_Joint_Committee_report_on_the_functionning_of_the_securitisation_regulation.pdf accessed 3 June 2026..

[15] Ibid., 5–6.

[16] S&P, 'European Structured Finance Outlook 2025 – Up In The Air' (2025) 3-4 https://www.spglobal.com/_assets/documents/ratings/research/101610216.pdf accessed 10 December 2025.

Moreover, the future outlook of the market for 'sustainable' debt instruments is highly uncertain. The gas energy shocks following Russia's full-scale invasion of Ukraine have brought new awareness among citizens, but also companies, financial institutions and investors, on the need to accelerate the transition towards alternative and 'greener' forms of energy.[17] Recurring weather events and the catastrophic consequences of climate change are a constant reminder of the urgency of meeting the Paris Agreement goals. In this context, financing techniques leveraging capital markets could act as crucial 'enablers' of the transition and of sustainability goals; in this sense, the further development of the 'sustainable securitisation' market appears to be a beacon of hope. Authoritative views by the Banque de France's governor and by the European Economic and Social Committee (EESC) have also underscored its pivotal importance.[18] At the same time, the rapidly evolving geopolitical circumstances have simultaneously acted as barriers to 'sustainable' investments, which have recently experienced a contraction. This could not only deter investments in these securities, but also the structuring of these transactions in the first place.[19]

Within the fluctuating market context, the present book aims to prompt a timely reflection, particularly at a time when regulators in the EU and globally have devoted significant attention to designing optimal policy measures to tackle climate change and to steer financial capital towards sustainability goals. Yet, the insufficiency and fragmentation of policy initiatives targeting 'sustainable securitisations' remain evident, despite

[17] See also a speech by a member of the ECB's Executive Board, Isabel Schnabel, 'Green transition' (8 January 2022) American Finance Association 2022 Virtual Annual Meeting https://www.ecb.europa.eu/press/key/date/2022/html/ecb.sp220108~0425a24eb7.en.html accessed 10 December 2025.

[18] François Villeroy de Galhau, *Financing the climate transition: building bridges between needs and resources,* 2024, https://www.banque-france.fr/en/governors-interventions/financing-climate-transition-building-bridges-between-needs-and-resources accessed 10 December 2025. See also an 'own-initiative' opinion by the European Economic and Social Committee, 'Review of the legal framework for securitisation in the EU' (2024) ECO/644, paras 1.3, 4.1.

[19] As also evidenced by 2025 data on the issuance of 'sustainable' debt instruments in Europe, which has experienced a significant contraction compared to 2024. Eg, green and sustainable bond issuance declined by 5.6% and 1% respectively. AFME, 'ESG finance report – Q3 2025' (2025) https://www.afme.eu/publications/data-research/afme-esg-finance-report-q3-2025/ accessed 3 June 2026.

the undeniable need to urgently mobilise substantially more capital to achieve the UN SDGs.[20]

This gap also transpires from the recent wave of initiatives to reboot the securitisation market in the EU, not least with the aim of accelerating the Capital Markets Union (CMU) project, recently rebranded as 'Savings and Investments Union'.[21]

As the European Commission assesses potential pathways to increase the EU's competitiveness globally, the revitalisation of the securitisation market has become a political priority.[22]

A review report on the Securitisation Regulation, issued by the European Supervisory Authorities in 2025, provides normative recommendations for streamlining the framework and improving supervisory practices,[23] but it fails to reflect on potential policy initiatives to boost the 'sustainable securitisation' market. Even a peer review conducted by ESMA to assess

[20] Between $5.4 and 6.4 trillion annually between 2023 and 2030, according to United Nations Conference on Trade and Development (*hereinafter:* UNCTAD), 'The costs of achieving the SDGs' (2023) https://unctad.org/sdg-costing accessed 10 December 2025.
[21] EU Commission, 'Communication from the Commission to the European Parliament, the European Council, the Council, the European Central Bank, the European Economic and Social Committee and the Committee of the Regions, 'Savings and Investments Union – –A Strategy to Foster Citizens' Wealth and Economic Competitiveness in the EU' (2025) 124.
[22] See Eurogroup, 'Statement of the Eurogroup on the future of Capital Markets Union', (2024) https://www.consilium.europa.eu/en/press/press-releases/2024/03/11/statement-of-the-eurogroup-in-inclusive-format-on-the-future-of-capital-markets-union/ accessed 10 December 2025. Authoritative 2024 reports by C. Noyer (chairing a group of experts), E. Letta and M. Draghi on the status of the Capital Markets Union and the competitiveness of the EU have all highlighted the underdeveloped potential of securitisation and the need to boost this market. Noyer et al, 'Developing European Capital Markets To Finance The Future – –Proposals for a Savings and Investments Union' (2024) 50 https://www.true-sale-international.de/fileadmin/tsi-gmbh/tsi_downloads/aktuelles/EN_-_Report_-_Developing_European_capital_markets.pdf accessed 10 December 2025; Enrico Letta, 'Much more than a market – –Speed, Security, Solidarity: Empowering the Single Market to deliver a sustainable future and prosperity for all EU Citizens' (2024) 32, 38 https://www.consilium.europa.eu/media/ny3j24sm/much-more-than-a-market-report-by-enrico-letta.pdf accessed 10 December 2025; Mario Draghi, 'The future of European competitiveness – –Part B | In-depth analysis and recommendations' (2024) 287, 292, 294 https://commission.europa.eu/document/download/ec1409c1-d4b4-4882-8bdd-3519f86bbb92_en?filename=The%20future%20of%20European%20competitiveness_%20In-depth%20analysis%20and%20recommendations_0.pdf accessed 10 December 2025.
[23] EBA, ESMA, EIOPA (n. 14).

supervisory practices over STS securitisations[24] offers no insights into the supervision of sustainability-related aspects of these deals.[25]

More importantly, a comprehensive review of the Securitisation Regulation[26] and of interrelated prudential provisions[27] is underway, based on proposals presented in June 2025, and spearheaded by a recent consultation.[28] Yet, amendments to the framework for 'sustainable securitisations' appear to remain off the table.

While not a panacea or a solution by itself, 'sustainable' securitisation brings several largely unexplored opportunities as a capital-raising mechanism to achieve the SDGs. The growth and development of the market, however, are contingent on the design of an adequate legal framework, providing for the right incentives, but also for the adequate safeguards against risks like 'greenwashing'. Thus, financial industry associations' efforts towards the standardisation of 'sustainable finance' transactions, as well as the ongoing review of the securitisation framework, represent crucial opportunities to reevaluate and dignify the role that 'sustainable securitisation' can play.

References

AFME, 'ESG finance report – Q3 2025' (2025) https://www.afme.eu/publications/data-research/afme-esg-finance-report-q3-2025/

[24] ESMA, 'Peer Review Report – –Peer Review on the implementation of the STS securitisation requirements' (2025) https://www.esma.europa.eu/sites/default/files/2025-03/ESMA42-2004696504-7945_STS_Securitisation_Peer_Review_Report.pdf accessed 10 December 2025.

[25] This gap is particularly striking given the existing disclosure requirements on sustainability matters, like the 'principal adverse impact' or energy efficiency of securitised assets. See Chap. 6, Sect. 6.4.1.

[26] Proposal for a Regulation of The European Parliament and of the Council amending Regulation (EU) 2017/2402 of the European Parliament and of the Council of 12 December 2017 laying down a general framework for securitisation and creating a specific framework for simple, transparent and standardised securitisation (2025) 826;

[27] Proposal for a Regulation of the European Parliament and of the Council amending Regulation (EU) No 575/2013 on prudential requirements for credit institutions as regards requirements for securitisation exposures (2025) 825.

[28] EU Commission, Targeted Consultation On The Functioning of the Eu Securitisation Framework (2024) https://finance.ec.europa.eu/document/download/fb451cdc-4e5b-4d74-9411-cb8bd0789090_en?filename=2024-eu-securitisation-framework-consultation-document_en.pdf accessed 10 December 2025.

Agostini F, 'One Step Forward, One Step Back': 'Sustainable Securitisations' Before and After the EU Green Bond Standard' in La Torre M, Leo S (eds.), *Contemporary Issues in Sustainable Finance* (2025) Palgrave Studies in Impact Finance.
Barmes D, *Securitisation is back, and green finance must stray far away* (Positive Money, September 2019) https://positivemoney.org/2019/09/securitisation-is-back-and-green-finance-must-stray-far-away/
Draghi M, 'The future of European competitiveness –Part B | In-depth analysis and recommendations' (2024) https://commission.europa.eu/document/download/ec1409c1-d4b4-4882-8bdd-3519f86bbb92_en?filename=The%20future%20of%20European%20competitiveness_%20In-depth%20analysis%20and%20recommendations_0.pdf
EBA, 'The role of environmental risks in the prudential framework' (2022) EBA Discussion Paper 2022/2, https://www.eba.europa.eu/sites/default/files/document_library/Publications/Discussions/2022/Discussion%20paper%20on%20the%20role%20of%20environmental%20risk%20in%20the%20prudential%20framework/1031947/Discussion%20paper%20on%20role%20of%20ESG%20risks%20in%20prudential%20framework.pdf accessed 3 June 2026.
EBA, ESMA, EIOPA, Joint Committee Report on the implementation and functioning of the Securitisation Regulation (Article 44) (2025) https://www.esma.europa.eu/sites/default/files/2025-03/JC_2025_14_Joint_Committee_report_on_the_functionning_of_the_securitisation_regulation.pdf
ESMA, 'Peer Review Report –Peer Review on the implementation of the STS securitisation requirements' (2025) https://www.esma.europa.eu/sites/default/files/2025-03/ESMA42-2004696504-7945_STS_Securitisation_Peer_Review_Report.pdf
EU Commission, 'Communication from the Commission to the European Parliament, the European Council, the Council, the European Central Bank, the European Economic and Social Committee and the Committee of the Regions, 'Savings and Investments Union –A Strategy to Foster Citizens' Wealth and Economic Competitiveness in the EU' (2025) 124.
European Economic and Social Committee, 'Review of the legal framework for securitisation in the EU' (2024) ECO/644
EU Commission, 'Omnibus Package' (1 April 2025) https://finance.ec.europa.eu/news/omnibus-package-2025-04-01_en
EU Commission, Targeted Consultation On The Functioning of the Eu Securitisation Framework (2024) https://finance.ec.europa.eu/document/download/fb451cdc-4e5b-4d74-9411-cb8bd0789090_en?filename=2024-eu-securitisation-framework-consultation-document_en.pdf
Eurogroup, 'Statement of the Eurogroup on the future of Capital Markets Union' (2024) https://www.consilium.europa.eu/en/press/press-releases/2024/03/11/statement-of-the-eurogroup-in-inclusive-format-on-the-future-of-capital-markets-union/

Gabor D, 'Securitisation for sustainability: does it help achieve the Sustainable development goals?' (2019) Heinrich Böll Stiftung https://us.boell.org/sites/default/files/gabor_finalised.pdf
IFRS, 'Sustainability Disclosure Standard –Climate-related disclosures' (2023) https://www.ifrs.org/content/dam/ifrs/publications/pdf-standards-issb/english/2023/issued/part-a/issb-2023-a-ifrs-s2-climate-related-disclosures.pdf?bypass=on
James K, Parker I, 'New Growth in the Green Securitisation Market' (2019) Simmons & Simmons, 1. www.simmons-simmons.com/-/media/files/articles/2019/capital%20markets/new%20growth%20in%20the%20green%20securitisation%20market%20january%202019.pdf
Letta E, 'Much more than a market –Speed, Security, Solidarity: Empowering the Single Market to deliver a sustainable future and prosperity for all EU Citizens' (2024) https://www.consilium.europa.eu/media/ny3j24sm/much-more-than-a-market-report-by-enrico-letta.pdf
MNB, 'Banks will soon have a wider variety of green loans to offer to companies and local governments' (2021) https://www.mnb.hu/letoltes/preferential-capital-requirements-for-green-corporate-and-municipal-financing-summary.pdf
Noyer Cet al, 'Developing European Capital Markets To Finance The Future – Proposals for a Savings and Investments Union' (2024) https://www.true-sale-international.de/fileadmin/tsi-gmbh/tsi_downloads/aktuelles/EN_-_Report_-_Developing_European_capital_markets.pdf
Petit C, Schlosser P, 'Rationale, Potential and Pitfalls of Green Securitisation' (2020) RSCAS Working Paper 35 https://cadmus.eui.eu/bitstream/handle/1814/67018/RSCAS%202020_35.pdf?sequence=1
Pistor K, 'A Legal Theory of Finance' (2013) 41 *Journal of Comparative Economics* 315.
S&P Ratings, 'European ABS Outlook 2023 –Under Pressure' (2023) 20 https://spratings.spglobal.com/ratings/en/regulatory/delegate/getPDF?articleId=2935594&type=COMMENTS&defaultFormat=PDF
S&P, 'European Structured Finance Outlook 2025 – Up In The Air' (2025) https://www.spglobal.com/_assets/documents/ratings/research/101610216.pdf
Schnabel I, 'Green transition' (8 January 2022) American Finance Association 2022 Virtual Annual Meeting https://www.ecb.europa.eu/press/key/date/2022/html/ecb.sp220108~0425a24eb7.en.html
Solana J, 'The Roles of Law in Finance' (2022) on file with author.
Tomadakis A, 'The EU is walking a fine line between simplification and deregulation' (2025) CEPS, https://www.ceps.eu/the-eu-is-walking-the-fine-line-between-simplification-and-deregulation/
United Nations Conference on Trade and Development (*hereinafter*UNCTAD), 'The costs of achieving the SDGs' (2023) https://unctad.org/sdg-costing

Villeroy de Galhau F, *Financing the climate transition: building bridges between needs and resources* (2024), https://www.banque-france.fr/en/governors-interventions/financing-climate-transition-building-bridges-between-needs-and-resources

Competing Interests

The author has no conflicts of interest to declare that are relevant to the content of this book. Chap. 4 of this work was partially funded by Fondazione Felice Gianani in the context of a one-year research fellowship, which supported my doctoral studies (a.a. 2021–2022).

F. Agostini, *The Law of Sustainable Securitisations*, EBI Studies in Banking and Capital Markets Law,
https://doi.org/10.1007/978-3-032-25046-9

Table of Cases

European Court of Justice

Case 436\03 European Parliament v Council [2006] I-03733.
Case C-377/98 Netherlands v Parliament and Council [2001] ECR I-7079.
Case C-376/98 Germany v Parliament and Council [2000] ECR I-7079.

UK

Barclays Bank Ltd v Quistclose Investments Ltd [1970] AC 567 (UK).
Citibank NA and MBIA Assurance SA v QVT Financial LP [2007] EWCA Civ 11 (UK).
Clough v Bond [1838] 3 My & Cr 490, 496-7 (UK).
Cowan v Scargill [1985] Ch 270 (UK).
Dearle v Hall [1828] 3 Russ 1; SS 4, 29, 32 (UK).
Electrim SA v Vivendi Holdings 1 Corp [2008] EWCA Civ 1178 (UK).
Equitix EEEF Biomass 2 Ltd v Fox [2021] EWHC 2531 (TCC) (UK).
Highberry Ltd v Colt Telecom Group (No 2) [2002] EWHC 2815 (UK).
Hong Kong Fir Shipping Ltd v Kisen Kaisha Ltd (1962) EWCA Civ 7 (UK).

F. Agostini, *The Law of Sustainable Securitisations*, EBI Studies in Banking and Capital Markets Law,
https://doi.org/10.1007/978-3-032-25046-9

Kelly v Coopers [1993] AC 205 (UK).
Knight v Knight [1855] 49 ER 58 (UK).
Learoyd v Whiteley [1887] UKHL 1 (UK).
Morice v Bishop of Durham [1805] EWHC Ch J80 (UK).
New Hearts v Cosmopolitan Investments [1997] BCLC 249 (UK).
Oscar Chess v Williams [1957] 1 WLR 370 (UK).
Roberts v Frohlich [2011] EWHC 257 (Ch) (UK).

US

Re Capstead Mortg. Corp. Securities Litigation [2003] 258 F.Supp.2d 533 ND Tex (US).
Katz v Oak Industries Inc, 508 A2d 873 (Del. Ch. 1986) (US).

Table of Legislation

EU

Commission Delegated Directive (EU) 2021/1270 of 21 April 2021 amending Directive 2010/43/EU as regards the sustainability risks and sustainability factors to be taken into account for Undertakings for Collective Investment in Transferable Securities (UCITS) [2021] OJ L 277/141.

Commission Delegated Regulation 2022/1214 of 9 March 2022 amending Delegated Regulation (EU) 2021/2139 as regards economic activities in certain energy sectors and Delegated Regulation (EU) 2021/2178 as regards specific public disclosures for those economic activities [2022] OJ L 188/1.

Commission Delegated Regulation (EU) 2023/2486 of 27 June 2023 supplementing Regulation (EU) 2020/852 of the European Parliament and of the Council by establishing the technical screening criteria for determining the conditions under which an economic activity qualifies as contributing substantially to the sustainable use and protection of water and marine resources, to the transition to a circular economy, to pollution prevention and control, or to the protection and restoration of biodiversity and ecosystems and for determining whether that economic activity causes no significant harm to any of the other environmental objectives and amending

F. Agostini, *The Law of Sustainable Securitisations*, EBI Studies in Banking and Capital Markets Law,
https://doi.org/10.1007/978-3-032-25046-9

Commission Delegated Regulation (EU) 2021/2178 as regards specific public disclosures for those economic activities [2023] C/2023/3851.

Commission Delegated Regulation 2017/565 as regards the integration of sustainability factors, risks and preferences into certain organisational requirements and operating conditions for investment firms [2021] OJ L 277/1.

Commission Delegated Regulation 2015/61 of 10 October 2014 to supplement Regulation (EU) No 575/2013 of the European Parliament and the Council with regard to liquidity coverage requirement for Credit Institutions [2014] OJ L 11/1.

Commission Delegated Regulation 2019/1851 of 28 May 2019 supplementing Regulation (EU) 2017/2402 of the European Parliament and of the Council with regard to regulatory technical standards on the homogeneity of the underlying exposures in securitisation [2019] OJ L 285/1.

Commission Delegated Regulation (EU) 2024/1700 of 5 March 2024 supplementing Regulation (EU) 2017/2402 of the European Parliament and of the Council with regard to regulatory technical standards specifying, for simple, transparent and standardised non-ABCP traditional securitisation, and for simple, transparent and standardised on-balance-sheet securitisation, the content, methodologies and presentation of information related to the principal adverse impacts of the assets financed by the underlying exposures on sustainability factors [2024] C/2024/1344.

Commission Delegated Regulation (EU) 2019/980 of 14 March 2019 supplementing Regulation (EU) 2017/1129 of the European Parliament and of the Council as regards the format, content, scrutiny and approval of the prospectus to be published when securities are offered to the public or admitted to trading on a regulated market, and repealing Commission Regulation ((EC) No 809/2004 [2019] OJ L 166/26.

Commission Delegated Regulation 2020/1224 of 16 October 2019 supplementing Regulation (EU) 2017/2402 of the European Parliament and of the Council with regard to regulatory technical standards specifying the information and the details of a securitisation to be made available by the originator, sponsor and SSPE [2020] OJ L 289/1.

Commission Delegated Regulation 2021/2139 of 4 June 2021 supplementing Regulation (EU) 2020/852 of the European Parliament and of the Council by establishing the technical screening criteria for determining the conditions under which an economic activity qualifies as contributing substantially to climate change mitigation or climate change adaptation and for determining whether that economic activity causes no significant harm to any of the other environmental objectives [2021] OJ L 442/1.

Commission Delegated Regulation (EU) 2023/2486 supplementing Regulation 2020/852 of the European Parliament and of the Council by establishing the technical screening criteria for determining the conditions under which an economic activity qualifies as contributing substantially to the sustainable use and protection of water and marine resources, to the transition to a circular economy, to pollution prevention and control, or to the protection and restoration of biodiversity and ecosystems and for determining whether that economic activity causes no significant harm to any of the other environmental objectives and amending Delegated Regulation (EU) 2021/2178 as regards specific public disclosures for those economic activities [2023] OJ L 2023/2486.

Commission Directive 2010/43/EU of 1 July 2010 implementing Directive 2009/65/EC of the European Parliament and of the Council as regards organisational requirements, conflicts of interest, conduct of business, risk management and content of the agreement between a depositary and a management company [2010] OJ 176\42.

Commission Implementing Regulation 2020/1225 of 29 October 2019 laying down implementing technical standards with regard to the format and standardised templates for making available the information and details of a securitisation by the originator [2020] OJ 289/217.

Consolidated Version of the Treaty on European Union [2016] OJ C 202/13.

Consolidated Version of the Treaty on the Functioning of the European Union [2012] C 326/49.

Directive 2003/41/EC of the European Parliament and of The Council of 3 June 2003 on the activities and supervision of institutions for occupational retirement provision [2003] OJ L 235/10.

Directive 2009/65 of the European Parliament and of the Council of 13 July 2009 on the coordination of laws, regulations and administrative provisions relating to undertakings for collective investment in transferable securities [2009] OJ 302/32.

Directive (EU) 2019/2162 of the European Parliament and of the Council of 27 November 2019 on the issue of covered bonds and covered bond public supervision and amending Directives 2009/65/EC and 2014/59/EU [2009] OJ L 302/32.Directive 2019/771 of the European Parliament and of the Council of 20 May 2019 on certain aspects concerning contracts for the sale of goods, amending Regulation (EU) 2017/2394 and Directive 2009/22/EC, and repealing Directive 1999/44/EC [2019] OJ 136/28.

Directive 2016/2341 of the European Parliament and of the Council of 14 December 2016 on the activities and supervision of institutions for occupational retirement provision [2016] OJ L 354/37.

Directive 2013/34/EU of the European Parliament and of the Council of 26 June 2013 on the annual financial statements, consolidated financial statements and related reports [2013] OJ L 182/19.

Directive 2014/17/EU of the European Parliament and of the Council of 4 February 2014 on credit agreements for consumers relating to residential immovable property and amending Directives 2008/48/EC and 2013/36/EU and Regulation (EU) No 1093/2010 [2014] OJ L 60/34.

Directive (EU) 2025/794 of the European Parliament and of the Council of 14 April 2025 amending Directives (EU) 2022/2464 and (EU) 2024/1760 as regards the dates from which Member States are to apply certain corporate sustainability reporting and due diligence requirements [2025] OJ 2025/794.

Directive 2017/828 of the European Parliament and of the Council of 17 May 2017 amending Directive 2007/36/EC as regards the encouragement of long-term shareholder engagement [2017] OJ L 132/1.

Directive (EU) 2022/2464 of the European Parliament and of the Council of 14 December 2022 amending Regulation (EU) No 537/2014, Directive 2004/109/EC, Directive 2006/43/EC and Directive 2013/34/EU, as regards corporate sustainability reporting.

Directive 2002/47/EC of the European Parliament and of the Council of 6 June 2002 on financial collateral arrangements [2002] OJ L 168/43.

Directive 2019/2162 of the European Parliament and of the Council of 27 November 2019 on the issue of covered bonds [2019] OJ L 328/29.

Guideline 2015/510 of the European Central Bank of 19 December 2014 on the implementation of the Eurosystem monetary policy framework (General Documentation Guideline) [2014] OJ L 91/5.
Guideline (EU) 2016/65 of the European Central Bank of 18 November 2015 on the valuation of haircuts applied in the implementation of the Eurosystem monetary policy framework [2015] OJ L 14/30.
Regulation 462/2013 of the European Parliament and of the Council of 21 May 2013 amending Regulation (EC) No 1060/2009 on credit rating agencies [2013] OJ L 146/1.
Regulation 2020/852 of the European Parliament and of the Council of 18 June 2020 on the establishment of a framework to facilitate sustainable investment, and amending Regulation (EU) 2019/2088 [2020] OJ L 198/13.
Regulation 2017/2402 of the European Parliament and of the Council of 12 December 2017 laying down a general framework for securitisation and creating a specific framework for simple, transparent and standardised securitisation, and amending Directives 2009/65/EC, 2009/138/EC and 2011/61/EU and Regulations (EC) No 1060/2009 and (EU) No 648/2012 [2017] OJ L 347/35.
Regulation 2021/557 of the European Parliament and of the Council of 31 March 2021 amending Regulation (EU) 2017/2402 laying down a general framework for securitisation and creating a specific framework for simple, transparent and standardised securitisation to help the recovery from the COVID-19 crisis [2021] OJ 116/1.
Regulation (EU) 2023/2631 on European green bonds and optional disclosures for bonds marketed as environmentally sustainable and for sustainability-linked bonds [2023] PE/27/2023/REV/1.
Regulation (EU) 2024/1623 of the European Parliament and of the Council of 31 May 2024 amending Regulation (EU) No 575/2013 as regards requirements for credit risk, credit valuation adjustment risk, operational risk, market risk and the output floor [2024] PE/80/2023/INIT.
Regulation 2017/1129 of the European Parliament and of the Council of 14 June 2017 on the prospectus to be published when securities are offered to the public or admitted to trading on a regulated market, and repealing Directive 2003/71/EC [2017] OJ L 168/12.
Regulation 2019/2088 of the European Parliament and of the Council of 27 November 2019 on sustainability-related disclosures in the financial services sector [2019] OJ L 317/1.

International Legal Instruments

BCBS, Basel III framework [2023].
Paris Agreement to the United Nations Framework Convention on Climate Change, [2015] TIAS 16-1104.
UN, Transforming our world: The 2030 Agenda for Sustainable Development [2015] RES 70/1.

National Jurisdictions

France

Code Monétaire et financier (FR).

Italy

Codice Civile, Regio Decreto 16 marzo 1942, n. 262.
L 130/1999.

Luxembourg

Loi du 22 mars 2004 relative à la titrisation.
Loi du 25 février 2022 portant 1° modification de la loi.
modifiée du 22 mars 2004 relative à la titrisation.

Spain

Ley 5/2015 de fomento de la financiación empresarial [2015].

UK

Companies Act 2006.
Land Registration Act 2002.
Law of Property Act 1925.
Securitisation Regulations 2024.

Index

F. Agostini, *The Law of Sustainable Securitisations*, EBI Studies in Banking and Capital Markets Law,
https://doi.org/10.1007/978-3-032-25046-9

R

S

T

U

V